Physical Chemistry

→ A Short Course ←

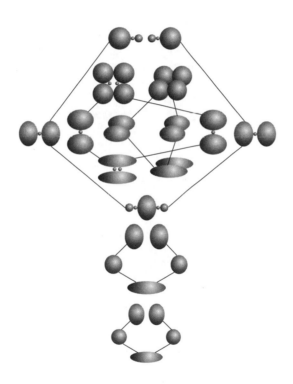

Physical Chemistry

→ A Short Course ←

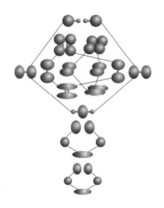

Wayne E. Wentworth

Department of Chemistry
University of Houston
Houston, Texas

**Blackwell
Science**

©2000 by Blackwell Science, Inc.

Editorial Offices:
Commerce Place, 350 Main Street, Malden, Massachusetts 02148, USA
Osney Mead, Oxford OX2 0EL, England
25 John Street, London WC1N 2BL, England
23 Ainslie Place, Edinburgh EH3 6AJ, Scotland
54 University Street, Carlton, Victoria 3053, Australia
Other Editorial Offices:
Blackwell Wissenschafts-Verlag GmbH, Kurfürstendamm 57, 10707 Berlin, Germany
Blackwell Science KK, MG Kodenmacho Building, 7-10 Kodenmacho Nihombashi, Chuo-ku, Tokyo 104, Japan

Distributors:
USA

Blackwell Science, Inc.
Commerce Place
350 Main Street
Malden, Massachusetts 02148
(Telephone orders: 800-215-1000 or 781-388-8250; fax orders: 781-388-8270)

Canada

Login Brothers Book Company
324 Saulteaux Crescent
Winnipeg, Manitoba, R3J 3T2
(Telephone orders: 204-224-4068)

Australia

Blackwell Science Pty, Ltd.
54 University Street
Carlton, Victoria 3053
(Telephone orders: 03-9347-0300; fax orders: 03-9349-3016)

Outside North America and Australia

Blackwell Science, Ltd.
c/o Marston Book Services, Ltd.
P.O. Box 269
Abingdon
Oxon OX14 4YN
England
(Telephone orders: 44-01235-465500; fax orders: 44-01235-465555)

Acquisitions: Nancy Hill-Whilton
Development: Jill Connor
Production: Louis C. Bruno, Jr.
Manufacturing: Lisa Flanagan
Cover design by Hannus Design Associates
Cover art by Loraine Crutchfield
Typeset by Pre-Press Company, Inc.
Printed and bound by Sheridan Books/Ann Arbor
Cover printed by Plymouth Color

Printed in the United States of America
00 01 02 03 5 4 3 2 1

Library of Congress Cataloging-in-Publication Data

Wentworth, Wayne Ernest.
    Physical chemistry: a short course / Wayne E. Wentworth.
        p. cm.
    ISBN 0-632-04329-6
    1. Chemistry, Physical and theoretical. I. Title
    QD453.2.W464   2000
    541—dc21   99-16543
                                                          CIP

To my son Rick, Charles Batten, and Al Hildebrandt,
all of whom meant so much to me personally and professionally

Contents

Preface

The original purpose for writing this text was to give secondary education majors in chemistry a background in the fundamental principles of physical chemistry. High schools frequently offer a second course in chemistry that is frequently referred to as an advanced placement course that is given college credit at most institutions. This course is equivalent to the freshman general chemistry course offered at most colleges and universities and commonly uses the same college-level text. This general chemistry course is generally based upon fundamental physical chemistry principles. Consequently, the secondary education teacher for this course must have a good understanding of physical chemistry. The curriculum for secondary education majors usually does not have room for the traditional two- or three-semester physical chemistry course for BS chemistry majors. Thus, the need for a one-semester course that covers the basic areas of thermodynamics, chemical kinetics, and molecular structure and energies. This was the original objective of this textbook.

From the inception of this course, however, the need for such a course was recognized for other disciplines such as biochemistry and engineering and for students preparing for professional careers who will be attending medical, dental, optometry, or pharmacy schools. With this large influx of students in mind, the text has been revised to include additional topics and exercises to illustrate the *practical applications* of the basic physical chemistry principles. Topics have been added at the suggestions of reviewers who teach such a course. As a result of these revisions, the text is more comprehensive than originally intended, which allows for optional use of the text to satisfy different objectives.

Organization of the Text

As one can readily see, the development of this text is calculus-based in order to understand the topics based on first principles. Since many of the students taking this course have limited backgrounds and aptitudes for mathematics, an extensive review of the basic principles required for this text are presented in Appendices A, B, and C. I recommend that all students, regardless of their mathematical preparation, first cover this material before starting the main text. The material in Appendices A, B, and C not only reviews the necessary principles of calculus but also demonstrates how these principles are applied to physical chemistry. The transition from the x, y, and z variables to lnK, 1/T, and P is often confusing, and the appendices attempt to bridge this gap. The time allotted to covering Appendices A, B, and C will vary depending on the mathematical background and aptitude of the students, but one week should suffice in most cases.

After the mathematical review, the topics in physical chemistry can be covered in Chapters 1 through 7. Chapter 1 is devoted principally to the *gaseous state* and introduces certain basic thermodynamic concepts such as work and its dependence on path. *Thermodynamics* is covered in Chapters 2, 3, and 4. All of the basic laws and application to gas phase reactions are in Chapters 2 and 3, and these should be covered. The application to *solution thermodynamics* has been separated in Chapter 4, and coverage of this material is optional. To devote sufficient time to molecular structure, Chapter 4 is skipped over in my course at the University of Houston. If the text is being used primarily for students in biochemistry, however, the material in Chapter 4 may be more beneficial than a more thorough treatment of molecular structure.

Chemical kinetics follows in Chapter 5. The kinetics of a reaction is independent of thermodynamics; however, the relationship between change in internal energy and activation energies shows the slight relationship. Also the concepts of bond dissociation energy and equilibrium constants are used so it would be preferable to cover Chapters 1 through 3 before Chapter 5.

Molecular structure and energies are covered in Chapters 6 and 7. The energies associated with nuclear motion (translational, rotational, and vibrational) are covered in Chapter 6. Quantum mechanics is introduced here using the simple particle in the box, which can be understood with minimal mathematical complexity. The Bohr theory is outlined primarily for its contribution to the development of quantum mechanics. Calculations of energy levels and transitions in hydrogenic systems are based on the Bohr theory. This introduces the concept of electronic energies. *Molecular structure and electronic energies* are considered in more detail in Chapter 7.

Course Content

The text contains more material than can be covered in a typical one-semester, three-credit-hour course; i.e., 3 hrs/week for 14 weeks. Consequently, there are alternative selections of the material. In the course at the University of Houston the students come from a variety of disciplines, and our choice of material includes the more popular molecular structure. This is also most important for secondary education majors. In order to devote more time to molecular structure we completely bypass Chapter 4 and much of the discussion of enzyme kinetics in Chapter 5. This allows us to cover in sequence: Appendices A, B, C, Chapters 1 to 3, Chapter 5 (up to section E), Chapter 6, and at least half of Chapter 7 (through section D). This gives the course the greatest breadth of topics.

On the other hand, if most of the students have majors in the biological sciences, it may be beneficial to cover Chapter 4 on *solution thermodynamics,* including the section on *bioenergetics,* and the section on *enzyme kinetics* in Chapter 5. Coverage of this material would preclude Chapter 7 and possibly some or all of Chapter 6. Not all of the material in Chapter 6 needs to be covered before going to Chapter 7. Only sections A through E and the one-dimensional particle in the box in section F need to be covered prior to Chapter 7. This would allow further coverage of the material in Chapter 7.

Pedagogical Features

Problems

Upon examining the text you can see that there are problems located at the end of most sections *within* the chapter. These problems pertain exclusively to the material just presented in this section. The purpose of these problems is to enlighten the student about the principal concepts in that section. Frequently a basic concept can not be fully understood unless there is a problem that requires the student to think through an application of the concept. This aspect of the book allows the comprehension of somewhat difficult material even for the student who is less adept mathematically. Furthermore, this aspect of the text allows the subject to be somewhat self-taught, in the absence of a lecture reinforcement of the material. All problems within the chapter being covered are automatically assigned at the beginning of the semester. Some class exam questions may be related directly to the problem assignments, and this further encourages the working of the problems.

Exercises

The *exercises* are placed at the end of the chapter and are not to be worked until the conclusion of the chapter. The exercises may include more than a single concept and frequently emphasize application to a practical problem. There is no specific criterion for the order of the exercises, but generally they progress according to difficulty. Just before completion of the chapter the professor may select about 10 or so exercises. The selection varies from semester to semester.

Tabular Data

Most tables of data are given in Appendix D so that they will be readily accessible. Many of the thermodynamic tables are required at several points in the text and for this reason need a common location. The alternative location within a chapter is very inconvenient since it is difficult to recall where they are located. On only a few occasions is a table positioned within a chapter, and this is when the data will be used in that specific section.

The form of the tables in Appendix D follow those found in the *Handbook of Chemistry and Physics.* This is the most common reference source for chemists, and for this reason the student should be familiar with the style of the tables in this reference source.

Acknowledgments

The text has been very successful in the teaching of the one-semester physical chemistry course at the University of Houston, which is entitled "Survey of Physical Chemistry." Initially, the course had an enrollment of 15 to 20 students, and the course was offered once a year. Presently the enrollment has risen in excess of 60 students, and the course is offered twice a year. Since the enrollment is so high, we plan to offer it also in the summer semester.

The development of this text would not have been possible without the assistance of several people whom I would like to recognize at this time. Typing of the original manuscript was performed by Susan Norman, and the many revisions were carried out on the word processor by Beatriz Fitz. Most of the figures were composed by Loraine Crutchfield using computer graphics software. Dr. Jules Ladner, my previous co-author and good friend, contributed much of the material in Chapters 4 and 7 in his usual precise manner. Mr. Jerry Johnson carried out a careful technical review of Chapter 4. Drs. Tom Apple, Rensselaer Polytechnic Institute; F. G. Baglin, University of Nevada, Reno; Cecil Dybowski, University of Delaware; Barry Friedman, Southern Methodist University; K. W. Hipps, Washington State University; Wolfgang Jaeger, University of Alberta, Edmonton; Merlyn D. Schuh, Davidson College; Richard W. Schwenz, University of Northern Colorado; and Paul D. Siders, University of Minnesota, Duluth were the technical reviewers of the entire manuscript. I am deeply appreciative of each of their contributions, and the text could not have been developed without their assistance. Finally I would like to thank my students for the many suggestions that they have made regarding the pedagogy and the numerous errors that they have detected in preliminary renditions.

1

INTRODUCTION: PROPERTIES OF GASES

As will be explained later in this chapter, all compounds have a unique critical point which can be defined in terms of mathematics. At this critical point there is a critical temperature, above which it is impossible to condense the gas to a liquid or solid, regardless of the pressure applied. If the temperature and pressure are in excess of the critical temperature and critical pressure, we have what is called a Supercritical Fluid which is thought to have unusual solubility properties. In recent years there has been considerable use of this supercritical state to carry out extractions of high boiling compounds, many of which are of biological interest. The technique is appropriately called Supercritical Fluid (SCFE) Extraction. The supercritical state has also been used as the carrier gas in chromatography and is appropriately called Supercritical Fluid Chromatography (SCFC). The use of Supercritical Fluids has the advantage that the fluid can be readily vaporized from the high boiling components.

A. Physical Chemistry and the Scope of This Text

In order to avoid complex and time consuming mathematics, we have restricted much of the thermodynamics to the ideal gas equation of state. Most of the students will be working in disciplines where the pressures do not exceed atmospheric or slightly higher where the ideal gas is quite adequate. In general a decision as to cover a certain topic is based upon the importance versus the mathematical complexity. If the topic is exceedingly important then an effort has been made to present it in sufficient depth regardless of the mathematical complexity.

The text is obviously based upon calculus and the author does not believe a student can get a good concept of physical chemistry without a calculus base. Obviously other authors feel otherwise and their texts consist of formulas given to the student. Problem solving then consists of substitution into these formulas. Frequently the limitations and restrictions on the formulas are not appreciated since the calculus-based derivation is not carried out.

Finally it should be pointed out that this book is <u>not</u> specifically applied towards biochemistry. There are several texts that have been written for a short course in P. Chem, but the emphasis has been heavily towards biochemistry. Obviously these texts serve the needs of majors in the biological areas and in particular pre-medical and pre-dental students. In this text there is no effort made to apply physical chemistry to any specific area, but rather an attempt has been made to show a broader application. At the University of Houston the application to biochemistry is taught in the Biochemistry Department after the student has obtained a background of the basics from this textbook.

Aside from a brief intoduction to gaseous properties in Chapter 1, the text contains three general areas of Physical Chemistry: Thermodynamics, Kinetics, and Molecular Structure and Energy States. A fouth general area, called Statistical Thermodynamics, has been left out due to its complex mathematical basis and its limitation to somewhat simple molecular systems. About one half of the course has been devoted to Thermodynamics (Chapters 2–4), one fourth to Kinetics (Chapter 5), and one fourth to Molecular Structure and Energies (Chapters 6 & 7). In the author's opinion, this material will give you the strongest background with which to solve problems in your specific field of interest.

B. Definitions and Terminology

The following definitions are necessary to introduce the subject of thermodynamics, though not intended to be complete. Other definitions and conventions will be given throughout the text in the presentation of a given subject.

System: Some portion of the universe under investigation.

Boundaries: Real or imaginary surfaces or lines which define the confines of the system.

Surroundings: All objects which can interact with the system. These could be work and heat sources or sinks, necessary for the system to undergo the process under the specified conditions; e.g., isothermal (constant temperature) or constant pressure.

Equilibrium: State of no net change in the system and surroundings.

Reversibility: A process is said to be reversible, if the system undergoing change is in equilibrium with its surroundings at all times. A truly reversible process would require an infinite amount of time to be carried out. Since equilibrium implies no net change, in order for a process involving net change to occur the system must be displaced from equilibrium. Such a process can be approached by a repetition of steps, each involving a minute displacement from equilibrium, followed by re-equilibration of the system. As the magnitude of the displacements is made smaller and the number of the steps larger (necessary in order to accomplish the desired total change), the process approaches the reversible path.

In order to obtain a better understanding of these somewhat abstract definitions we will consider an easily visualized mechanical expansion of a gas. We will consider the gas contained in a cylinder with a weightless piston. The motion of the piston is perpendicular to the earth's surface so that when weights are placed on the weightless piston a force will be generated on the piston due to the gravitational attraction. The cylinder/piston are placed in a water bath so that the temperature of the gas can be controlled. A diagram of this arrangement is shown in Figure 1-B-1.

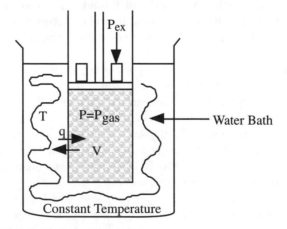

Figure 1-B-1. Gas Contained in a Cylinder/Piston
Temperature (T) controlled by water bath, pressure (P) controlled by weights
added to a weightless piston.

We can assign the above definitions to the apparatus in Figure 1-B-1 as follows:

System: Gas contained in the cylinder/piston.

Boundaries: Walls of the cylinder and piston.

Surroundings: Water bath, capable of absorbing or liberating heat, is used to control the temperature; weights on the piston are used to control the external pressure applied to the gas.

Equilibrium: $P_{ex} = P_{gas}$, no change in position of the piston; T = constant, heat change from cylinder to water bath and water bath to cylinder are equal.

Now in order to describe "reversibility" we must consider a *process* in which the piston undergoes some movement. For simplicity we will select an isothermal process where the temperature of the gas is held constant. We will allow this isothermal process to occur by removal of some of the weights placed on the piston. In Figure 1-B-2 we show three such processes. For simplicity we will not include the constant temperature bath but we will assume it is there to control the temperature of the gas. In Figure 1-B-2(a) one weight is removed in a single step and we place it at the level of the initial position of the piston. The piston then expands lifting the one remaining weight to its final position where the pressure of the gas is equal to the external pressure which is half the initial pressure. The work involved in this process is the work that the gas did on the surroundings by lifting one weight to its final position. We call this process an irreversible process since in order to reverse it we must lift the weight at the initial level of the piston to the upper level and place it on the piston. Compression will then occur back to the initial position. Since additional work was required to reverse the process, we refer to this as an *irreversible* process.

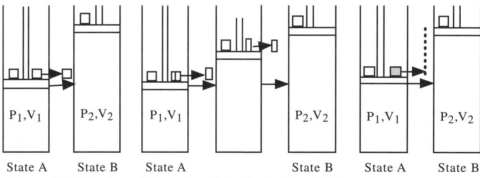

State A   State B   State A   State B   State A   State B
Figure 1-B-2. Isothermal Expansion of the Gas by Three Different Processes

In Figure 1-B-2(b) the expansion is depicted as a two step process. The weight is split in two and half of it is removed at the first level and the gas expands to an intermediate level corresponding to a pressure of 1 1/2 weights. The remaining 1/2 weight is then removed and the expansion leads to the same final state as in Figure 1-B-2(a). The work performed by the gas on the surroundings is greater for the expansion in (b) than for the expansion in (a). Note that not only is one of the weights lifted to the upper level, but 1/2 of a weight is lifted to an intermediate level. Again this is an irreversible expansion.

Finally in Figure 1-B-2(c) we describe a path in which only an infinitesimal amount of mass is removed at each step of the expansion. These are designated by a series of small dots running vertically between the lower level of the piston to the upper level. This process would give the maximum work done by the gas. This is obvious by the extrapolation from one → two → infinite number of steps. This is also the *reversible* path since the same path can be followed by compression of the gas by simply returning the infinitesimal weights to the piston, starting from the upper level. We conclude then that the reversible path occurs in infinitesimal steps and would require an infinite amount of time. This idealized reversible expansion obtains the *maximum* work done by the gas on the surroundings. The following definitions refer to the states of systems and we can apply these to our system of the gas in a cylinder/piston.

Properties: Quantities which can be used to describe a system, for example, density, color, physical state, index of refraction.

Many properties are variables which are quantitatively measurable, and these are of particular interest to us. These properties which have a definable magnitude can be further classified into two groups:

Intensive property: Magnitude is *independent* of the mass of the system, for example, index of refraction, temperature, concentration, pressure.

Extensive property: Magnitude is *dependent* on the mass, for example, volume, heat of reaction, number of moles.

State: A specific condition of a system which is completely and precisely described through its properties. Generally, only the *minimum* number of properties necessary to completely describe the state are specified. From a practical standpoint, the state of a system can also be viewed as the specification of certain variables so that any scientist could reproduce the state at any location on the earth. For example, if P and T for a certain number of moles of a gas are fixed, the state is fixed and consequently, the volume.

Referring to Figure 1-B-2 we note that the *initial state* can be defined by the *properties*: P_1, V_1, T_1. It is assumed that the gaseous species is known and, if we know the species and P_1, V_1, T_1, this is sufficient information to define this state. The *final state* can be defined similarly by the properties: P_2, V_2, T_1. If we let A represent the initial state and B the final state, the process of expansion would be

$$A (P_1,V_1,T_1) \rightarrow B (P_2,V_2,T_1) \qquad (1\text{-}B\text{-}1)$$

Of course the path taken in the expansion would have to be defined in order to evaluate the work. The following definitions are also important and their application will be shown later.

State function: (Also called state variable or thermodynamic function.) A variable which is dependent only on the state of the system. Therefore, when a change in the system occurs, the change in such a function is dependent only on the initial and final states, and independent of the path traveled in going from one state to the other. Since this type of function is of great significance in thermodynamics, it is frequently called a thermodynamic function. An example of such a function is volume, which depends only on the initial and final states.

Equation of state: A mathematical relationship between the state variables, P, V, T, and n, the number of moles of the substance. The relationship or equation may also contain constants which are characteristic of the substance.

Heat and Temperature: Heat and temperature can be defined in a phenomenological manner through the equation

$$Q = mc(t_2 - t_1) \qquad (1\text{-}B\text{-}2)$$

where t_1 and t_2 are two temperatures (deg) which are not too widely different, m is the mass of the substance (g), c is the specific heat capacity (cal or joules/g-deg), and Q is the heat absorbed or lost (cal or joules). The equation gives the heat, Q, required to raise or lower the temperature from t_1 to t_2 of a substance of mass, m. The specific heat capacity c, is characteristic of the specific substance. (An alternate definition of temperature will be presented in the next section.)

C. The Ideal Gas

Volume-Temperature Relationship—Ideal Gas Thermometer

In the previous section we referred to temperature, implying that it was the intensive property which governs the direction of flow of heat. This heat flow was expressed quantitatively in equation (1-B-1). Temperature can be measured quantitatively by thermometry and expressed on some arbitrary temperature scale, such as the centigrade scale. Although thermometers and the measurement of temperature are familiar to almost everyone, the exact nature and meaning of temperature remains a mystery to many. This is not surprising since temperature is, in fact, a man-made dimension designed to measure a property of matter which can be perceived only through the sense of touch.

The awareness of warmth or lack of warmth (cold) in material objects is ageless in the history of man's knowledge. However, not until it was noted that an increase in warmth was accompanied by an increase in the spatial extent or volume of a material object did it become possible to associate a quantitative dimension or scale with this property. As a result of this observation, it became apparent that the intensity of the warmth of a material object, referred to as its temperature, could be arbitrarily defined by relating a change in the temperature to a change in the volume of the substance being cooled or warmed.

Thermometers function by means of the thermal expansion of a liquid such as alcohol or mercury. Generally, two points are fixed by measurements at predetermined temperatures, such as freezing and boiling points of water. Actually the volume of a liquid does not vary precisely in a linear manner, and a certain error could be introduced, depending on the substance used and the temperature range. Thus, the nonlinear expansion of the liquid with temperature is taken into account in the construction of modern thermometers.

Fortunately we have another thermometer, which may not be as convenient as the sealed liquid thermometer, but has the advantage of being very precise. This is the so-called *ideal gas thermometer*. Its precision does not mean better experimental techniques, but refers to the precise linear relationship between volume and temperature. From introductory chemistry courses, we know that several gases approximate an ideal gas at low pressures, on the order of one atmosphere or less. The specific thermal expansion of an ideal gas is known as **Charles' Law** and a graph of the relationship is shown in Figure 1-C-1. The solid line represents a straight line drawn through the experimental data graphed as volume versus temperature in $^{\circ}C$, with pressure and the number of moles held constant. The extrapolations to zero volume, as shown by the broken lines in Figure 1-1, establish the **absolute temperature scale**. The intercept occurs at t = -273.16$^{\circ}C$.

Actually no real gas is exactly an ideal gas, and an accurate determination of the intercept is obtained in a different manner. The extrapolation to V = 0 is carried out with data at various pressures, such as P_1 and P_2 in Figure 1-1. The extrapolations to the abscissa for each of the pressures are then plotted versus pressure and an extrapolation to zero pressure is carried out. As the pressure approaches zero, the real gas approaches an ideal gas and the extrapolated value in this limit can be taken as the true absolute zero. In this manner the intercept with the t$^{\circ}C$ axis has been precisely determined as -273.16. Charles' Law can then be expressed in terms of absolute temperature: *The volume of a gas of a given number of moles is directly proportional to absolute temperature under the condition of constant pressure.*

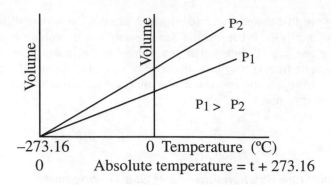

Figure 1-C-1. Charles' Law Plots

$$V \propto T \text{ (constant P and n)} \quad \textbf{Charles' Law} \qquad \text{(1-C-1)}$$

Actually, Charles' Law can be thought of as a restatement of the definition of temperature in terms of the volume.

Charles' Law can be presented in a more useful form by expressing the proportionality as an equality of ratios of volume to temperature.

$$\frac{V_1}{T_1} = \frac{V_2}{T_2} = \frac{V}{T} = \text{constant} \qquad \text{(1-C-2)}$$

The meaning of equation (1-C-2) is that the volume of the gas divided by the absolute temperature is always equal to the same constant value, provided the pressure and number of moles remain constant.

Pressure-Volume Relationship

The behavior of the ideal gas, under the condition of constant temperature, is described by **Boyle's Law**, which states that the *volume of a given number of moles of gas at constant temperature is inversely proportional to the pressure.*

$$V \propto \frac{1}{P} \text{ (constant T and n)} \quad \textbf{Boyle's Law} \qquad \text{(1-C-3)}$$

Boyle's Law can also be expressed in a more useful mathematical form for pressure-volume calculations.

$$P_1V_1 = P_2V_2 = PV = \text{constant} \qquad \text{(1-C-4)}$$

Equation (1-C-4) states that the product of the volume and the pressure of a gas is always equal to the same constant value, as long as the temperature and number of moles remain constant.

Volume-Moles Relationship

An additional relationship known as **Avogadro's Law** must be presented before we can develop the **Ideal Gas Equation of State**. Avogadro's Law states that *if both the temperature and the pressure of a gas remain constant, the total volume of the gas is directly proportional to the number of moles*, n.

$$V \propto n \text{ (constant T and P)} \quad \textbf{Avogadro's Law} \qquad \text{(1-C-5)}$$

Thus, according to Avogadro's Law, if the number of moles of a gas is doubled at the same temperature and pressure, the volume of the gas will also be doubled.

The Ideal Gas Equation of State

The Ideal Gas Law can be derived in a number of ways that are based on the various relationships between pairs of variables discussed previously. In equations (1-C-1), (1-C-3), and (1-C-5) we note that V is directly proportional to T and n and inversely proportional to P. Although this is not a rigorous derivation, one might anticipate that the combination of these laws will yield

$$V \alpha \frac{nT}{P}$$

If we assign the proportionality constant to be the gas constant, R, then

$$V = R \frac{nT}{P} \tag{1-C-7}$$

Rearranging equation (1-C-7) we obtain what is referred to as the Ideal Gas Equation of State

$$\textbf{PV = nRT} \quad \textbf{Ideal Gas Equation of State} \tag{1-C-8}$$

In Section B we defined an *equation of state* as a mathematical relationship between the state variables, P, V, T, n. Obviously the Ideal Gas Law is also an *equation of* state and thus called the Ideal Gas Equation of State. Equation (1-C-8) completely describes any state of an ideal gas. If the values of any three of the four quantities P, V, n, and T are known, the value of the fourth can be calculated from equation (1-C-8).

The gas constant R can be expressed in several units. The value of 0.08205 L-atm/K-mol is the most convenient to use in evaluating the ideal gas equation of state when P is expressed in atmospheres and V in liters, the units most commonly used by chemists. Other units commonly used for pressure and their equivalents are

$$1 \text{ atm} = 760 \text{ mm Hg} = 760 \text{ torr} = 7.60 \times 10^5 \text{ microns (of Hg)} = 14.7 \text{ lb/in}^2$$
$$= 1.01325 \text{ bar} = 0.101325 \text{ MPa}.$$

The use of these equalities in converting units is readily accomplished with dimensional analysis, as illustrated in the following example.

Example 1-C-1. Suppose we have a gas sample, NO, of volume 800 mL, pressure 2.00 microns (μ), and temperature 25°C. How many grams of material do we have?

First, expressing the number of moles in the Ideal Gas Equation of State in terms of mass and molecular weight, we have

$$PV = \frac{m}{M} RT \text{ or } m = \frac{PVM}{RT}$$

where m = mass and M = molecular weight. Substituting in the known quantities, (M = (14.0 + 16.0) = 30.0 g/mole)

$$m = \frac{(2.00 \text{ }\mu)(800 \text{ mL})(30.0 \text{ g/mol})}{(.08205 \text{ liter - atm/deg - mol}(298 \text{ deg}))}$$

The units cancel, except for the pressure and volume. Using the equivalent expressions and solving for the ratios

$$1.00 \text{ atm} = 7.60 \times 10^5 \mu, \qquad \frac{1.00 \text{ atm}}{7.60 \times 10^5 \mu} = 1$$

and

$$1.00 \text{ liter} = 1000 \text{ mL}, \qquad \frac{1.00 \text{ liter}}{1000 \text{ mL}} = 1$$

Since the ratios are unity, we can multiply by them to cancel the pressure and volume units thusly:

$$m = \frac{(2.00 \text{ μ}) \left(\dfrac{1.00 \text{ atm}}{7.60 \times 10^5 \text{ μ}} \right) (800 \text{ mL}) \left(\dfrac{1.00 \text{ liter}}{1000 \text{ mL}} \right) (30.0 \text{ g})}{(0.08205 \text{ liter-atm})(298)}$$

$$m = 2.58 \times 10^{-6} \text{ g}$$

The gas constant can also be expressed in terms of ergs or joules and in a convenient unit of heat, the calorie. The calorie is *approximately* the heat required to raise the temperature of one gram of water one degree centigrade. It is precisely defined in terms of the erg or joule. Though it is not apparent at this time, the product of liters times atmospheres is a measure of energy. Liter-atmospheres are readily converted to ergs, if one recalls that the pressure, expressed in terms of the height of mercury, arises from the force (f) of the mercury per unit area. If A represents the cross-sectional area, then

$$P = \frac{f}{A}$$

The force (f), in turn, is equal to the mass (m) times the acceleration of gravity (g). Thus,

$$P = \frac{mg}{A}$$

If we consider a column of mercury with a constant cross-sectional area (A), the mass of the mercury can be related to the volume, which then can be related to the height (h) of the mercury.

$$m = \rho V = \rho h A$$

where ρ(rho) = density of mercury, and v = volume of mercury. Substituting in the expression for pressure,

$$P = \frac{(\rho h A)g}{A} = \rho h g$$

The PV product, expressed in terms of liter-atmosphere, can be converted to erg units by recalling that 1 atm supports 760 mm Hg.

$$PV = \rho h g V$$

$$1 \text{ L-atm} = (13.55 \text{ g/cm}^3)(76.0 \text{ cm})(980.6 \text{ cm/sec}^2)(1 \text{ L})(1000 \text{ cm}^3)/1 \text{ L}$$
$$= 1.013 \times 10^9 \text{ g-cm}^2/\text{sec}^2$$

Since 1 erg = 1 g-cm^2/sec^2 = 10^{-7} joules (see Table 1, Appendix D)

$$1 \text{ liter-atm} = 1.013 \times 10^9 \text{ g-cm}^2/\text{sec}^2 = 1.013 \times 10^9 \text{ erg} = 101.3 \text{ joules.}$$

The calorie is defined in terms of the joule (absolute) as:

$$1 \text{ calorie} = 4.1840 \text{ joules.}$$

Thus, the gas constant can be expressed in these various units:

$$R = 0.08205 \text{ liter-atm/K-mol}$$
$$= 8.3142 \times 10^7 \text{ erg/K-mol}$$
$$= 8.3142 \text{ joules/K-mol}$$
$$= 1.987 \text{ cal/K-mol}$$

Problem 1-C-1. Nitrogen is sold compressed in cylinders. If a cylinder is filled to 2000 lb/in^2 pressure at 27°C, how many kilograms of nitrogen does it contain if the volume of the cylinder is 4.1 ft^3? How many liters would this occupy at one atmosphere pressure at 27°C? Assume the gas obeys the Ideal Gas Equation of State.
Answer: 18.0 kg, 1.58x10^4 L.

D. Real Gases

Deviations from Ideality

Under conditions of relatively high temperature and low pressure, the behavior of many real gases is adequately described by the Ideal Gas Law. However, modern equipment and techniques have made possible the study of the behavior of gases over wide temperature and pressure ranges. A dramatic deviation from that predicted on the basis of the Ideal Gas Law has been observed as the temperature is decreased and/or the pressure increased to any great extent. Moreover, the character and degree of the deviation from the ideal gas (ideality) is different for each particular gas. The nature of the deviation from ideality is not only dependent on the particular gas, but also on the temperature range and pressure range being considered. In other words, the variation of the volume of a given gas as a function of the pressure at constant temperature, will, in general, differ with each temperature chosen.

The behavior of real gases can be compared to ideal behavior in a number of ways. A convenient method of comparison is a graphic representation of pressure-volume data at constant temperature, in which the experimentally determined values of $(P\overline{V}/RT)$ are plotted as a function of the pressure. $\overline{V}$ is defined as the molar volume = V/n. Figure 1-D-1 illustrates this graphic technique for several common gases.

It is advantageous to plot the values of the quotient $(P\overline{V}/RT)$ for several reasons. Since $P\overline{V} = RT$ for an ideal gas, $P\overline{V}/RT$ must be equal to unity for all values of P and T, as represented by the broken line in Figure 1-D-1. Hence, the degree of deviation of the plots for *real gases* is immediately obvious from the graph. In addition, the value of $P\overline{V}/RT$ at any point on one of the curves constitutes a direct comparison of the *actual* molar volume of the gas at the particular pressure and temperature, to the molar volume of an *ideal* gas at the same temperature and pressure.

Since we are concerned with experimentally determined values of $P\overline{V}/RT$, we may indicate this by using the subscripted variables, V_{act}, where the subscript "act" means "actual."

$$\frac{P\overline{V}}{RT} = \frac{P\overline{V}_{act}}{RT} = \frac{\overline{V}_{act}}{RT/P}$$

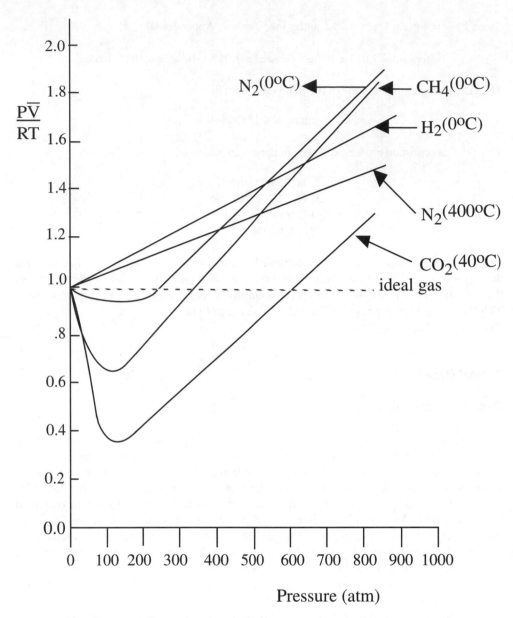

Figure 1-D-1. Pressure-Volume Product for Real Gases (Constant Temperatures)

However, since the product RT divided by P is equal to the molar volume the gas would have if it were ideal, then,

$$\frac{P\overline{V}}{RT} = \frac{\overline{V}_{act}}{\overline{V}_{ideal}}$$

The quantity ($P\overline{V}/RT$) is usually referred to as the *compressibility factor* and given the symbol Z.

$$Z = \frac{P\overline{V}}{RT} = \frac{\overline{V}_{act}}{\overline{V}_{ideal}} \qquad \textbf{Compressibility Factor} \qquad (1\text{-}D\text{-}2)$$

Thus, the value of Z, compressibilty factor, at any pressure and temperature gives the ratio of the actual molar volume of the gas to the molar volume of an ideal gas at that pressure and temperature. When the plot of Z versus P lies below one, the actual molar volume of the gas is smaller than the ideal molar volume.

Figure 1-D-1 also illustrates the fact that the nature of the deviation of real gases from ideal behavior is dependent both on the particular gas and on the temperature. In the case of hydrogen at $0^{\circ}C$, the curve is almost a straight line lying above the ideal gas line. On the other hand, the curve for carbon dioxide at $40^{\circ}C$, illustrating an extreme case, first dips far below the ideal gas line to exhibit a minimum before finally rising above unity. It is important to note that the complexity of the plots of Z versus P correlates in a sense with the **critical temperature** of the gas, that is, the temperature above which liquefaction cannot occur, regardless of how great the applied pressure on the gas. The closer the temperature of the gas is to its critical temperature, the more dramatic is the deviation of its Z curve from $Z = 1$ (ideal gas). In the case of several gases at the same temperature, such as those illustrated in Figure 1-D-1, the gas with the highest critical temperature will exhibit the greatest deviation from ideal gas behavior. The most important aspect concerning the deviation of the isothermal Z graph from $Z = 1$ appears to be the increase in the temperature of the gas above its critical temperature.

Other Equations of State

It may be noted in Figure 1-D-1 that the graphs of Z versus P show a linear region in the vicinity of $P = 0$. Obviously this initial linear region has a slope which is a function of temperature. For a given isotherm the linear equation can be represented by

$$Z = \frac{P\overline{V}}{RT} = B'P + 1 \qquad (1\text{-}D\text{-}3)$$

where B' = slope and the intercept is one. Since the slope changes with temperature we must remember that B' is a function of temperature, B'(T). For each gas B' will be different and the temperature dependence will also be different. Equation 1-D-3 can be arranged to a form that shows similarity to the ideal gas equation of state. Multiplying by RT

$$P\overline{V} = (B'RT)\,P + RT$$

Since B' is a function of temperature we can define a new B which is also a function of T

$$B = B'RT$$

and

$$P\overline{V} = RT + BP \qquad (1\text{-}D\text{-}4)$$

If we want to include n as a variable

$$P\frac{V}{n} = RT + BP$$

$$\mathbf{PV = nRT + nBP} \qquad (1\text{-}D\text{-}5)$$

Note the similarity of equation (1-D-5) with the ideal gas equation of state. The n BP is a single "correction" term to the ideal gas equation of state.

Many equations of state have been proposed which describe the behavior of real gases more precisely than the Ideal Gas Law. Some of these have been obtained by the empirical analysis of specific data such as equation (1-D-5), and, hence, are limited in their usefulness. Others, which find more general application, have been suggested by arguments based on various theoretical ideas. One of the better-known, useful equations of state is

$$\left(P + \frac{n^2 a}{V^2}\right)(V - nb) = nRT \qquad \textbf{Van der Waals' Equation of State} \qquad (1\text{-}D\text{-}6)$$

The constants a and b in equation (1-D-6) have unique values for each and every substance, hence, they must either be known or determined for a gas before van der Waals' Equation can be employed for P-V-T calculations.

The units of the constants in equation (1-D-6) can be determined through dimensional analysis. Since each factor on the left side of the equals sign in equation (1-D-6) is a sum of terms, each of the two terms in either of the factors must have the same units. Thus, the term n^2a/V^2 must have the units of pressure, and the term nb must have the units of volume. If the pressure is taken in atmospheres and the volume in liters, then

$$\frac{n^2a}{V^2} = \frac{(moles)^2(a)}{(liters)^2} = atmospheres$$

or

$$a = \frac{(liters)^2(atmospheres)}{(moles)^2}$$

and

$$nb = (moles)(b) = liters$$

or

$$b = \frac{liters}{mole}$$

Comparison of Van der Waals' Equation and the Ideal Gas Law

Although van der Waals' Equation is far from the ultimate equation of state, there are several advantages for considering it:

1. Van der Waals' Equation provides a better description of the behavior of real gases than the Ideal Gas Law, while still in a relatively simple form that is not too difficult to apply.

2. The validity of van der Waals' Equation can be supported with plausible arguments based on realistic modifications on the Kinetic Molecular Theory of gases.

3. Van der Waals' Equation is reasonably consistent with experimental knowledge of the critical state of substances. To a limited extent, it is capable of describing the liquid state near the critical point. We will consider this point shortly.

Some elaboration on number two should be valuable in providing insight into the mathematical form of van der Waals' Equation. It can be shown that the Ideal Gas Law can be theoretically derived from the postulates of the basic Kinetic Molecular Theory. The interested reader should consult a more comprehensive textbook on physical chemistry for this derivation such as P. Atkins, Physical Chemistry, W.H. Freeman. The derivation of the Ideal Gas Law in the Kinetic Molecular Theory assumes that molecules are *mass-points*, that is, dimensionless masses which occupy no space themselves. This postulate, in effect, says that the entire volume of a container is available for each and every molecule to move about in. But, since molecules, in fact, do occupy space, the actual volume available to each molecule should be equal to the total volume of the container minus some quantity, say nb, representing the volume of space used up by the existence of n moles of molecules, where b is proportional to the volume of a single molecule multiplied by Avogadro's number. Thus, the volume in the Ideal Gas Law is changed to the actual volume available to the n moles of molecules by replacing V with (V - nb).

The basic Kinetic Molecular Theory further postulates that molecules do not interact with one another, except during collisions, and hence, does not admit the possibility of the existence of intermolecular forces of attraction. There is an abundance of information which suggests that these forces do exist, so that molecules also influence each other over a distance, as well as during collisions. Analysis of the motion of molecules shows that the pressure exerted on the walls of a container due to the collisions of these molecules would be reduced as a direct result of the intermolecular attraction between molecules. It can be argued that the decrease in pressure, from the ideal situation, should be proportional to $1/\overline{V}^2$ or equal to $a(n^2/V^2)$, where a is a *constant of proportionality*. Consequently, the pressure given in the Ideal Gas Law would be greater than the actual pressure, all else being constant. To compensate, P in the Ideal Gas Law is replaced by the term

$$P + \frac{n^2 a}{V^2}$$

The preceding discussion is not intended as a proof of the van der Waals' Equation, but only to show the reader that it can be viewed as a modification of the Ideal Gas Law that is more precise.

Problem 1-D-1. Using the number of moles of N_2 obtained in Problem 1-C-1, calculate the pressure in the tank, according to the van der Waals' Equation of State. See Table 1-D-1 for the constants a and b. Answer: 132 atm.

Problem 1-D-2. Expand the left side of van der Waals' Equation (1-D-6), transposing all but the PV term to the right side of the equals sign. Then, show that as the pressure approaches zero and the temperature becomes infinitely large, van der Waals' Equation reduces to the Ideal Gas Law. Hint: Assume the ideal gas equation in the correction terms since they are small in magnitude.

Critical Temperature, Pressure, Molar Volume

When sufficiently high pressures are applied to gases, the molecules come so close together that the attractive forces increase to the point where the gas condenses to a liquid or a solid. However, this phenomenon does not occur if the temperature of the gas is sufficiently high. Generally increasing the temperature diminishes the attractive forces between molecules and this tends to inhibit condensation to a liquid or a solid. For all gases there is a unique temperature above which the gas cannot be condensed regardless of the pressure applied. This is called the *critical temperature*, T_c. There is also a *critical pressure*, P_c, and *critical molar volume*, V_c, which along with T_c define the *critical point*.

The *critical point* can best be described by examination of a graph of pressure versus volume for a gas at various temperatures (isotherms). In Figure 1-D-2 an isotherm at T_1 is shown where the $T_1 > T_c$. Note that the isotherm at T_1 shows no discontinuities or points at which the slope goes to zero. Along the isotherm T_1 there is only the gas phase. At T_3 the isotherm shows two discontinuities. If one follows the T_3 isotherm from low pressure, high volume back towards high pressure we first note a discontinuity at point A where the slope becomes zero. At this point A the gas begins to condense to a liquid and the pressure is the vapor pressure of the liquid at this temperature T_3. As the volume decreases the pressure remains constant since the pressure of the gas is constant, equal to the vapor pressure. In the region from A to B both liquid and gas are present, only the relative amounts change. At point B all of the gas has been condensed to liquid and the volume is now

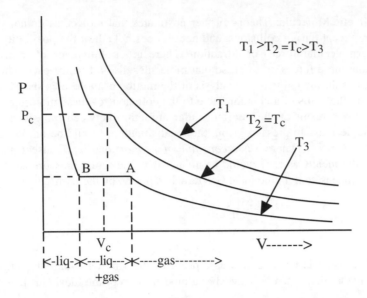

Figure 1-D-2. Graphs of P versus V at Various Isotherms Including the Critical Temperature T_c

decreased by increasing the pressure on the liquid. Since liquids are very incompressible, the pressure rises rapidly for a small change in volume.

At an intermediate temperature T_2 there is a *unique* isotherm which shows a *point of inflection*. Along this isotherm T_2 there are no discontinuities as observed along T_3, thus the gas does *not* condense to a liquid. At any temperature $< T_2$, the isotherm would show discontinuities, such as noted at T_3. The temperature of the isotherm which shows the point of inflection is called the *critical temperature*, T_c. The point of inflection defines the critical pressure, P_c, and the critical volume, V_c, as noted in Figure 1-D-2. From Figure 1-D-2 it is obvious that the definition of the critical temperature is "The temperature above which a gas cannot be condensed to a liquid regardless of the pressure applied."

Every gas has its unique critical values and the values of T_c range from very low values, well below room temperature where gases remain as gases, to very high temperatures where the gas is condensed to a liquid even at room temperature and P = 1 atm. Take for example H_2 and H_2O which have quite extreme values of T_c. The critical temperature of H_2 is 33.2 K = -239.9°C and that for H_2O is 647.2 K = 374.1°C. In order to liquefy H_2 the temperature must be less than 33.2 K and obviously H_2 will remain a gas at ambient temperatures since the temperature would never get below T_c = 33.2 K. On the other hand, H_2O has a very high T_c = 647.2 K and at ambient conditions it would exist as a liquid in equilibrium with the gas providing $P = P_{vap}$. If the $P < P_{vap}$, the liquid water would vaporize to the gas phase. In order to assure H_2O being in the gas phase regardless of pressure the temperature must be increased so that $T > T_c$ = 647.2 K.

As mentioned previously, the van der Waals' equation of state represents reasonably well the experimental P, $\overline{V}$, T relationship in the vicinity of the critical point. A point of inflection is characterized mathematically by the fact that both the first and second derivatives are zero. Van der Waals' equation can be differentiated to find the first and second derivatives of P with respect to $\overline{V}$ along the critical temperature isotherm. If one evaluates these derivatives at the critical point where they are zero (point of inflection), then these two equations can be combined with van der Waals' equation itself evaluated at the critical point to give three equations.

$$\left(P_c + \frac{a}{\overline{V}_c^2}\right)(\overline{V}_c - b) = RT_c$$

$$0 = \left(\frac{\partial P}{\partial V}\right)_T \text{evaluated at Pc,Vc, Tc}$$

$$0 = \left(\frac{\partial 2P}{\partial V2}\right)_T \text{evaluated at Pc, Vc, Tc}$$

These three equations can be solved simultaneously to obtain the relationship between P_c, V_c, T_c and the three constants in van der Waals' equation a, b, R. These relationships are:

$$T_c = \frac{8a}{27Rb} \qquad \overline{V}_c = 3b \qquad P_c = \frac{a}{27b^2} \qquad (1\text{-}D\text{-}7)$$

Solving for a and b,

$$b = \frac{\overline{V}_c}{3} \qquad a = \frac{9}{8}R\,\overline{V}_c\,T_c \text{ or } 3\,\overline{V}_c^2 P_c \qquad (1\text{-}D\text{-}8)$$

Thus, the van der Waals' constants for any gas can be calculated from equations (1-D-8), provided its critical constants (T_c, V_c, P_c) are known. Table 1-D-1 contains values of van der Waals' constants for several common substances, along with its critical values.

Table 1-D-1 Van der Waals' Constants and Experimentally Determined Critical Values

Substance	a(bar L^2/mol^2)	b(L/mol)X10^2	T_c(K)	P_c(MPa)	V_c (cm^3/mol)
He	0.0346	2.38	5.19	0.227	57
Ne	0.208	1.672	44.4	2.76	42
Ar	1.355	3.201	150.87	4.898	75
Kr	2.325	3.96	209.4	5.50	91
Xe	4.192	5.156	289.73	5.84	118
H$_2$	0.2453	2.651	32.97	1.293	65
N$_2$	1.370	3.87	126.21	3.39	90
O$_2$	1.382	3.186	154.59	5.043	73
CO$_2$	3.656	4.286	304.14	7.375	94
H$_2$O	5.537	3.049	647.14	22.06	56
SO$_2$	6.865	5.679	430.8	7.884	122
CH$_4$	2.300	4.301	190.53	4.604	99
C$_2$H$_6$	5.570	6.499	305.4	4.884	148
C$_3$H$_8$	9.385	9.044	369.82	4.250	203
C$_4$H$_{10}$	13.93	11.68	425.14	3.784	255

The units shown for van der Waals' constants and the critical values are those used in the 74th Edition CRC Handbook of Chemistry and Physics, 1993-94. The following conversion factors can be used to convert pressure to atmospheres: 1 atm = 1.01325 bar = 0.101325 MPa.

Note in Table 1-D-1 that van der Waals' constants, both a and b, increase with increasing size of the molecules or atoms. This is especially apparent when one examines the series of inert gases: He, Ne, ... or the hydrocarbons: CH$_4$, C$_2$H$_6$, ... The only exception for the inert gases is the unusually large value of b for He and low value for Ne. One would expect the magnitude of "a" to increase with size since the number of electrons/molecule or atom increases and, consequently, the induced dipole-induced dipole (London) forces would increase. Polar molecules such as H_2O would be expected to have unusually large "a" values compared to a non-polar molecule such as CO_2. The values given in Table 1-D-1 are experimental and are subject to error. Furthermore, the values of a and b are those that give a good representation of the P, V, T data according to equation (1-D-6), not necessarily the data at the critical point. Consequently, calculations using equations (1-D-7) and (1-D-8) would not be expected to give precise agreement with the experimental results in Table 1-D-1. Also, since R is taken as the fixed gas constant, the three equations in (1-D-8) relating T_c, $\overline{V}_c$, P_c to the constants "a" and "b" is over determined and the two equations for "a" will not necessarily give the same values.

Problem 1-D-3. Calculate T_c, $\overline{V}_c$, P_c for N_2 from van der Waals' constants using equations (1-D-7). Answer: $T_c = 126$ K, $\overline{V}_c = 116.1$ cm^3/mol, $P_c = 33.4$ atm = 3.38 MPa.

Problem 1-D-4. Calculate van der Waals' constants for CO_2 from the critical values using equations (1-D-8). Answer: $a = \dfrac{9}{8} R \overline{V}_c T_c = 2.674$ bar L^2/mol^2, $a = 3 \overline{V}_c^2 P_c = 1.955 \dfrac{barL^2}{mol^2}$, $b = 3.13 \times 10^{-2}$ L/mol.

E. Work

When force is exerted through a distance to bring about a change, we say that **work** is done. Mathematically, work is defined as a force times a displacement—in other words, that quantity necessary to cause a displacement against a given force. We usually think of work as mechanical work, where the displacement is an actual displacement in distance. However, there can be different forms of work, such as electrical work, where the force is the electrical force, ε, and the displacement is a transfer of charge, Δq. The differential of work, in this case for the displacement of an infinitesimal amount of charge dq, is

$$dW_{elec} = -\varepsilon dq \qquad (1\text{-}E\text{-}1)$$

where W_{elec} is the electrical work.

In general, thermodynamics is concerned with all types of work associated with a system; however, in some cases only a particular type can be considered. For example, there is an expression for magnetic work which involves the displacement of magnetization in a magnetic field of given strength. However, if no magnetic field is present, there cannot be any work of this nature.

At present we are concerned only with mechanical work—actually, a *restricted* type of mechanical work. *Mechanical work* is the product of force times displacement, with the restriction that *the force is in the direction of the displacement*. The expansion of a gas against an external pressure is an important example of mechanical work under the above restriction. Figure 1-B-2 previously showed the expansion of a gas in a cylinder, against a frictionless piston which supplies a given resisting force against such expansion. The initial state in Figure 1-B-2(a) illustrates the force of a piston resulting from the two weights placed on top of the piston. When the piston is stationary, the gas exerts an internal pressure which balances the pressure resulting from the piston and weights. The mechanical work (dW) associated with the displacement ($d\ell$) *in the direction of the force, f,* is given by

$$dW = -f \, d$$

(1-E-2)

Since pressure is force/unit area, the external pressure (P_{ex}) from the piston is given by

$$P_{ex} = \frac{f}{A} \text{ or } f = P_{ex}A$$

(1-E-3)

where A = cross-sectional area of the cylinder. Substituting (1-E-3) into (1-E-2),

$$dW = - P_{ex}A \, d$$

The displacement d can be expressed in terms of a change in volume (dV), since V = A and the cross-sectional area A is constant. Thus,

$$dV = A \, d$$

(1-E-4)

or

$$d = \frac{dV}{A}$$

and our expression for the work that results from a small change in volume becomes

$$\mathbf{dW = - P_{ex}dV}$$

(1-E-5)

Note that the sign for work is negative when there is an expansion, since dV is positive. A negative sign for work infers that the system is doing work on the srroundings, and correspondingly a positive value of work infers that work is done on the system by the surroundings.

Figure 1-B-2(a) describes two states of the gas at the same temperature: an initial state A, shown with an external pressure, P_1, corresponding to the two weights shown, and the final B, with the external pressure reduced to P_2 by the removal of one of the weights. We might ask, What is the work associated with the expansion from (P_1, V_1) to (P_2, V_2)? Graphically we can plot (V_1, P_1) and (V_2, P_2), as shown in Figure 1-E-1; however, the work cannot be calculated, until we have specified the curve, P_{ex}, connecting the points (V_1, P_1) and (V_2, P_2). In other words, the *work is dependent upon the path* going from state 1 to state 2. Let us visualize various paths that could be followed in going from one state to the other. Note that an expansion gives a negative work which designates work done by the system on the surroundings.

Expansion against a constant pressure: $P_{ex} = P_2$. In this case we visualize the immediate removal of one of the weights so that P_1 suddenly changes to P_2, which then remains constant throughout the expansion as shown in Figure 1-B-1(a).

Graphically this is shown as Path #1 in Figure 1-E-1. The work for this path is

$$W = - P_{ex} \int_{V_1}^{V_2} dV = - P_2(V_2 - V_1)$$

(1-E-6)

In Figure 1-E-1, it is represented by the darker-dotted area of the rectangle.

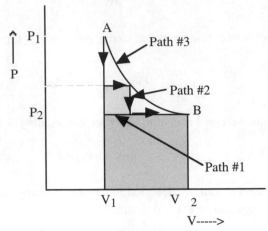

Figure 1-E-1. Pressure versus Volume for Expansion of Gas by Three Paths:
Path #1: $P_{ex} = P_2$: one step $P_{ex} = P_2$
Path #2: two steps: $P_{ex} = (P_1 + P_2)/2$ and $P_{ex} = P_2$
Path #3: infinite number of steps: $P_{ex} = P$

Expansion by removal of one weight in two steps, 1/2 weight each time: One of the
weights is separated into two equal parts. The expansion is carried out by removing these
two weights one at a time as depicted in Figure 1-B-1(b). Graphically it is represented by
Path #2 in Figure 1-E-1. The work for this process is the area under the two step curve #2.

Expansion by successive removal of infinitesimal weights: This process is depicted in
Figure 1-B-1(c) where extremely small weights are taken off at each step so that the path
approaches the smooth curve from (V_1, P_1) to (V_2, P_2). The work in this process is the
total area under the curve from (V_1, P_1) to (V_2, P_2) in Figure 1-E-1. It is obviously bigger
than either of the previous cases, and is, in fact, the *maximum work* that can be done.

This last process, a most important one, is a **reversible process**, that is, one in which
the system is in equilibrium with its surroundings at all times. Indeed, if we remove only
an *infinitesimal* weight, the system will remain in equilibrium and the external and internal
pressures will be equal, thus

$$P_{ex} = P_{gas} = P$$

Since the area under a curve between limits is defined by the definite integral over the
limits, we can write

$$W = \int dw = - \int_{V_1}^{V_2} P_{ex}dV = - \text{ area under the curve of } P_{ex} = f(V) \qquad (1\text{-E-}7)$$

$$\text{from } V_1 \text{ to } V_2$$

Equation (1-E-7) is applicable to any path, $P = f(V)$, but it gives the maximum
possible work only when the path is reversible (in other words, only when $P = f(V)$ is the
locus of points corresponding to equilibrium states of the system). Generally, an equation
of state is necessary in order to calculate the maximum work by equation (1-E-7). For
example, if the gas behaves like an ideal gas, then, $PV = nRT$ is the desired relationship.
Solving for P gives

$$P = \frac{nRT}{V}$$

Substituting into (1-E-7),

$$W = - \int_{V_1}^{V_2} \frac{nRT}{V} dV$$

For an isothermal process (T = constant), we thus obtain:

Isothermal Reversible Work for an Ideal Gas

$$W = - nRT \int_{V_1}^{V_2} \frac{dV}{V} = - nRT \ln \frac{V_2}{V_1} \qquad (1\text{-}E\text{-}8)$$

This expression for work gives the area under the smooth curve in Figure 1-E-1, path #3, provided the gas obeys the ideal gas equation of state, in which case the curve is a plot, of P = nRT/V versus V. The concept of reversibility is an important one, and the reversible expansion of a gas is a good example of mechanical reversibility. Later we will consider chemical reactions which are carried out reversibly.

Equation 1-E-7 gives a general definition of work, regardless of the path taken, whereas equation (1-E-8) was derived for the reversible path P = nRT/V. The expansion under constant pressure ($P_{ex} = P_2$) could similarly be followed, and equation (1-E-7) becomes:

Work of Expansion against a Constant Pressure ($P_{ex} = P_2$)

$$W = - \int_{V_1}^{V_2} P_{ex} dV = - \int_{V_1}^{V_2} P_2 dV = - P_2 \int_{V_1}^{V_2} dV$$
$$W = - P_2(V_2 - V_1) \qquad (P_{ex} = P_2) \qquad (1\text{-}E\text{-}6)$$

This is identical to equation (1-E-6) previously derived for this constant pressure expansion.

Problem 1-E-1.

(a) Consider one mole of an ideal gas at constant t = 200°C. Show, from a P versus V plot, how the reversible work can be calculated for the compression of the gas from P = 1.0 atm to 10 atm. Evaluate the work numerically.

(b) Show that the work involved in an isothermal, reversible compression of n moles of an ideal gas is

$$W = - nRT \ln \frac{V_2}{V_1} = - nRT \ln \frac{P_1}{P_2}$$

Evaluate the work for the compression of one mole from 1.0 atm to 10 atm and compare with (a). Answer: (a) and (b) 89.4 L-atm = 2.16 kcal = 9.057 kJ.

Problem 1-E-2. Calculate the work that would result if the ideal gas in Problem 1-E-1 were compressed irreversibly by *suddenly* changing the initial pressure from P = 1.0 atm to P = 10 atm, and allowing the gas to compress at P = 10 atm. Answer: 349.2 L-atm = 8.46 kcal = 35.4 kJ.

F. Heat

We have previously defined heat as that quantity required to raise the temperature of a substance from t_1 to t_2, but we can also express heat change as a differential quantity, dQ. Considering temperature, T, as the only independent variable, dQ can be expressed as

$$dQ = \frac{dQ}{dT}\ dT = mc\ dT = nC\ dT \qquad (1\text{-}F\text{-}1)$$

where c = specific heat capacity (cal or joules/g-deg), C = molar heat capacity (cal or joules/deg-mol), m = mass in grams, and n = number of moles. Since dQ/dT will vary with temperature, the heat capacities, c and C, will also vary with temperature.

As for dW, the dQ in equation (1-F-1) is an *inexact* differential and the integral of dQ is generally dependent on the path. In Chapter 2 the molar heat capacity will be defined with certain restrictions on the process (namely, constants P and V) which will make dQ an exact differential. In general, it should be remembered that when an integral is dependent on the path, the differential is an inexact differential.

G. Summary

From your study of this chapter you should have a good understanding of the following principles.

1. Using the Ideal Gas equation of State, $PV = nRT$, you should be able to calculate the value of any variable P, V, T, or n providing the other three variables are specified.

2. Understand how the Compressibility Factor $Z = \dfrac{PV}{RT}$ varies with pressure and the reason for the deviation from the ideal gas equation.

3. Define the critical state based upon $\left(\dfrac{\partial P}{\partial V}\right)_T = 0$ and $\left(\dfrac{\partial^2 P}{\partial V^2}\right)_T = 0$.

4. Identify the Supercritical fluid state.

5. Calculate work, W, for a reversible process $(P = P_{ex})$ and an irreversible process $(P \neq P_{ex})$ from the definition of mechanical work, $dW = -P_{ex}dV$.

Exercises

1. (a) A one liter flask is filled with helium to a pressure of one atmosphere at room temperature of 25^oC. Assuming helium obeys the ideal gas equation of state, calculate the moles of helium in the flask.
   (b) Evaluate the partial derivative
   $$\left(\frac{\partial n}{\partial T}\right)_{P,\,V}$$
   for an ideal gas at : P = 1 atm, t = 25^oC, V = 1 liter.
   (c) If the temperature in part (a) was in error by 0.1^oC, what error would there be in the calculation of the number of moles? <u>Use a differential expression</u>.
   Answer: (a) 0.0409 moles (b) $-\dfrac{PV}{RT^2} = -0.000137$ mol/K (c) -1.37×10^{-5} moles.

2. Sketch the isotherm (P versus V) at the critical temperature (T_c) for CO_2 ($T_c = 304.2$ K, $P_c = 72.8$ atm, $V_c = 0.094$ L/mole). Point out the mathematical significance of the isotherm and identify the critical pressure (P_c) and critical volume (V_c).

3. The work required to stretch a rubber band is given directly by equation (1-E-2). The force, f, is the restoring force, which is in the direction *opposite* to the direction of displacement. For this reason, the force is negative for an increase in length, L (stretch). The force generally increases with increasing length, and in this problem we will assume that the force increases linearly with displacement (L - L_0).

$$f = -kT(L - L_o)$$

where L_o is the length of the unstretched rubber band ($f = 0$) and T is the absolute temperature. Derive an expression for the work required to stretch the rubber band reversibly under isothermal conditions from L_1 to L_2. Note the sign of W when work is being done on the system ($L_2 > L_1 > L_0$). Then, on an f versus L graph, show the work required for this expansion.

Answer: $W = \dfrac{kT(L_2^2 - L_1^2)}{2} - kTL_0(L_2 - L_1)$.

4. (a) If the stretching of the rubber band in Exercise 3 was done under a *constant* force, so that

$$f = -kT(L_2 - L_o)$$

calculate the work required for the process from L_1 to L_2.
(b) Compare the result from (a) with the reversible work found in Exercise 3. Is the reversible work a minimum for this process? Show this on a graph of f versus L.
Answer: (a) $W = kT(L_2 - L_o)(L_2 - L_1)$ (b) Work required to stretch the rubber band under a constant force is greater than that required for the reversible process in Exercise 3.

5. The average velocity of a molecule is given by the formula

$$\bar{c} = \left(\frac{8RT}{\pi M}\right)^{1/2}$$

where R = gas constant, M = molecular weight. Using the value for R in ergs/deg-mol, calculate $\bar{c}$ for benzene at 25°C in cm/sec.
Answer: $\bar{c} = 2.8 \times 10^4$ cm/sec.

6. The rate at which molecules can effuse through a small opening is given by

$$\frac{dn}{dt} = \frac{\bar{c}A}{4}\left(\frac{n}{V}\right)$$

where A = area of the opening, $\bar{c}$ = average speed, V = volume of the container, n = moles. Calculate the rate at which benzene molecules effuse through an opening 1 μ^2, when the pressure of benzene is equal to its vapor pressure at 25°C ($P_{vap} = 95$ mm). Assume that benzene obeys the Ideal Gas Equation of State and use this to calculate the concentration (n/V) of benzene vapor. Express the rate in mol/sec.
Answer: 3.6×10^{-10} mol/sec.

7. Calculate the work involved in expanding an ideal gas from an initial state ($P_1 = 10$ atm, T = 1000K), under a constant pressure of 1 atm. Assume that the process is isothermal and that there are 2 moles of gas.
Answer: $W = -3.58 \times 10^3$ cal $= -1.50 \times 10^4$ J.

8. Suppose that a gas at P = 1 atm and V = 10 L is compressed isothermally by suddenly increasing the external pressure to 10 atm (path A), so that after compression the volume is 1 L. Then, the gas is allowed to expand isothermally against a constant pressure of 1.0 atm until it reaches a final volume of 10 liters (path B), thus returning to its initial state. Calculate the net work.

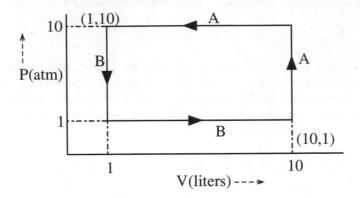

Answer: 81 L-atm = 1.96 kcal = 8.21 kJ.

9. A simple nonideal gas equation of state is given by

$$PV = nRT + nbP.$$

(a) Show that the isothermal, reversible work is given by

$$W = -nRT \ln \frac{(V_2 - nb)}{(V_1 - nb)}$$

(b) If b = .050 L/mole, evaluate the work for the compression in Problem 1-E-1 from 1.0 to 10 atm, at t = 200°C, and compare answers.

(c) Calculate the reversible work of expansion from 0.5 liter to 5.0 liters at 200°C for both 1 mole of an ideal gas and the nonideal gas given by the above equation of state. Compare these results. Answer: (b) The work will be the same, W = 9.06 kJ; however, the initial and final volumes will be different. (c) W = -9.43 kJ for nonideal gas, W = 9.05 kJ for ideal gas.

10. A gas obeys the equation of state

$$PV = nRT + nbP$$

where b is a function of temperature.

(a) Derive an expression for

$$\left(\frac{\partial P}{\partial V} \right)_T$$

(b) Derive an expression for

$$\left(\frac{\partial P}{\partial T} \right)_V$$

(c) If the temperature is increased by a small increment, ΔT, derive an expression for the corresponding incremental change in pressure, ΔP.

Answer: (a) $- \dfrac{nRT}{(V - nb)^2}$ (b) $\dfrac{nR}{V - nb} + \dfrac{n^2 RT}{(V - nb)^2} \dfrac{db}{dT}$

(c) $\Delta P = \left(\dfrac{nR}{V - nb} + \dfrac{n^2 RT}{(V - nb)^2} \dfrac{db}{dT} \right) \Delta T.$

11. A real gas is expanded isothermally and the following P-V data were observed:

P (atm)	V (L)
91.0	1.00
46.0	2.00
23.5	4.00
16.0	6.00
11.5	8.00
10.0	10.00

Show how you would calculate the *maximum* work performed by the gas in expanding from 1.00 L to 10.00 L. Hint: Use the trapezoidal approximation, described in Appendix C, to obtain the area under the curve. Answer: Graph P versus V and evaluate the area under the curve from V= 1.00 L to V = 10.00 L. Using the trapezoidal approximation W~ - 226.5 L-atm = - 5.485 kcal = - 22.95 kJ.

12. Consider the following equation of state, where "a" is a constant:

$$\left(P + \frac{a}{V^2}\right) V = RT \quad \text{(one mole)}$$

(a) Evaluate $(\partial P/\partial V)_T$ for this equation of state.
(b) Show mathematically that this function has a maximum in a graph of P versus V.
(c) Derive an expression for W for an isothermal, reversible expansion for a gas with this equation of state.
(d) Graph P versus V in the region V = 0 to V = 5 L. Take P = - 20 atm as the lower limit for P.

   Answer: (a) $(\partial P/\partial V)_T = - RT/V^2 + 2a/V^3$ (b) $(\partial P/\partial V)_T = 0$ and $(\partial P^2/\partial V^2)_T < 0$
(c) $= - RT \ln(V_2/V_1) - a(1/V_2 - 1/V_1)$ (d) maximum at V = 0.288 L.

2

THE FIRST LAW OF THERMODYNAMICS

*In the author's opinion the greatest single problem facing civilization in the long term is the lack of a renewable energy source. This, coupled with the increasing population, could lead to disaster in regards to supplying the basic needs of food, comfort, and transportation. As everyone knows, food production is highly energy dependent and with the diminishing supply of fossil fuels, the disasterous outcome is obvious. In Chapters 2 and 3 the basic principles of thermodynamics are presented which lay the background for predicting the efficiency of using fuels to generate electricity or produce work. For example, it is well known that fuel cells have a high efficiency for converting fuels to electrical energy. However, only fuel cells that use hydrogen have been developed and hydrogen is not a readily available fuel. Procedures to produce hydrogen efficiently from other raw materials have not been found. The use of fossil fuels in fuel cells would greatly prolong our existence based on fossil fuels, but the technology in this effort has not been worked out. Again the gain in efficiency using fossil-fuel based fuel cells is well understood based on thermodynamic principles. However, in the long term we must find a renewable source such as **solar energy** or **nuclear fusion**. The technology for the practical use of these sources is an arduous task, but one that should be addressed immediately.*

A. The Objectives and Scope of Thermodynamics

In general, thermodynamics is concerned with the energetics of various processes—energy in many different forms, such as work, heat, and free energy. The subject deals with essentially all processes; however, our discussions will be restricted to chemical processes and other physical processes which are helpful in the study of chemical reactions. A more appropriate title for our subject would be *chemical thermodynamics*.

In a sense, the word thermodynamics is misleading, in that dynamics is generally associated with dynamic systems, or systems involving moving objects. However, this is *not* true of thermodynamics. On the contrary, thermodynamics is associated with *static* systems or states, and is concerned only with the *possibility* of the process occurring between these states and the manner in which this is related to the energetics of the processes involved. It is not usually concerned with the *actual* process. For example, thermodynamic parameters can be calculated, and predictions for a given chemical reaction can be made, although in practice, the reaction may not even proceed at all under the present conditions. It is important to remember this point, since it differentiates two distinct subjects, namely chemical thermodynamics and chemical kinetics.

There are two *primary objectives of chemical thermodynamics:*

Establishment of criterion for spontaneity. Here spontaneity refers to a process that can occur without external aid, that can proceed in *some* period of time. Therefore, as this objective, we wish to establish some quantitative criterion to judge whether a given process is spontaneous.

Calculation of equilibrium yield. In a given chemical reaction, an equilibrium between reactants and products is eventually established. As the second objective of chemical thermodynamics, we wish to establish a quantitative relationship between a thermodynamic parameter and this equilibrium, such that we can predict the equilibrium yield. The *equilibrium yield* is the *maximum* yield of products that can be attained. The actual yield depends on the extent of the reaction at the time of measurement. Since the processes discussed here include not only chemical, but physical transformations, such as phase changes, thermodynamic principles can also be used to predict spontaneity and equilibrium conditions between phases—for example, vapor pressure as a function of temperature.

To put chemical thermodynamics in its proper perspective, we should point out its *limitations*:

Time of process. As suggested in the discussion on spontaneity, chemical thermodynamics is unable even to consider the time required for a process to occur.

Mechanism of process. Thermodynamics is unable to determine the mechanism of a process, since the prediction of spontaneity depends only on the initial and final states, not on the actual reaction path. This prediction only states that the process may go if the criterion is satisfied, and does not even hint as to *how* the process can actually go.

Another aspect of thermodynamics, which is neither an objective nor a limitation, is that it is concerned only with *macroscopic* objects, in contrast to features of molecular motions and energies. Thermodynamics is based upon laws formulated from astute observations of phenomena associated with macroscopic objects and, therefore, must be restricted to this domain. In fact, both thermodynamics and chemical thermodynamics could be studied without reference to molecules or molecular structure. Only the basic chemical reactions and identification of the chemical substances need be known.

Although a study of thermodynamics does not require a knowledge of molecular motions and energies, the values of the thermodynamic parameters are directly dependent upon the molecular motions, bond energies, bond lengths, resonance energies, the associated energy levels, and molecular interactions. In Chapter 5 we will obtain an understanding of molecular energies and structures for small molecules. In order to use these molecular energies in the calculation of thermodynamic properties, it is necessary to utilize the principles of another subject called *statistical thermodynamics*. As mentioned in Chapter 1 this topic will not be covered in this text. For the present, we will look at thermodynamics as a subject in itself, in order to see the great value it has in predicting chemical phenomena. At the same time we can see its limitations or boundaries, with respect to fundamental chemical knowledge.

B. Importance of Energy

Anyone who has been exposed to even the most elementary science course has heard the word energy. In fact, we are so familiar with its use and so readily accept its importance that we might be hard pressed to answer questions like, Why is energy such an important parameter? Why is energy defined in the precise manner that it is? Couldn't kinetic energy be defined as $\frac{1}{2} mv^3$ (instead of $\frac{1}{2} mv^2$), with the units g-cm^3/sec^3? The key to answering the latter questions is that energy, as defined, is conserved. The definition suggested above simply doesn't work. Exactly why energy is a specific function can be shown from classical mechanics, starting with Newton's laws of motion, with the following critical restriction: *The potential energy is a function of position only*. This defines a conservative system, in which there is no heat transferred or work done, wherein the energy remains constant (as, for example, in planetary motion). The First Law of Thermodynamics is also a statement of the conservation of energy, but it applies specifically to systems in which there is a change in heat and work.

The previous discussion indicates that the importance of energy is not always obvious; rather, we accept it primarily because of familiarity. Later in the chapter, we will encounter other thermodynamic functions which will be unfamiliar, whose uses will not be obvious.

However, each function will be defined in a unique manner, and we will soon acquire a familiarity with how it can be used to satisfy or accomplish a given objective.

C. First Law of Thermodynamics

The **First Law of Thermodynamics**, as is true with most laws, is simply a statement of fact based on observations of numerous processes, involving a variety of systems. It cannot be proved per se and its premise is based on the fact that it has worked on these systems in the past, so that we believe it to be a universal concept or fact. In the event we find that the law can be disobeyed, then it must be abandoned or modified.

The First Law of Thermodynamics states that conservation of energy occurs where there is a possibility of heat exchanged with and work done by or on the surroundings. It can be expressed very succinctly in terms of a simple mathematical differential relationship. For a differential change in the state of a system,

$$dU = dQ + dW \qquad \text{(2-C-1)}$$

where dQ, dW are differential amounts of heat and work as defined in Chapter 1, and dU is the differential change in a new function called **internal energy**. Despite the fact that internal energy includes the term energy, it is as new and different as kinetic or potential energy. To the question, What is it? the only answer is that it is defined in terms of equation (2-C-1), just as kinetic energy in mechanical systems is defined as $\frac{1}{2}(mv^2)$. Admittedly, one can visualize velocity or movement of an object, but it is difficult to visualize $\frac{1}{2}(mv^2)$. Later, we shall show how U can be described in terms of molecular energies for some simple systems; however, for the present, it is quite satisfactory to define U through equation (2-C-1) in terms of Q and W, which can be determined experimentally. If we wish, equation (2-C-1) can be integrated to give

$$\int_{U_1}^{U_2} dU = \int dQ + \int dW$$

$$U_2 - U_1 = \Delta U = Q + W \qquad \text{(2-C-2)}$$

where ΔU is the *change* in internal energy for the system as it is carried through some finite process. Since Q and W refer only to processes and hence, only to changes in the system or surroundings, there is no need to affix a Δ to these quantities to infer a change.

The differentials in equation (2-C-1) are written in an identical manner and have the general properties of differentials (see Appendix B). However, in this case the differentials are of a different nature; namely, dQ and dW are said to be *inexact* differentials, whereas dU is an *exact* differential. The physical feature which distinguishes these two types of differentials is that the integral of the differential (that is, the finite change in the quantities ΔU, Q, and W) is *dependent on the path for an inexact differential* and *independent of the path for an exact differential*. There are formal mathematical tests that can be used to examine a differential expression to determine precisely whether it is an exact or inexact differential, however, the physical essence of an exact or inexact differential will suffice for this course. Since dU is exact, a change in U depends only on the initial and final states, therefore, U is a state function or variable.

Let us pursue this aspect of exactness and inexactness further, since the consequences are extremely important to thermodynamics. Figure 2-C-1 depicts three general processes or paths going from state A to state B. The significance of the First Law, as it applies to paths (1), (2), and (3) in going from A to B, can be summarized in the following statements:

$$Q_1 \neq Q_2 \neq Q_3$$

$$W_1 \neq W_2 \neq W_3$$

$$\Delta U_1 = \Delta U_2 = \Delta U_3 = (U_B - U_A)$$

or

$$Q_1 - W_1 = Q_2 - W_2 = Q_3 - W_3 = (U_B - U_A)$$

To use equation (2-C-2), we must adopt a convention of sign for the magnitude of Q and W:

Q positive = heat absorbed by the system from the surroundings
Q negative = heat liberated by the system to the surroundings
W positive = work done on the system by the surroundings
W negative = work done by the system on the surroundings

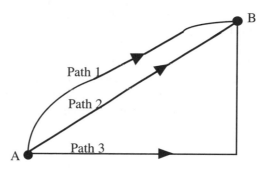

Figure 2-C-1. Different Paths from State A to State B

We can graphically represent these quantities in relation to possible processes or paths followed in going from state A to state B by Figure 2-C-2. It is apparent from

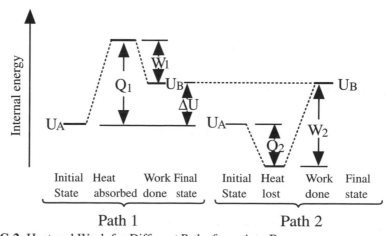

Figure 2-C-2. Heat and Work for Different Paths from A to B

Figure 2-C-2, in regard to the heat gained or lost and the work done on or by the system, that there are an infinite number of paths through which the system may be changed from state A(U_A) to state B(U_B).

Problem 2-C-1. An engine undergoes a change from state A to state B. (i) When the process is carried out reversibly 100 kcal of heat are absorbed and the engine performs - 65 kcal of work. Calculate the change in internal energy in going from state A to state B. (ii) When the process going from the same initial state A to the same final state B is carried out irreversibly only 80 kcal of heat are absorbed. Calculate the work performed by the engine for this irreversible process. Answer: (i) 35 kcal (ii) - 45 kcal.

To further emphasize the significance of an exact differential or state function, let us consider the evaluation of volume change, ΔV, and work, for two different paths between the initial and final states. These paths and processes are shown in Figure 2-C-3.

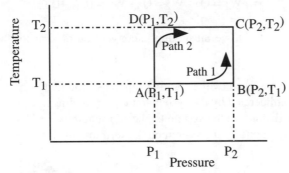

Figure 2-C-3. Temperature-Pressure Diagram

Consider the paths designated by (1): A $\rightarrow$ B $\rightarrow$ C and (2): A $\rightarrow$ D $\rightarrow$ C. We shall calculate ΔV and W for these two paths traveled by an ideal gas in reversible steps. Because of the very nature of volume, the change in volume should be independent of the path. However, in accordance with our statement of the First Law, W should be dependent on the path.

Path (1):   A $\rightarrow$ B $\rightarrow$ C

The first step, A $\rightarrow$ B, is an isothermal, reversible compression of an ideal gas ($P_2 > P_1$ so $V_2 < V_1$).

$$PV = nRT$$

$$\Delta V_{AB} = V_B - V_A = \frac{nRT_1}{P_2} - \frac{nRT_1}{P_1} = nRT_1 \left(\frac{1}{P_2} - \frac{1}{P_1} \right)$$

Utilizing equation (1-E-8),

$$W = - nRT \ln \frac{V_2}{V_1} \qquad\qquad (1\text{-E-8})$$

$$W_{AB} = - nRT_1 \ln \frac{V_2}{V_1} = - nRT_1 \ln \frac{P_1}{P_2}$$

The next step, B $\rightarrow$ C, is a constant pressure expansion of an ideal gas.

$$\Delta V_{BC} = V_C - V_B = \frac{nRT_2}{P_2} - \frac{nRT_1}{P_2}$$

Using equation (1-E-6),

$$W = -P_2(V_2 - V_1) \qquad\qquad (1\text{-E-6})$$

$$W_{BC} = -P_2(V_C - V_B) = -P_2\left(\frac{nRT_2}{P_2} - \frac{nRT_1}{P_2}\right) = -nR(T_2 - T_1)$$

Therefore,

$$\Delta V_1 = \Delta V_{AB} + \Delta V_{BC} = nRT_1\left(\frac{1}{P_2} - \frac{1}{P_1}\right) + \frac{nRT_2}{P_2} - \frac{nRT_1}{P_2}$$

$$\Delta V_1 = nR\left(\frac{T_2}{P_2} - \frac{T_1}{P_1}\right)$$

$$W_1 = W_{AB} + W_{BC} = -nRT_1 \ln\frac{P_1}{P_2} - nR(T_2 - T_1)$$

Path (2): $\qquad\qquad A \rightarrow D \rightarrow C$

In this case, the first step, $A \rightarrow D$, is a constant pressure expansion of an ideal gas.

$$\Delta V_{AD} = (V_D - V_A) = \frac{nRT_2}{P_1} - \frac{nRT_1}{P_1} = nR\left(\frac{T_2}{P_1} - \frac{T_1}{P_1}\right)$$

From equation (1-E-6)

$$W_{AD} = -P_1 (V_D - V_A) = -nR(T_2 - T_1)$$

The next step, $D \rightarrow C$, is an isothermal, reversible compression of an ideal gas.

$$\Delta V_{DC} = (V_C - V_D) = \frac{nRT_2}{P_2} - \frac{nRT_2}{P_1} = nR\left(\frac{T_2}{P_2} - \frac{T_2}{P_1}\right)$$

Using equation (1-E-8),

$$W_{DC} = -nRT_2 \ln\frac{V_2}{V_1} = -nRT_2 \ln\frac{P_1}{P_2}$$

Therefore,

$$\Delta V_2 = \Delta V_{AD} + \Delta V_{DC} = nR\left(\frac{T_2}{P_1} - \frac{T_1}{P_1}\right) + nR\left(\frac{T_2}{P_2} - \frac{T_2}{P_1}\right)$$

$$\Delta V_2 = nR\left(\frac{T_2}{P_2} - \frac{T_1}{P_1}\right)$$

$$W_2 = W_{AD} + W_{DC} = -nR(T_2 - T_1) - nRT_2\ln\frac{P_1}{P_2}$$

Comparing ΔV_1 and ΔV_2,

$$\Delta V_1 = \Delta V_2$$

However, comparing W_1 and W_2,

$$W_1 \neq W_2$$

$$- nRT_1 \ln \frac{P_1}{P_2} - nR(T_2 - T_1) \neq - nR(T_2 - T_1) - nRT_2 \ln \frac{P_1}{P_2}$$

since $T_1 \neq T_2$. Again, this confirms that W is *dependent* on the path, but ΔV is independent of the path.

Problem 2-C-2. How would the ΔU along path #1 in Figure 2-C-3 compare to that along path #2? Answer: $\Delta U_1 = \Delta U_2$.

Problem 2-C-3. How would the Q along path #1 in Figure 2-C-3 compare to that along path #2? Hint: Consider the relative magnitudes of W_1 and W_2 calculated above and the result of Problem 2-C-2. Answer: $Q_1 \neq Q_2$.

D. Enthalpy and Heat Capacity

If we consider a process carried out under *constant pressure* P, in which there is a volume change from V_1 to V_2, the First Law of Thermodynamics states

$$\Delta U = Q + W$$

Using equation (1-E-6), and restricting the process to PV work,

$$(U_2 - U_1) = Q - P (V_2 - V_1)$$

Rearranging gives

$$(U_2 + PV_2) - (U_1 + PV_1) = Q \qquad\qquad (2\text{-}D\text{-}1)$$

We note that the change in a new function $U_i + PV_i$, gives the heat change for the process. Furthermore, we note that $U_i + PV_i$ is a state variable, since it is composed of only state variables U, P, and V. Because this function is so useful, we define a new thermodynamic variable, H, which has the properties of the function shown in (2-D-1), so that $\Delta H = Q$, when the pressure is held constant and the system is restricted to PV work only.

$$\mathbf{H = U + PV} \qquad\qquad (2\text{-}D\text{-}2)$$

H is called **enthalpy** and it is also a state function, since both U and the PV product are state functions. Obviously, for a constant pressure process in which only PV work is performed,

$$\mathbf{dH = dU + PdV \text{ or } \Delta H = \Delta U + P\Delta V} \quad \text{(constant pressure).}$$

From equation (2-C-1), the First Law of Thermodynamics,

$$dH = dQ_P + dW + PdV = dQ_P - P_{ex}dV + PdV$$

but,

$$P_{ex} = P = constant$$

hence,

$$dH = dQ_P \qquad (2\text{-}D\text{-}3)$$

For a finite process,

$$\mathbf{\Delta H = Q_P} \qquad (2\text{-}D\text{-}4)$$

where the pressure is constant, and only mechanical work ($dW = - P_{ex}dV$) is done. (The subscript P specifies that the pressure is being held constant.)

A relationship similar to (2-D-4)) can be obtained for the change in internal energy when the volume is held constant. Using the First Law, when only PV work is done,

$$dU = dQ + dW = dQ - P_{ex}dV$$

If V = constant, then dV = 0 and

$$dU = dQ_V \qquad (2\text{-}D\text{-}5)$$

Integrating the differential,

$$\mathbf{\Delta U = Q_V} \qquad (2\text{-}D\text{-}6)$$

where the volume is constant, and only PV work is done.

Using equations (2-D-3) and (2-D-5) as a basis, we can define heat capacity more completely. In Chapter 1, heat capacity was defined as the derivative of Q with respect to temperature:

$$C = \frac{dQ}{dt}$$

Since the absolute temperature (Kelvin) is a simple displacement from the centigrade temperature,

$$T = t + 273.16$$

$$dT = dt$$

and

$$C = \frac{dQ}{dT} \qquad (2\text{-}D\text{-}7)$$

In (2-D-7), as in Chapter 1, we did not specify whether the heat change with temperature was carried out at constant pressure or constant volume. This is not critical in the example with water in Appendix B, since the values are about the same. However, for gases, it is important, so we define:

Heat capacity at constant pressure:

$$C_P = \left(\frac{\partial Q}{\partial T}\right)_P \qquad\qquad (2\text{-D-}8)$$

In light of equation (2-D-3),

$$\mathbf{C_P} = \left(\frac{\boldsymbol{\partial H}}{\boldsymbol{\partial T}}\right)_{\mathbf{P}} \qquad \text{(constant pressure)} \qquad (2\text{-D-}9)$$

Heat capacity at constant volume:

$$C_V = \left(\frac{\partial Q}{\partial T}\right)_V \qquad\qquad (2\text{-D-}10)$$

In view of equation (2-D-5),

$$\mathbf{C_V} = \left(\frac{\boldsymbol{\partial U}}{\boldsymbol{\partial T}}\right)_{\mathbf{V}} \qquad \text{(constant volume)} \qquad (2\text{-D-}11)$$

It should be noted that C_P and C_V are molar quantities and have units of kJ/K-mol. The units have mol in the denominator, and this infers per mol. As we defined intensive properties in Chapter 1, molar heat capacities would be intensive quantities, independent of the mass of material.

Equations (2-D-9) and (2-D-11) can be used to calculate the change in H and U for processes in which P and V are held constant. At constant P,

$$dH = \left(\frac{\partial H}{\partial T}\right)_P dT = C_P\, dT \qquad \text{(one mole)}$$

For n moles

$$\int dH = \int nC_P dT$$

$$H_2 - H_1 = \Delta H = n \int_{T_1}^{T_2} C_P dT \qquad\qquad (2\text{-D-}12)$$

At constant V,

$$dU = \left(\frac{\partial U}{\partial T}\right)_V dT = C_V\, dT \qquad \text{(one mole)}$$

For n moles

$$\int dU = \int nC_V dT$$

$$U_2 - U_1 = \Delta U = n \int_{T_1}^{T_2} C_V dT \qquad \text{(2-D-13)}$$

A relationship between C_P and C_V can be derived using the Second Law of Thermodynamics

$$C_P = C_V + \left[P + \left(\frac{\partial U}{\partial V} \right)_T \right] \left(\frac{\partial V}{\partial T} \right)_P \qquad \text{(2-D-14)}$$

Furthermore, an expression for $(\partial U / \partial V)_T$ can be derived using the Second Law of Thermodynamics, which will be introduced subsequently. These equations are generally applicable without any restrictions. This expression for $(\partial U / \partial V)_T$ can be evaluated from the equation of state for the substance and the equation

$$\left(\frac{\partial U}{\partial T} \right)_T = T \left(\frac{\partial P}{\partial T} \right)_V - P \qquad \text{(2-D-15)}$$

With an expression for $(\partial U / \partial V)_T$, the difference in heat capacities can be obtained with equation (2-D-14). Since C_P and C_V are molar quantities, $(\partial U / \partial V)_T$ should be evaluated for one mole.

At this time it should be noted that the equations developed in this section are general thermodynamic expressions which are applicable to any phase—liquid, solid, gas, or solution. Many of the later applications will be restricted to ideal gases, but equations (2-D-2) through (2-D-15) are generally applicable.

Problem 2-D-1. The C_P for CCl_4 liquid is 130.7 J/K-mol. Calculate the change in enthalpy if the temperature of 2.5 moles of CCl_4 is raised from 15°C to 75°C. Assume C_P remains constant, independent of temperature. Answer: 19.6 kJ.

Problem 2-D-2. The heat capacity, C_P, for benzene(g) is given in the following table as a function of temperature.

T(K)	Cp(J/K-mol)
300	83.020
400	113.510
500	139.340
600	160.090
700	176.790
800	190.460
900	201.840
1000	211.430
1100	219.580
1200	226.540
1300	232.520
1400	237.680
1500	242.140

Using this data calculate the heat required to raise the temperature of one mole of benzene gas at constant pressure (standard) from 300 K to 1500 K. Answer: 227.25 kJ.

E. Calculation of Q, W, ΔU, ΔH

Unfortunately, there is no simple procedure that can be followed in evaluating the various thermodynamic quantities of interest in every case. The basic laws and definitions must be rigorously adhered to, and their application to a specific process should be made with care. The best instruction that can be given in this regard, is through examples and the assignment of numerous problems. Whenever possible, the equations needed for the calculation should be derived, rather than attempting to memorize all the possible equations for all conditions, which would constitute literally hundreds. The calculations shown in this section may seem trivial, but in later sections, their value and necessity will become apparent.

Properties of an Ideal Gas

The inherent thermodynamic properties of an ideal gas can be calculated from the equation of state and the expressions developed earlier. We will show that in this special case, *the internal energy is dependent only on the temperature of the system.* From equation (2-D-15)

$$\left(\frac{\partial U}{\partial V}\right)_T = T\left(\frac{\partial P}{\partial T}\right)_V - P$$

In order to evaluate the partial derivative, solve the equation of state for P.

$$PV = nRT$$

$$P = \frac{nRT}{V}$$

$$\left(\frac{\partial P}{\partial T}\right)_V = \frac{nR}{V}$$

and,

$$\left(\frac{\partial U}{\partial V}\right)_T = T\left(\frac{nR}{V}\right) - P = \frac{nRT}{V} - P = P - P$$

$$\left(\frac{\partial U}{\partial V}\right)_T = 0$$

Therefore, for an ideal gas, *U is a function of T only*, and ΔU can be calculated from equation (2-D-13). Also, from the definition of enthalpy,

$$H = U + PV$$

$$dH = dU + d(PV)$$

But, for a constant T process (isothermal),

$$d(PV) = d(RT) = 0$$

Hence,

$$dH = dU = 0$$

Therefore, *for an ideal gas, H is a function only of T.*

It can also be shown that C_P and C_V, for an ideal gas, are related by a very simple equation. From equation (2-D-14), the heat capacity relationship for an ideal gas is

$$C_P = C_V + [P + 0] \left(\frac{\partial V}{\partial T}\right)_P$$

Since C_P and C_V are molar quantities, we must evaluate the derivative for one mole.

$$\overline{V} = \frac{RT}{P}$$

$$\left(\frac{\partial \overline{V}}{\partial T}\right)_P = \frac{R}{P}$$

Then,

$$C_P = C_V + P\left(\frac{R}{P}\right)$$

$$\mathbf{C_P = C_V + R} \tag{2-E-1}$$

Therefore, if C_V is specified for an ideal gas, C_P can be readily calculated, and vice versa.

The equations developed in Chapters 1 and 2, based upon the First Law of Thermodynamics and basic definitions, can be applied to the following processes.

Isothermal Process

An *isothermal process* is a constant T process (dT = 0) and for an ideal gas

$$\Delta U = 0$$

since U is a function of T only. Also for an ideal gas H is a function of T only and for the isothermal process

$$\Delta H = 0$$

From the First Law of Thermodynamics and the definition of mechanical work

$$0 = dU = dQ + dW$$

$$- dQ = dW = - P_{ex}dV$$

The evaluation of work for reversible and irreversible processes was discussed in Chapter 1, Section D.

For non-ideal gases the change in U must be evaluated from the differential

$$\Delta U = \int dU = \int \left(\frac{\partial U}{\partial V}\right)_T dV$$

using the derivative in equation (2-D-15)

$$\left(\frac{\partial U}{\partial V}\right)_T = T\left(\frac{\partial P}{\partial T}\right)_V - P \tag{2-D-15}$$

It is convenient if an equation of state is known so that $(\partial U/\partial V)_T$ can be evaluated.

ΔH can be evaluated from

$$\Delta H = \Delta U + \Delta(PV)$$

where $\Delta(PV)$ can be evaluated from the equation of state. W can be evaluated by integrating

$$dW = -P_{ex}dV$$

if the process of reversibility or non-reversibility is defined and the equation of state is given. Finally, the Q can be evaluated from the First Law

$$Q = \Delta U - W$$

Problem 2-E-1. In problem 1-E-1 the work was calculated for the reversible compression of an ideal gas at $200^{\circ}C$ from P = 1.0 atm to 10 atm. Calculate the Q, ΔU, ΔH for this process. Answer: $\Delta U = \Delta H = 0$, Q = - 2.16 kcal = - 9.04 kJ, W = 2.16 kcal = 9.04 kJ.

Problem 2-E-2. In problem 1-E-2 the work was calculated for the compression when the pressure of the ideal gas was suddenly increased from P = 1.0 atm to P = 10 atm. Calculate Q, ΔU, ΔH for this process. Answer: $\Delta U = \Delta H = 0$, Q = - 8.46 kcal = - 35.4 kJ, W = 8.46 kcal = 35.4 kJ.

Isobaric Process

An *isobaric process* is one in which the pressure is held constant (dP = 0) at *all* times during the process. For a single phase of a substance then the temperature and volume are changing. According to equation (2-D-4)

$$\Delta H = Q_P \qquad (2\text{-}D\text{-}4)$$

The heat capacity, C_P, needs to be given in order to evaluate ΔH from equation (2-D-12) for n moles of the substance

$$H_2 - H_1 = \Delta H = n \int_{T_1}^{T_2} C_P dT$$

The work would be given by equation (1-E-5) where $P_{ex} = P = $ constant

$$W = -P(V_2 - V_1) \qquad (2\text{-}E\text{-}5)$$

The V_1 and V_2 need to be supplied or an equation of state must be given and the volumes evaluated from the known P, T, n. Finally the ΔU would be calculated from the First Law.

$$\Delta U = Q + W = \Delta H + W$$

The calculations of Q, W, ΔU, and ΔH for an isobaric process apply to any substance regardless of the physical state: gas, liquid, or solid.

If we have an ideal gas the equation of state obviously would be known and C_P could be calculated if C_V were given

$$C_P = C_V + R \qquad (2\text{-}E\text{-}1)$$

Since U for an ideal gas is a function of T only, the ΔU can be calculated from

$$\Delta U = n \int_{T_1}^{T_2} C_V dT \qquad (2\text{-E-6})$$

Problem 2-E-3. In problem 2-D-1 the ΔH was calculated for heating CCl_4 liquid from $15^\circ C$ to $75^\circ C$. If we assume the expansion of the liquid CCl_4 is negligible over this small temperature change, calculate Q, W, ΔU for the process.
Answer: $Q = \Delta H = 19.6$ kJ, $W = 0$, $\Delta U = Q$.

Problem 2-E-4. The C_P for $N_{2(g)}$ can be approximated by the equation

$$C_P = 27.296 \text{ J/K-mol} + (5.23 \times 10^{-3} \text{ J/K}^2\text{-mol}) \ T$$

Calculate Q, W, ΔU, ΔH for cooling one mole of $N_{2(g)}$ from 1,000K to 298K at P = 1.0 atmosphere. Assume $N_{2(g)}$ obeys the ideal gas equation of state. Answer: $Q = \Delta H = -21.545$ kJ, $\Delta U = -15.708$ kJ, $W = +5.837$ kJ.

Isochoric Process

An *isochoric process* is one in which the volume V is held constant (dV = 0) at all times during the process. Again for a single phase of a substance the temperature and pressure are changing. According to equation (2-D-6)

$$U_2 - U_1 = \Delta U = Q_V$$

$$\Delta U = n \int_{T_1}^{T_2} C_V dT \qquad (2\text{-E-6})$$

where n = number of moles of the substance. Since there is no volume change

$$W = 0$$

ΔH can be evaluated from the basic definition of H

$$\Delta H = \Delta U + \Delta(PV) = \Delta U + V(\Delta P) \qquad (2\text{-E-7})$$

In order to evaluate ΔP the initial and final pressures P_1 and P_2 must be given

$$\Delta P = P_2 - P_1$$

or evaluated from an equation of state and known V, T, n.
For an ideal gas ΔH can be evaluated from

$$\Delta H = n \int_{T_1}^{T_2} C_P dT$$

Or alternatively, P_1 and P_2 can be calculated from the ideal gas equation and $\Delta P = P_2 - P_1$ can be substituted into equation (2-E-7).

Problem 2-E-5. If the $N_{2(g)}$ in Problem 2-E-4 were cooled at *constant volume* from 1000K to 298K, calculate Q, W, ΔU, and ΔH for the process. Assume $N_{2(g)}$ obeys the ideal gas equation of state. Answer: W = 0, Q = ΔU = - 15.708 kJ, ΔH = - 21.545 kJ.

Problem 2-E-6. In Section C a temperature-pressure diagram, Figure 2-C-3, was given in which two paths were described in going from State A to State C. The work was calculated along paths #1 and #2. The gas was assumed to be an ideal gas with n moles. Calculate Q, ΔU, ΔH for these two paths and compare the values; i.e., are they equal or different. Account for the fact that they are equal or different.

$$\text{Answer: } Q_1 = n \int_{T_1}^{T_2} C_V dT + nRT_1 \ln \frac{P_1}{P_2} + nR(T_2 - T_1), \quad \Delta U_1 = \Delta U_2 = n \int_{T_1}^{T_2} C_V dT$$

$$Q_2 = n \int_{T_1}^{T_2} C_V dT + nR(T_2 - T_1) + nRT_2 \ln \frac{P_1}{P_2}, \quad Q_1 \neq Q_2, \quad \Delta H_1 = \Delta H_2 = n \int_{T_1}^{T_2} C_P dT$$

ΔH and ΔU are the same along the two paths since H and U are state functions.

Problem 2-E-7. Consider the two paths followed in going from A to C in the following diagram:

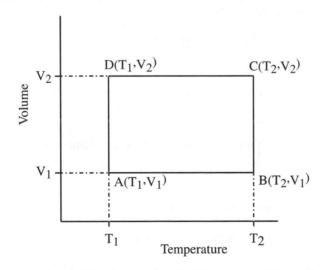

(a) Derive expressions for Q, W, ΔU, ΔH for the path ABC for one mole of an ideal gas. Assume reversibility.
(b) Derive expressions for Q, W, ΔU, ΔH for the path ADC for one mole of an ideal gas. Assume reversibility.
(c) Compare the expressions for Q, W, ΔU, and ΔH for the two paths. Can you reconcile the results with respect to the First Law of Thermodynamics? Explain.

$$\text{Answer: (a) } Q_{ABC} = n \int_{T_1}^{T_2} C_V dT + nRT_2 \ln \frac{V_2}{V_1}; W_{ABC} = -nRT_2 \ln \frac{V_2}{V_1}, \quad \Delta U_{ABC} = n \int_{T_1}^{T_2} C_V dT,$$

$$\Delta H_{ABC} = n \int_{T_1}^{T_2} C_P dT \text{ ; (b) } Q_{ADC} = nRT_1 \ln \frac{V_2}{V_1} + n \int_{T_1}^{T_2} C_V dT , \quad W_{ADC} = - nRT_1 \ln \frac{V_2}{V_1}$$

$$\Delta U_{ADC} = n \int_{T_1}^{T_2} C_V dT \; , \; \Delta H_{ADC} = n \int_{T_1}^{T_2} C_P dT \; \text{(c)} \; Q_{ABC} \neq Q_{ADC}, \; W_{ABC} \neq W_{ADC},$$

$\Delta U_{ABC} = \Delta U_{ADC}$, $\Delta H_{ABC} = \Delta H_{ADC}$. U and H are state functions, thus ΔU and ΔH must be independent of the path.

F. Thermochemistry

Thermochemistry is that phase of thermodynamics which deals with the measurement or determination of heats of reaction. From equations (2-D-6) and (2-D-4), we know that the heat of the reaction is a most important quantity, since it permits evaluation of ΔU and ΔH for the reaction. The values of ΔU and ΔH are important, not only for thermodynamic predictions, but they can also be important parameters in the understanding of molecular structure. Discussions on the latter topic will be deferred to Chapter 6.

The important fact that U and H are state functions is brought out again in the subject of thermochemistry. We can depict a general reaction as

$$aA + bB = cC + dD \tag{2-F-1}$$

The ΔU and ΔH for the conversion of a moles of A and b moles of B are unique values, since ΔU and ΔH are independent of the path; that is, ΔU and ΔH are the differences in U and H for the products minus the reactants, as written in (2-F-1). If U and H were not state functions, ΔU and ΔH for a given reaction would not be very significant, since there frequently are alternate mechanisms or paths which give the desired products. If that were the case, ΔU and ΔH would be different for each path.

Standard States/Reference State

We note that in a reaction only the *change* in internal energy and enthalpy can be measured, not absolute values. Similarly, from the First Law of Thermodynamics, only ΔU can be found, not U. Nowhere do we find an expression for an absolute value of U, and since enthalpy is defined in terms of U, it too cannot be determined absolutely. This does not actually present a problem, since we are only interested in ΔU and ΔH for a given chemical reaction or process, and for this, absolute values of U and H are not required. Although relative values are satisfactory for our purpose, we still must decide on some convention to which we can conform for the relative values to be meaningful. It is also convenient to define standard states, so that values can be compared to a table of standard values at corresponding conditions. If values for ΔU and ΔH are desired at other conditions, the change in the standard values can be calculated using principles discussed in the previous section.

The conventions that have been adopted for enthalpy are:

Standard states (designated by a superscript zero):
liquid or *solid* - pure material under one bar pressure* at the specified temperature;
        gas - pure material with properties of an ideal gas at one bar pressure and
             specified temperature;

Reference states: the enthalpies of the elements in their most stable form in the
             respective standard state are arbitrarily assigned a value zero.

* 1 bar = 0.986923 atmospheres

The **standard heat of formation** (ΔH_f^o) of a compound is defined as the enthalpy change associated with the chemical reaction forming one mole of the compound of interest from the elements, all species being in their standard states. For example, the standard heat of formation of nitrobenzene in the liquid state is the ΔH^o for the reaction

$$6\ C_{(graphite)} + O_{2(g)} + \frac{1}{2}\ N_{2(g)} + \frac{5}{2}\ H_{2(g)} = C_6H_5NO_{2(\ell)}$$

For an ideal gas the enthalpy is independent of pressure; recall that $(\partial H/\partial P)_T = 0$ for an ideal gas. Consequently, any pressure can be used as the standard state for an ideal gas. For consistency with the liquid and solid states we use one bar pressure as the standard state. For real gases, H is a function of pressure; however, at one bar pressure real gases closely behave like an ideal gas and one bar pressure can also be used as the standard state for real gases. For a rigorous treatment real gases only obey the ideal gas equation of state at zero pressure and this should be taken as the standard state for real gases.

The term "most stable form," used in the definition of reference states, may need clarification. For some elements there could be more than one form or phase possible at one atmosphere pressure; for example, carbon can exist as graphite or diamond. But, since graphite is the most stable at one atmosphere, it is assigned the value zero. Also, different phases are used as reference states for similar type compounds, according to which is the most stable form; for example, at 25°C:

$$\Delta H^o_{f,I_{2(s)}} = 0,\ \Delta H^o_{f,Br_{2(\ )}} = 0,\ \Delta H^o_{f,Cl_{2(g)}} = 0,\ \Delta H^o_{f,F_{2(g)}} = 0$$

Thus, iodine solid, bromine liquid, and chlorine and fluorine gas are assigned values of zero in their standard states at 25°C.

Finally, a comment on the term "*specified temperature*" is appropriate. The temperature specified for the reaction under consideration is called the specified temperature. Since the standard state is at a specified temperature, the reference state of the most stable form of the elements in their standard state must be zero at any specified temperature. The specified temperature may be different for different reactions; nevertheless, the enthalpy of formation of the elements is taken as zero at *all* specified temperatures. The heat of formation of a substance at any specified temperature is thus, the heat of reaction necessary to produce the compound from the elements in their reference state, where the substance and the elements are all at the specified temperature.

Heat of Reaction

The *heat of reaction,* the enthalpy change for the reaction, is defined as the difference in enthalpies of formation of the products and the enthalpies of formation of the reactants. Referring to the general reaction given previously, the heat of reaction is expressed as

$$\Delta H = c\Delta H_{f,C} + d\Delta H_{f,D} - a\Delta H_{f,A} - b\Delta H_{f,B} \qquad (2\text{-}F\text{-}2)$$

A glance at the reaction scheme and enthalpy diagram in Figure 2-F-1 may clarify the need for this relationship. If the standard heat of reaction (ΔH^o) is desired, then the standard heats of formation (ΔH^o_f) should be used. Some selected values of (ΔH^o_f) can be found in Appendix D, Table 2 for non-carbon containing compounds (most inorganics) and Table 3 for carbon containing compounds. For example, since CO, CO_2, and carbonates contain carbon, they are found in Table 3 even though they are classified as inorganic compounds.

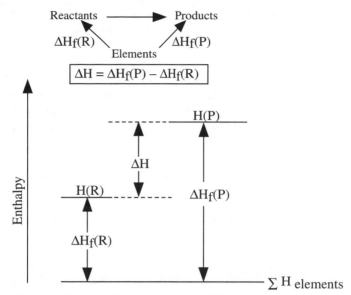

Figure 2-F-1. Heat of Reaction Related to Heats of Formation

Heats of Formation

The *heats of formation* of compounds are determined from experimentally measured heats of reaction. In some cases, the heat of formation can be found directly from the heat of reaction of the elements to produce the compound. For example, carbon could be combined with O_2 to give CO_2, and H_2 could be readily oxidized to H_2O.

$$C_{(s)} + O_{2(g)} = CO_{2(g)} \qquad \Delta H^o = \Delta H^o_{f,CO_{2(g)}}$$

$$H_{2(g)} + \frac{1}{2} O_{2(g)} = H_2O_{(g)} \qquad \Delta H^o = \Delta H^o_{f,H_2O_{(g)}}$$

Unfortunately most compounds cannot be readily made from their elements, as, for example, the reaction previously given, showing the formation of nitrobenzene from its elements. Any chemist knows that this reaction is not a feasible method for preparing nitrobenzene. In these cases, other reactions must be considered, such that by combination, they will give the proper formation of the compound from the elements in their standard states. Many times some devious reaction paths must be carried out, in order to determine the ΔH^o_f precisely. Although this is not meant to be an experimental text, a few criteria for a satisfactory reaction step might help you have a better appreciation for some rather complicated reaction schemes.

1. The reaction must go in a reasonable length of time in order that extraneous heat losses, which accumulate with time, will not greatly affect the precision of the experimental measurement.

2. Preferably, the reaction should go to completion or near completion. In any event, the extent of the reaction must be known.

3. The reaction products must be known quantitatively, with no unknown side reactions occurring. This can be an extreme problem with some compounds because, for example, mixed oxides such as NO and NO_2 can be produced in almost unpredictable quantities.

Satisfying all these criteria for a given reaction is not always an easy task. The following examples show two approaches to the determination of heats of formation.

Example 2-F-1. The first, a rather conventional technique which is applicable to many compounds, involves the burning or combustion of the sample completely, and measuring the associated heat of combustion. A table of selected heats of combustion can be found in Appendix D, Table 5. For C, H, O containing compounds, an excess of O_2 will generally give complete conversion to CO_2 and H_2O. Consider the determination of ΔH_f^o for benzoic acid.

$$C_6H_5COOH_{(s)} + \frac{15}{2} O_{2(g)} = 7 CO_{2(g)} + 3 H_2O_{(l)} \quad \Delta H^o = -3,226.9 \text{ kJ}$$

This result for the heat of combustion can be combined with the heats of formation of $CO_{2(g)}$ and $H_2O_{(l)}$, to give ΔH_f^o for benzoic acid.

$$\Delta H^o = 7 \Delta H_{f,CO_{2(g)}}^o + 3 \Delta H_{f,H_2O_{(l)}}^o - \Delta H_{f,C_6H_5COOH_{(s)}}^o - \frac{15}{2} \Delta H_{f,O_{2(g)}}^o$$

$$-3,226.9\text{kJ} = 7(-393.5 \text{ kJ}) + 3(-285.8\text{kJ}) - \Delta H_{f,C_6H_5COOH_{(s)}}^o - \frac{15}{2}(0)$$

$$\Delta H_{f,C_6H_5COOH_{(s)}}^o = -385.0 \text{ kJ/mole}$$

Example 2-F-2. The heat of formation of $NH_{3(g)}$ can be determined from the heat of reaction with oxalic acid in solution, along with the heats of formation of the other reactants and products in the reaction. The reaction is

$$(COOH)_2 \cdot (H_2O)_{2(aq)} + 2 NH_{3(g)} = (COONH_4)_2 \cdot H_2O_{(aq)} + H_2O_{(l)} \quad \Delta H^o = -180.33\text{kJ}$$

In order to determine the heats of formation of the oxalic acid and ammonium oxalate in solution, we need the heat of solution of the solids and the heat of formation of the solids. We can obtain the latter from heats of combustion. The heats of solution are:

$$(COOH)_2 \cdot (H_2O)_{2(s)} = (COOH)_2 \cdot (H_2O)_{2(aq)} \qquad\qquad \Delta H^o = 36.0 \text{ kJ}$$

$$(COONH_4)_2 \cdot (H_2O)_{(s)} = (COONH_4)_2 \cdot (H_2O)_{(aq)} \qquad\qquad \Delta H^o = 48.1 \text{ kJ}$$

The heats of combustion are:

$$(COOH)_2 \cdot (H_2O)_{2(s)} + \frac{1}{2} O_{2(g)} = 2 CO_{2(g)} + 3H_2O_{(l)} \qquad\qquad \Delta H^o = -222.2 \text{ kJ}$$

$$(COONH_4)_2 \cdot (H_2O)_{(s)} + 2O_{(2g)} = N_{2(g)} + 2 CO_{2(g)} + 5 H_2O_{(l)} \qquad \Delta H^o = -794.5 \text{ kJ}$$

In addition, we need the heat of formation of $H_2O_{(l)}$

$$H_{2(g)} + \frac{1}{2}O_{2(g)} = H_2O_{(l)} \qquad\qquad\qquad\qquad \Delta H^o = -285.8 \text{ kJ}$$

These reactions can be combined in the following manner to get the desired heat of formation of $NH_{3(g)}$:

The $\Delta H^o = -180.3$ kJ for the reaction of NH_3 with oxalic acid in solution, given by

$$\Delta H^o = \Delta H^o_{f,(COONH_4)_2 \cdot H_2O_{(aq)}})) + \Delta H^o_{f,H_2O_{(l)}} - \Delta H^o_{f,(COOH)_2 \cdot H_2O_{(aq)}} - 2\Delta H^o_{f,NH_{3(g)}}$$

The $\Delta H^o_{f,H_2O_{(\ell)}}$ is given as $- 285.8$ kJ. The $\Delta H^o_{f,NH_{3(g)}}$ can be evaluated if the two remaining $\Delta H^o_{f,i}$ are known. First we determine the $\Delta H^o_{f,(COOH)_2 \cdot H_2O_{(aq)}}$ from the heat of solution of the solid and the heat of combustion of the solid.

$$(COOH)_2 \cdot (H_2O)_{2(s)} = (COOH)_2 \cdot (H_2O)_{2(aq)} \qquad\qquad \Delta H^o = 36.0 \text{ kJ}$$

$$2CO_{2(g)} + 3H_2O(\ell) = (COOH)_2 \cdot (H_2O)_{2(s)} + \frac{1}{2} O_{2(g)} \qquad\qquad \Delta H^o = 222.2 \text{ kJ}$$

$$2CO_{2(g)} + 3H_2O(\ell) = (COOH)_2 \cdot (H_2O)_{2(aq)} + \frac{1}{2} O_{2(g)} \qquad\qquad \Delta H^o = 258.2 \text{ kJ}$$

$$\Delta H^o = \Delta H^o_{f,(COOH)_2 \cdot (H_2O)_{2(aq)}} + \frac{1}{2} \Delta H^o_{f,O_{2(g)}} - 2\Delta H^o_{f,CO_{2(g)}} - 3\Delta H^o_{f,H_2O_{(l)}}$$

$$\Delta H^o_{f,(COOH)_2 \cdot (H_2O)_{2(aq)}} = \Delta H^o - \frac{1}{2} \Delta H^o_{f,O_{2(g)}} + 2\Delta H^o_{f,CO_{2(g)}} + 3\Delta H^o_{f,H_2O_{(l)}}$$

$$\Delta H^o_{f,(COOH)_2 \cdot (H_2O)_{2(aq)}} = 258.2 \text{ kJ} - \frac{1}{2} (0) + 2(- 393.5 \text{ kJ}) + 3(- 285.8 \text{ kJ})$$

$$\Delta H^o_{f,(COOH)_2 \cdot (H_2O)_{2(aq)}} = - 1386.4 \text{ kJmol}$$

Similarly the $\Delta H^o_{f,(COONH_4)_2 \cdot (H_2O)_{2(aq)}}$ can be found from the heat of solution of the solid and the heat of formation of the solid. The result of this calculation is

$$\Delta H^o_{f,(COONH_4)_2 \cdot (H_2O)_{2(aq)}} = - 1,373.6 \text{ kJ/mol}$$

The $\Delta H^o_{f,NH_{3(g)}}$ is then found from the ΔH^o for the original reaction between oxalic acid in solution and ammonia.

$$\Delta H^o_{f,NH_{3(g)}} \frac{1}{2}\left[-\Delta H^o - \Delta H^o_{f,(COONH_4)_2 \cdot H_2O_{(aq)}} + \Delta H^o_{f,H_2O_{(\ell)}} - \Delta H^o_{f,(COOH)_2 \cdot H_2O_{(aq)}} \right]$$

$$\Delta H^o_{f,NH_{3(g)}} = \frac{1}{2} [180.3 - 1,373.6 - 285.8 + 1,386.4] = - 46.0 \text{ kJ/mol}$$

Calculation of Heat of Reaction

If the standard heats of formation are known for all reactants and products in a reaction, then the ΔH^o for reaction can be calculated from equation (2-F-2). For example, if we wanted the ΔH^o for the reaction

$$NH_{3(g)} + \frac{7}{4} O_{2(g)} = NO_{2(g)} + \frac{3}{2} H_2O_{(g)}$$

we find from Table 2 the ΔH_f^o 's for all reactants and products

$$\Delta H^o = \Delta H_{f,NO_2(g)}^o + \frac{3}{2} \Delta H_{f,H_2O(g)}^o - \Delta H_{f,NH_3(g)}^o - \frac{7}{4} \Delta H_{f,O_2(g)}^o$$

$$\Delta H^o = 33.2 + \frac{3}{2}(-241.8) - (-45.9) - \frac{7}{4}(0) = -283.6 \text{ kJ/mole.}$$

Thus the burning of $NH_{3(g)}$ to $NO_{2(g)}$ is exothermic by 283.6 kJ/mol.

Problem 2-F-1. From the heats of formation given in the Appendix, calculate the standard heats of reaction for

$$NH_{3(g)} + HCl_{(g)} = NH_4Cl_{(s)}$$

$$C_6H_{6(g)} + HNO_{2(g)} = C_6H_5NO_{2(g)} + H_{2(g)}$$

Answer: - 176.2 kJ and 64.4 kJ.

Problem 2-F-2. Calculate the heat of formation for ethane from the heat of combustion found in the Appendix. Answer: - 83.7 kJ/mol.

Problem 2-F-3. From the following known heats of reaction (all at 298K), calculate the heat of formation of the $CH_3CO_{(g)}^{\cdot}$ radical. The heat of formation of $CH_3CHO_{(g)}$ = - 166.2 kJ/mol is needed.)

	$\Delta H^o(kJ)$
$I_{(g)}^{\cdot} + CH_3COI_{(g)} = CH_3CO_{(g)}^{\cdot} + I_{2(g)}$	64.4
$CH_3CHO_{(g)} + I_{2(g)} = CH_3COI_{(g)} + HI_{(g)}$	3.1
$I_{2(s)} = I_{2(g)}$	62.4
$H_{2(g)} + I_{2(s)} = 2 HI_{(g)}$	52.7
$I_{2(g)} = 2 I_{(g)}^{\cdot}$	148.8

Answer: - 19.4 kJ/mol.

Bond Dissociation Energy/Bond Strength

Two terms are associated with the energy required to break chemical bonds—**bond dissociation energy** and **bond energy** or **normal covalent bond energy**. Although these terms are quite different, they are frequently confused with each other. Therefore, specific definitions are given, along with examples.

Bond dissociation energy is the energy required to break a specific bond in a specific gaseous compound. The bond to be broken is designated by a bond (line) joining the two radical fragments. The reaction is assumed to be carried out at 0 K where

$$D = \text{bond dissociation energy} = \Delta E^o_{0\ K} = \Delta H^o_{0\ K}$$

Since bond dissociation energies are frequently used to calculate heats of formation and heats of reaction at 25°C, many tables give the ΔH_{298}^o (*bond strength*) for the bond dissociation process. Such tables of bond strength (bond enthalpies) at 25°C or 298K are given in Appendix D, Tables 6 and 9. As we will see in in Chapter 5, the bond dissociation is always greater than the bond strength, since bond dissociation has all molecules in their lowest rotational and vibrational states at 0 K whereas at 25°C some of the molecules are in excited rotational and vibrational states and it requires less energy to dissociate molecule when some of the molecules occupy excited states.

Heats of Formation of Radicals

Since radicals are generally formed when a bond is broken, the bond dissociation energy can be used to calculate the heat of formation of the radicals. The bond strength is the ΔH^o_{298} for the reaction to break the specific bond. If the ΔH^o_f's are known for the other species in the process, then the ΔH^o_f for the desired radical can be found. For example, the bond strength of H_2 would give the ΔH^o_f for the hydrogen atom, H:

$$H_{2(g)} = 2H^{\cdot}_{(g)} \qquad\qquad D_{H\text{-}H} = \Delta H^o = 435.990 \text{ kJ}$$

Since the ΔH^o of reaction is the sum of heats of formation of products minus the sum of heat of formation of reactants

$$\Delta H^o = 2\Delta H^o_{f,H(g)} - \Delta H^o_{f,H2(g)}$$

$$435.990 \text{ kJ} = 2\Delta H^o_{f,H(g)} - 0$$

$$\Delta H^o_{f,H(g)} = 217.995 \text{ kJ}$$

From the bond strength of C-H in CH_4 and using the $\Delta H^o_{f,H(g)}$ from above, we can determine the $\Delta H^o_{f,CH3(g)}$

$$CH_{4(g)} = CH^{\cdot}_{3(g)} + H^{\cdot}_{(g)} \qquad\qquad \Delta H^o = D_{H3C\text{-}H} = 438.5 \text{ kJ}$$

$$D_{H3C\text{-}H} = \Delta H^o_{f,CH3(g)} + \Delta H^o_{f,H(g)} - \Delta H^o_{f,CH4(g)}$$

$$438.5 \text{ kJ} = \Delta H^o_{f,CH3(g)} + 217.995 \text{ kJ} - (-74.4 \text{ kJ})$$

$$\Delta H^o_{f,CH3(g)} = 146.1 \text{ kJ}$$

The $\Delta H^o_{f,CH3(g)}$ can also be determined from the C-C bond strength in ethane.

$$H_3C\text{ - }CH_3 = 2\,CH^{\cdot}_3 \qquad\qquad \Delta H^o = D_{H3C\text{-}CH3} = 376.0 \text{ kJ}$$

$$D_{H3C\text{-}CH3(g)} = \Delta H^o = 2\Delta H^o_{f,CH3(g)} - \Delta H^o_{f,CH3CH3(g)}$$

$$376.0 \text{ kJ} = 2\Delta H^o_{f,CH3(g)} - (-83.8)$$

$$2\Delta H^o_{f,CH3(g)} = 292.2 \text{ kJ}$$

$$\Delta H^o_{f,CH3(g)} = 146.1 \text{ kJ}$$

The $\Delta H^o_{f,Cl(g)}$ can be determined from the bond strength in $Cl_{2(g)}$.

$$Cl_{2(g)} = 2Cl^{\cdot}_{(g)} \qquad\qquad \Delta H^o = D_{Cl\text{-}Cl} = 242.580 \text{ kJ}$$

$$D_{Cl\text{-}Cl} = \Delta H^o = 2\Delta H^o_{f,Cl(g)} - \Delta H^o_{f,Cl2(g)}$$

$$242.580 \text{ kJ} = 2\Delta H^o_{f,Cl(g)} - 0$$

$$\Delta H^o_{f,Cl(g)} = 121.29 \text{ kJ}$$

It is obvious that a continued use of bond strengths will give numerous $\Delta H^o_{f,radicals}$.

Problem 2-F-4. Using the bond strength for F_2 calculate the $\Delta H^o_{f,F(g)}$. Answer: 79.39 kJ.

Problem 2-F-5. Using the bond strength for CH_3 - CN calculate the $\Delta H^o_{f,CN(g)}$. Answer: 428 kJ.

Average Bond Energy or Normal Covalent Bond Energy

Bond Energy or Normal Covalent Bond Energy is an average of the bond dissociation energies for a given type of bond in many different compounds. The value may vary in different tabulations, since values from different compounds may have been used in the average. This is only an approximation of the actual bond dissociation energy, which is a specific quantity, and discrepancies of 20 - 60 kJ/mole can be found. We will use the symbol ε for bond energy.

The heat of formation can be *estimated* from bond energy values; however, remember that this is only an approximation. The heat of formation of 2-methyl-2-propanol (tertiary butyl alcohol) can be used as an example.

Example 2-F-3. Each of the atomic radicals must first be formed from the elements in their respective standard states. Then the bonds are formed and ΔH for the reaction is assumed to be the algebraic sum of the bond energies. When bonds are broken, energy must be absorbed, and the sign of ΔH is positive; when bonds are formed, energy is lost, and the sign of ΔH is negative.

Reaction Step	ΔH (kJ)
$4\,C_{(graphite)} \longrightarrow 4\,C_{(g)}$	$4\,\Delta H_{f,C\cdot(g)} = 4(716.68)$
$\frac{1}{2}\,O_{2(g)} \longrightarrow O_{(g)}$	$\Delta H_{f,O\cdot(g)} = 249.170$
$5\,H_{2(g)} \longrightarrow 10\,H_{(g)}$	$10\,\Delta H_{f,H\cdot(g)} = 10(217.995)$
$4\,C_{(g)} \longrightarrow C-C-C_{(g)}$ (with C below middle)	$-3\varepsilon_{C\text{-}C} = -3(334)$
$C-C-C_{(g)} + O_{(g)} \longrightarrow C-C-C_{(g)}$ (with O and C)	$-\varepsilon_{C\text{-}O} = -1(338)$
$9H_{(g)} + C-C-C_{(g)} \longrightarrow HC-C-CH_{(g)}$	$9\,\varepsilon_{C\text{-}H} = -9(414)$
$H_{(g)} + HC-C-CH_{(g)} \longrightarrow HC-C-CH_{(g)}$	$-\varepsilon_{O\text{-}H} = -1(464)$

$$4\,C_{(graphite)} + \frac{1}{2}\,O_{2(g)} + 5\,H_{2(g)} \longrightarrow HC-C-CH_{(g)} \quad \Delta H_f^o = -234$$

Note that in steps 4 through 7 bonds are formed and ΔH is negative.

Problem 2-F-6.
(a) Estimate the heat of formation of ethylmethylether, using bond strengths given in Appendices D-6 and D-10 and heats of formation given in Appendix D-7.
(b) Calculate the heat of formation of ethylmethylether, using the heat of combustion found in the Appendix. Compare with the estimated value found in part (a).
Answer: (a) - 179 kJ (b) - 216 kJ.

Temperature Dependence of ΔH Reaction

The dependence of the heat of reaction on temperature is of value in predicting the extent of reaction as a function of temperature, as we shall see later. This relationship between ΔH and T can be derived by considering the following general reactions carried out at T_1 and T_2.

$$\Delta H_1$$
$$T_1 \quad aA + bB = cC + dD$$
$$\uparrow \qquad\qquad \downarrow$$
$$\Delta H_2$$
$$T_2 \quad aA + bB = cC + dD$$

ΔH for the reaction at T_2 can be found by considering the path indicated by the arrows. Since H is a state function, ΔH_2 can be represented by the sum of the ΔH,s for the three steps.

$$\Delta H_2 = \int_{T_2}^{T_1}\left(a\,C_{P,A} + b\,C_{P,B}\right)dT + \Delta H_1 + \int_{T_1}^{T_2}\left(c\,C_{P,C} + d\,C_{P,D}\right)dT$$

reactants $T_2 \to T_1$ Heat of Reaction   products $T_1 \to T_2$

Reversing the limits of integration simply changes the sign of the integral. Hence,

$$\Delta H_2 = -\int_{T_1}^{T_2}\left(a\,C_{P,A} + b\,C_{P,B}\right)dT + \Delta H_1 + \int_{T_1}^{T_2}\left(c\,C_{P,C} + d\,C_{P,D}\right)dT$$

$$\Delta H_2 = \Delta H_1 + \int_{T_1}^{T_2}\Delta C_P dT \qquad (2\text{-}F\text{-}3)$$

where

$$\Delta C_P = cC_{P,C} + dC_{P,D} - aC_{P,A} - bC_{P,B}$$

Equation (2-F-3) can be used to evaluate ΔH at some temperature T_2, from known ΔC_P and ΔH at T_1.

Example 2-F-4. To illustrate the use of equation (2-F-3), let us calculate ΔH^o for the reaction

$$\tfrac{1}{2} N_{2(g)} + \tfrac{3}{2} H_{2(g)} = NH_{3(g)}$$

at 500°C. From Example 2-F-2, ΔH^o at 25°C is $\Delta H^o_{f,NH_{3(g)}}$ = - 46.0 kJ. The heat capacities at 25°C are:

$$C_{P,N_{2(g)}} = 28.857 \text{ J/K-mol,}$$

$$C_{P,H_{2(g)}} = 28.995 \text{ J/K-mol,}$$

$$C_{P,NH_{3(g)}} = 37.216 \text{ J/K-mol.}$$

Thus

$$\Delta C_P = C_{P,NH_{3(g)}} - \frac{1}{2} C_{P,N_{2(g)}} - \frac{3}{2} C_{P,H_{2(g)}} \,,$$

$$\Delta C_P = 37.216 - \frac{1}{2}(28.995) - \frac{3}{2}(28.857) \,,$$

$$\Delta C_P = - 20.567 \text{ J/K}$$

If we assume that the C_P's remain constant over the temperature range 25°C to 500°C, then ΔC_P can be considered independent of temperature and equation (2-F-3) becomes

$$\Delta H^o_{T_2} = \Delta H^o_{T_1} + \Delta C_P \int_{T_1}^{T_2} dT \,,$$

$$\Delta H^o_{T_2} = \Delta H^o_{T_1} + \Delta C_P(T_2 - T_1),$$

$$\Delta H^o_{773K} = \Delta H^o_{298K} + \Delta C_P(773K - 298K),$$

$$\Delta H^o_{773K} = - 46.0 \text{ kJ} - 20.567 (475) \text{ J} = -55.8 \text{ kJ}$$

In general, C_P is not constant, but is in fact a function of the temperature. A more accurate calculation must take into account the temperature dependence of the C_P and consequently that of ΔC_P. In the case of the reaction considered here, a more accurate expression for the heat capacities, as determined experimentally, are

$$C_{P,N_{2(g)}} = 27.296 + 5.23 \times 10^{-3} \text{ T (J/K-mol),}$$

$$C_{P,H_{2(g)}} = 27.065 - 0.8363 \times 10^{-3} \text{ T (J/K-mol),}$$

$$C_{P,NH_{3(g)}} = 29.748 + 25.104 \times 10^{-3} \text{ T (J/K-mol).}$$

Thus

$$\Delta C_P = C_{P,NH_{3(g)}} - \frac{1}{2} C_{P,N_{2(g)}} - \frac{3}{2} C_{P,H_{2(g)}}$$

$$\Delta C_P = (29.748 + 25.104 \times 10^{-3} \text{ T}) - \frac{1}{2}(27.296 + 5.23 \times 10^{-3} \text{ T}) - \frac{3}{2}(27.065 - 0.8363 \times 10^{-3} \text{ T})$$

$$\Delta C_P = - 27.49 + 21.1 \times 10^{-3} \text{ T}$$

Hence, ΔH at $T = 500 + 273 = 773$ K is

$$\Delta H_{773} = \Delta H_{298} + \int_{298}^{773} (-27.49 + 21.1 \times 10^{-3} T) dT$$

$$\Delta H_{773} = -46.0 \text{ kJ} + \left[-27.49 T + \frac{21.1 \times 10^{-3} T^2}{2} \right]_{298}^{773}$$

$$\Delta H_{773} = -46.0 \text{ kJ} - 13,054 \text{ J} + 5,355 \text{ J} = -53.7 \text{ kJ}$$

Problem 2-F-7. Calculate ΔH_{500K}^{o} for the reaction

$$C_2H_{2(g)} + \frac{5}{2} O_{2(g)} = 2CO_{2(g)} + H_2O_{(g)}$$

using constant heat capacities at 298K. Use ΔH_f and heat capacities from Appendix D, Tables 2 and 3. Answer: - 1,258.9 kJ.

The integration in equation (2-F-3) can also be expressed as an indefinite integral, where the constant of integration ΔH_0 refers to ΔH at $T = 0$.

$$\Delta H = \Delta H_0 + \int \Delta C_P dT \qquad (2\text{-}F\text{-}4)$$

The form of the integral will depend upon the form for ΔC_P. Let us assume

$$\Delta C_P = \Delta a + \Delta b T$$

Then,

$$\Delta H = \Delta H_0 + \int \Delta a \, dT + \int \Delta b T \, dT$$

$$\Delta H = \Delta H_0 + \Delta a T + \frac{\Delta b}{2} T^2 \qquad (2\text{-}F\text{-}5)$$

Any value of ΔH at some T can be used to evaluate ΔH_0 since all terms in equation (2-F-5) are known except ΔH_0. Formally, ΔH_0 in equation (2-F-5) is equal to the ΔH at $T = 0$ K. However, unless ΔH data near absolute zero is used to evaluate ΔH_0, it is a rather poor estimate of the true ΔH_0.

G. Summary

The following basic principles were covered in this chapter:

1. Objectives of Thermodynamics: (a) Prediction of Spontaneity for a process or chemical reaction (b) Calculation of the equilibrium yield for the reaction.

2. First Law of Thermodynamics: $dU = dQ + dW$ where dU is an exact differential (U is a state function called Internal Energy) whereas dQ and dW are inexact differentials.

3. Enthalpy: $H = U + PV$ for processes involving only mechanical work, $dH = dQ$ at constant pressure.

4. For an ideal gas U and H are functions of T only and $C_P = C_V + R$.

5. For non-ideal gases $(\partial U/\partial V)_T$ must be evaluated to calculate ΔU. Also the relationship between C_P and C_V must be evaluated.

6. Bond Dissociation Energy (D_{A-B}) is the energy (H) required to break the A-B bond at $0°K$. Bond Stength is a comparable quantity but at $25°C$.

7. The ΔH for a reaction at some temperature T can be evaluated from a known ΔH_1 at some T_1 (such as 298K) and the ΔC_P for the reaction

$$\Delta H = \Delta H_1 + \int_{T_1}^{T} \Delta C_P dT$$

If ΔC_P can be assumed constant, then

$$\Delta H = \Delta H_1 + \Delta C_P(T - T_1)$$

Exercises

1. Consider the change from state A to state C , shown in the following diagram:

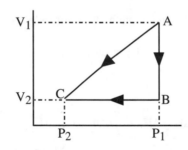

(a) How would Q, W, ΔH, and ΔU for the path AC compare with the path ABC? That is, would the quantities be necessarily equal, unequal, greater, or otherwise? Would these relationships between paths AC , and ABC, be restricted to ideal gases?

(b) Obtain expresions for $\Delta U = U_C \longrightarrow U_A$, and $\Delta H = H_C \longrightarrow H_A$, for an ideal gas. Let T_A = absolute temperature for state A, and T_C = absolute temperature for state C. Let C_V = heat capacity at constant volume. (C_V is not necessarily a constant.) How many different ways can ΔU and ΔH be calculated?

Answer: (a) $\Delta U_{ABC} = \Delta U_{AC}$, $\Delta H_{ABC} = \Delta H_{AC}$, $Q_{ABC} \neq Q_{AC}$, $W_{ABC} \neq W_{AC}$. These relationships would not be restricted to ideal gases.

$$(b)\ \Delta U_{AC} = n \int_{T_A}^{T_C} C_V dT \ , \Delta H_{AC} = n \int_{T_A}^{T_C} C_P dT$$

2. Criticize the statement: "An investigation has been initiated to find a new reaction sequence so that the ΔU for the reaction could be lowered."

3. If the reaction between $H_2 + Br_2$ ocurred along paths A and B,

Path A

$$Br_2 \longrightarrow 2\,Br^{\cdot}$$

$$Br^{\cdot} + H_2 \longrightarrow HBr + H^{\cdot}$$

$$H^{\cdot} + Br^{\cdot} \longrightarrow HBr$$

Path B

$$H_2 + Br_2 \longrightarrow 2HBr$$

show that ΔH is the same for both paths.

4. Calculate ΔU, ΔH, Q, and W for the vaporization of one mole of NH_3 at - 10.0°C. The vapor pressure of $NH_3(\ell)$ at - 10°C is 2.966 kg/cm². The heat of vaporization = 1,297 J/g at - 10°C. Assume the vapor obeys the ideal gas equation of state. Neglect the volume of liquid compared to the volume of vapor. (Hint: at constant P the work is $W = - P(V_g - V = - PV_g)$.) Answer: $\Delta H = Q = 22.1$ kJ, $W = - 2.19$ kJ, $\Delta U = 19.9$ kJ.

5. Assume a gas obeys the equation of state

$$PV = RT + bP \quad \text{(one mole)}$$

where b = constant.
(a) Using equation (2-D-15), show that

$$\left(\frac{\partial U}{\partial V} \right)_T = 0$$

for this equation of state.
(b) Derive expressions for Q,W, ΔU, and ΔH for an isothermal, reversible expansion from V_1 to V_2 for one mole of a gas obeying this equation of state.
Answer: $\Delta U = 0$, $Q = - W = RT\ln \dfrac{V_2 - b}{V_1 - b}$; $\Delta H = b(P_2 - P_1)$

6. 0.0400 moles of a real gas (non-ideal) is heated at a constant pressure of one atmosphere from a volume of 0.250 liters at 25°C to a volume of 0.550 liters at 325°C. The heat capacity of the gas is $C_p = 29.29$ J/mole-K. Calculate Q, W, ΔU, ΔH for this process. Answer: $Q = \Delta H = 351.5$ J, $W = - 30.4$ J, $\Delta U = 321.1$ J.

7. Ideal rubber, similar to an ideal gas, is defined as one whose molar internal energy (U) is a function of temperature only. Derive expressions for ΔU, Q, W, and ΔH for the reversible isothermal expansion described in Exercise 3 at the end of Chapter 1. The enthalpy for stretching and contraction is defined as

$$H = U + fL$$

Answer: $\Delta U = 0$, $Q = - W = - kT\dfrac{\left(L_2^2 - L_1^2 \right)}{2} + kTL_0(L_2 - L_1)$,
$\Delta H = - kT(L_2 - L_1)(L_2 + L_1 - L_0)$

8. Derive expressions for ΔU, Q, and ΔH for the isothermal expansion described in exercise 4 at the end of Chapter 1. Assume the rubber is ideal.
Answer: $\Delta U = 0$, $Q = - W = - kT(L_2 - L_0)(L_2 - L_1)$, $\Delta H = - kT(L_2 - L_1)$
x $(L_2 + L_1 - L_0)$

9. (a) Derive expressions for Q, W, ΔU, and ΔH for heating one mole of ideal rubber from T_1 to T_2 at constant length. Let C_L represent the heat capacity at constant length.

 (b) Show that

$$\Delta H = Q$$

 for heating or cooling of ideal rubber at constant f.

 (c) Calculate ΔH, ΔU, Q, and W, for heating at constant f. Let C_f represent the molar heat capacity at constant force.

 Answer: (a) $Q = \Delta U = \int_{T_1}^{T_2} C_L dT$; $W = 0$, $\Delta H = \int_{T_1}^{T_2} C_L dT + kL(L - L_o)(T_2 - T_1)$

 (c) $Q = \Delta H = \int_{T_1}^{T_2} C_f dT$, $W = - f(L_2 - L_1)$, $\Delta U = \int_{T_1}^{T_2} C_f dT - f(L_2 - L_1)$

10. Estimate the heat of formation of

 using heats of formation of the elements and average bond energies from the Appendices. Answer: - 208 kJ.

11. From the bond dissociation energies at 25°C (bond strengths) in Appendix D, Table 9, calculate an average bond energy for N - H. Answer: ε_{N-H} = 397 kJ.

12. Calculate ΔH_f^o in the gas phase for *o*-methylaniline and *m*-methylaniline, using the heats of combustion found in Table 5 in the Appendix, Section D. Then, compare $\Delta H_{f,(g)}^o$ for these isomers. Answer: $\Delta H_{f,o\text{-methylaniline}}^o$ = 56.6 kJ/mol and $\Delta H_{f,m\text{-methylaniline}}^o$ = 55.3 kJ/mol. The greater steric hindrance expected in the ortho isomer increases the ΔH_f^o only slightly over that for the meta isomer.

13. If methane and propane cost the same price per pound, which of the two would be the most economical fuel for heating? Assume complete combustion in both cases. Answer: Methane but the difference is small.

14. (a) Calculate the bond dissociation energies for $CH_3 - NO_2$ and $C_6H_5 - NO_2$, using data given below.

 (b) Calculate the heat of reaction at 25°C for

$$CH_3NO_2(g) + 3\ H_2(g) = CH_3NH_2(g) + 2\ H_2O(g)$$

Standard Heats of Formation at 25°C are

Species	ΔH_f^o(kcal/mol)	Specie	ΔH_f^o(kcal/mol)
$CO_2(g)$	- 94.052	$C_6H_5NO_2(g)$	16.8
$H_2O(g)$	- 57.798	$N_2O_4(g)$	2.31
$H_2O(l)$	- 68.32	$CH_3(g)$	34.
$CH_4(g)$	- 17.889	$C_2H_5(g)$	25.7
$C_2H_6(g)$	- 20.236	$C_6H_5(g)$	71.
$C_3H_8(g)$	- 24.820	$H_2(g)$	0.0
$C_6H_6(g)$	19.820	C(graphite)	0.0
$CH_3NH_2(g)$	- 6.7	H	52.10
$CH_3NO_2(l)$	- 21.28		

Bond Dissociation Energy: $D_{O_2N-NO_2}$ = 13 kcal/mol

Heat of Vaporization: $\Delta H_{vap,CH_3NO_2}$ = 9.2 kcal/mol

Answer: (a) $D_{CH_3-NO_2}$ = 53.7 kcal/mole $D_{C_6H_5-NO_2}$ = 61.9 kcal/mole,
(b) - 110 kcal.

15. (a) Estimate the heat of formation of

$$NH_2 - CH_2 - CH_2 - OH$$

using average bond energies.
(b) Estimate $\Delta H_{f,NH_2CH_2CH_2OH}^o$ using ΔH_f^o for C_2H_5OH, $C_2H_5NH_2$, and C_2H_6.
Assume ε_{C-C} is the same for all compounds. Compare this estimate with that from
part (a). Answer: (a) - 80 kJ (b) -199 kJ.

16. (a) Calculate the heat of hydrogenation of benzene (gas) to cyclohexane (gas). Use
ΔH_f^o's in Appendix D, Table 3.
(b) Calculate the heat of hydrogenation of the following compounds, all species in
the gas phase.

ethylene$_{(g)}$ $\longrightarrow$ ethane$_{(g)}$

propene$_{(g)}$ $\longrightarrow$ propane$_{(g)}$

1-butene$_{(g)}$ $\longrightarrow$ butane$_{(g)}$

cis-2-butene$_{(g)}$ $\longrightarrow$ butane$_{(g)}$

trans-2-butene$_{(g)}$ $\longrightarrow$ butane$_{(g)}$

(c) From the data in (b), obtain an estimate of the heat of hydrogenation for
    H   H
cis or *trans* R - C = C - R. Compare with the value for *trans*-2-pentene. Account for
the difference in heat of hydrogenation for *cis* and *trans*-2-butenes.
(d) From the result in (c), estimate the heat of hydrogenation of benzene if the 3
double bonds were independent. Account for the discrepancy between this estimate
and the true heat of hydrogenation obtained in (a). (Hint: Consider resonance or de-
localization energy.) Answer: (a) -206.0 kJ (b) -136.5 kJ, -124.7 kJ, -125.7 kJ, -118.6
kJ, -114.5 kJ (c) average ~- 124 kJ, which is a larger heat of hydrogenation than that
for *trans*-2-pentene (-115.0 kJ). *Cis*-2-butene would have a higher ΔH_f than *trans*-2-
butene due to steric hindrance. Consequently, the $\Delta H_{hydrogention}$ for *cis*-2-butene
should be higher than that for *trans*-2-butene (more heat liberated). (d) - 372 kJ; this
is greater in magnitude than $\Delta H_{hydrogenation}$ of benzene (-206.5 kJ). The difference
can be attributed to the resonance stabilization of benzene which should be ~ 166 kJ.

17. From $\Delta H^o_{f,elements}$, $\Delta H^o_{f,radicals}$, and ΔH^o_f for compounds found in Appendix D (Tables 3, 7, and 8), calculate the bond dissociation energies in $C_6H_5 - H$, $CH_3 - H$, and $C_6H_5CH_2 - H$. Can you account for the relative values? Answer: 464.3 kJ, 438.4 kJ, 367.6 kJ. Apparently the H bonded to the benzene ring has some resonance stabilization. The H bonded in the CH_3 of toluene has a lower bond dissociation energy due to the resonance stabilization of the remaining CH_2 group with the phenyl ring.

18. Calculate the heat of formation of $CH_3CO^{\cdot}_{(g)}$ from the ΔH for the reaction

$$I^{\cdot} + CH_3CHO = CH_3CO^{\cdot} + HI \qquad \Delta H^o_{298K} = 61.1 \pm 5.0 \text{ kJ}$$

and the heats of formation: $\Delta H^o_{f,I(g)} = 106.767$ kJ/mol, $\Delta H^o_{f,HI(g)} = 26.4$ kJ/mol. Compare the results with those in Problem 2-F-3. Should they be the same value? Answer: - 24.7 kJ; they should be the same.

19. From your knowledge of organic chemistry, devise a scheme for the determination of the heat of formation of nitrobenzene.

20. Using reactions other than those in Example 2-C-2, show how the heat of formation of NH_3 could be determined.

21. The reaction

$$\underset{\text{(forsterite)} \quad \text{(quartz)}}{Mg_2SiO_4 + SiO_2} \longrightarrow \underset{\text{(enstatite)}}{2MgSiO_3}$$

is virtually impossible to carry out rapidly enough to determine ΔH directly from calorimetric measurements. For this reason, ΔH must be determined indirectly from heats of solution. Use the following heats of solution in HF to evaluate ΔH^o for the above reaction.

	ΔH^o(kcal/mol)
$Mg_2SiO_4 + \infty HF \rightarrow$ solution I	- 95.4
$SiO_2 + \infty HF \rightarrow$ solution II	- 33.0
Solution I + Solution II $\rightarrow$ solution III	0.0
$MgSiO_3 + \infty HF \rightarrow$ solution III	- 62.9

Answer: - 2.6 kcal.

22. Calculate $\Delta H^o_{f,Mg_2SiO_4}$ from the following data:

$\Delta H^o_{f,SiO_2} = -205.4$ kcal/mol

$\Delta H^o_{f,MgO} = -143.8$ kcal/mol

$$\underset{\text{(quartz)} \quad \text{(pericalse)}}{SiO_2 + 2MgO} \longrightarrow \underset{\text{(fosterite)}}{Mg_2SiO_4} \qquad \Delta H^o = -15.1 \text{ kcal/mol}$$

Answer: - 508.1 kcal.

23. Suggest a procedure to determine ΔH for the process

$$\underset{\text{calcite}}{CaCO_3} \rightarrow \underset{\text{aragonite}}{CaCO_3}$$

An indirect process must be used since the reaction is slow. (Hint: $CaCO_3$ is soluble in acid solution.)

24. Calculate the standard heat of formation of NH_3 at $500^\circ C$, assuming the heat capacities, Cp, for ammonia and the elements are constant. The heat capacities can be found in Table 2 in Appendix D. Answer: - 56.7 kJ.

25. Calculate Q, W, ΔU, and ΔH for heating 100 g of water from $25^\circ C$ to $100^\circ C$ at a constant pressure of 1 atm. The density of water at $25^\circ C$ is 1.10534 g/mL and at $100^\circ C$ is 1.06346 g/mL. Assume the molar heat capacity is constant, $C_P = 18.0$ cal/mol-K Answer: $Q = \Delta H = 31.4$ kJ, $W = - 3.6 \times 10^{-4}$ kJ, $\Delta U \sim Q = \Delta H$.

26. 0.3 moles of an ideal gas is in an initial state of P = 1 atm, V = 10 L. Calculate the ΔH or the change to a final state of P = 3 atm and V = 3 L. Assume $C_P = 7/2$ R. Answer: -355 J.

27. Consider the reaction

$$CH_{4\,(g)} + H_2O_{\,(g)} = CO_{\,(g)} + 3H_{2\,(g)}$$

(a) Calculate ΔH^O at $25^\circ C$ using data from the Appendices.
(b) If you were given a table of the heat capacities as a function of temperature from $25^\circ C$ to $900^\circ C$, show how you would calculate ΔH^O at $900^\circ C$.

Answer: (a) 205.7 kJ (b) $\Delta H^o_{1173} = 205.7$ kJ $+ \Delta a\,(T_2 - T_1) + \dfrac{\Delta b}{2}(T_2^2 - T_1^2) +$

$\dfrac{\Delta c}{3}(T_2^3 - T_1^3)$, $T_1 = 298$ and $T_2 = 1173$.

28. An ideal gas undergoes a change along the path A → B → C as described in the following graph

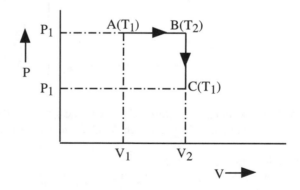

The initial state A has $P_1 = 4$ atm, $V_1 = 2$ L, $T_1 = 300$ K and the final state C has $P_2 = 2$ atm, $V_2 = 4$ L, $T_1 = 300$ K. The intermediate state B has a temperature T_2. Calculate Q, W, ΔU, and ΔH for the path A → B → C.
Answer: $\Delta U = \Delta H = 0$, $Q_{ABC} = - W_{ABC} = 810$ J.

29. Consider the equation of state

$$PV = RT + BP \qquad \text{(one mole)}$$

where B is a function of temperature. Derive an expression for the relationship between Cp and C_V. Answer: $C_P = C_V + \left[1 + \dfrac{1}{V - nb}\dfrac{dB}{dT}\right]\left[R + P\dfrac{dB}{dT}\right]$.

30. 0.288 moles of an ideal gas is expanded isothermally at t = 150°C along the path ABC. Describe on the following P versus V graph.

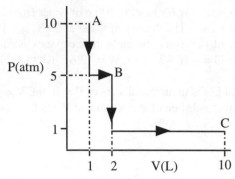

At points A and B there are sudden decreases in pressure to the final pressures of the expansion. Calculate q,w, ΔU, and ΔH for the process. Express the answers in kcal.
Answer: $\Delta U_{ABC} = \Delta H_{ABC} = 0$, $Q_{ABC} = -W_{ABC} = 0.315$ kcal. = 1.32 kJ.

31. Two moles of an ideal gas undergo the changes of state shown in the following diagram:

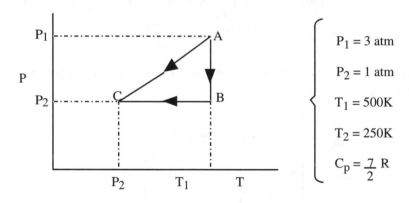

$P_1 = 3$ atm

$P_2 = 1$ atm

$T_1 = 500K$

$T_2 = 250K$

$C_p = \dfrac{7}{2} R$

(a) Calculate the work along the path ABC if the process is carried out reversibly.
(b) Calculate the heat change along the path ABC if the process is carried out reversibly.
(c) Calculate ΔU along the path AC.
Answer: (a) - 4.98 kJ (b) - 5.41 kJ (c) -10.39 kJ.

32. For the following reaction

$$HCN_{(g)} + 3H_{2(g)} = CH_{4(g)} + NH_{3(g)}$$

(a) Calculate ΔH^o at 298 K.
(b) Calculate ΔH^o at 500 K assuming the heat capacity of each gas is constant.
Answer: (a) -255.4 kJ (b) -265.9 kJ.

33. Helium is passed through a heated tube at a rate of 30 mL/min measured at 25°C. The outlet of the tube is exposed to the atmosphere, which we will assume is at one atmosphere pressure. At what rate (cal/min) must heat be supplied to the tube in order to increase the temperature of the helium from 25°C to 150°C? The heat capacity for helium is $C_V = 3/2R$. Assume helium obeys the ideal gas equation of state. Answer: 0.762 cal/min.

3

THE SECOND AND THIRD LAWS
OF THERMODYNAMICS

There are two well-known solid phases of carbon: graphite and diamond. In this chapter you will understand on the basis of thermodynamics the conversion from one form into another. However, in recent years other solid forms of carbon have been found which are classified as fullerenes. These solid phases consist of molecules containing 60 or more carbon atoms. In general they have cage-like structures in which molecular and ionic species can reside in the large opening in the center of the cage. They have been referred to as buckyballs since their outward appearance is a ball with planar rings bonded together on the surface. Because of their unique structures, the fullerenes are thought to possess unique properties that may catalyze various reactions. The thermodynamics of the formation and vaporization of the fullerenes and mixtures of the fullerenes has been carried out in order to better understand the formation and properties of these unique species. [C.K. Mathews, M.Sai Baba, and T.S. Lakshmi Norasimhan, "Thermodynamics of Fullerenes," pp. 459-475 in Recent Advances in the Chemistry and Physics of Fullerenes and Related Materials, *Ed. K.M. Kadish and R.S. Ruoff, The Electrochemical Society, Inc., Pennington, N.J., USA.] The importance of the discovery of the fullerenes was recognized by awarding the Nobel Prize in Chemistry to Drs. R.E. Smalley and R.F. Curl, professors at Rice University and Dr. H.W. Kroto, professor at Sussex University.*

A. ΔU, ΔH, and Spontaneity

As stated at the outset of Chapter 2, one of the principal objectives of thermodynamics is the establishment of a criterion for spontaneity. As defined earlier, spontaneity refers to the possibility that a process *may* take place. We cannot, however, be definite that a spontaneous process will proceed in a specific time, but only that it may proceed, given an infinite amount of time. On the other hand, if we can decide that a process is not spontaneous, it means we can *guarantee* that the process will *not* occur.

At this stage of our study of thermodynamics we might ask whether we have yet encountered a thermodynamic parameter which can be used as a criterion for spontaneity. We have been introduced to two energy terms, ΔU and ΔH, which are state functions and could possibly serve the purpose. Because of everyday experiences, we frequently associate a decrease in energy with a process that proceeds by itself. This probably arises from our association with classical mechanics. For example, suppose a boulder were situated on the edge of a cliff. If we asked even a small child what would happen if we were to give the boulder a shove over the edge, he would immediately reply that it would fall to the bottom; in other words, the boulder would fall of its own accord, or fall spontaneously, once it was freed of support. If the question, Why? were asked of an older person, his answer would probably be associated with the gravitational attraction to the earth. Although this is essentially correct, we could also express the answer in terms of potential energy which arises from the gravitational attraction. This potential energy is a function only of the height above the earth's surface and the mass of the object, and is independent of the velocity of the object if the atmosphere is neglected. Assuming that the object's initial velocity was zero, we can postulate that the boulder falls because there is a *decrease in potential energy*. If we noticed that this phenomenon occurred for numerous objects, we could then generalize the phenomena into a law, using a decrease in potential energy as the general criterion for the spontaneity of the process.

Now, are the signs of ΔU and ΔH useful criteria for spontaneity? In other words, can a process always occur if it is accompanied by a decrease in energy or enthalpy? We can show that the answer is *no* by giving a few simple examples, and only one exception proves the criterion is not generally valid.

Example 3-A-1. *Isothermal expansion of an ideal gas against a constant pressure.* This process has been discussed in Chapter 1, Section E. A graph of P versus V describes the path taken in Figure 1-E-1 (path [#]1). The work for the process was shown to be

$$W = - P_2(V_2 - V_1)$$

We know from Chapter 2 that U and H are a function of temperature only for an ideal gas. Since the process is isothermal, U and H must remain constant and

$$\Delta U = \Delta H = 0$$

It is obvious that this process of expansion against the final pressure P_2 is spontaneous, since the external pressure was reduced suddenly by removing the weights simultaneously. In this process we conclude that the process is spontaneous and $\Delta U = \Delta H = 0$.

Example 3-A-2. *Melting of ice at 10°C.* Experience has shown us another example which goes spontaneously. The change in enthalpy can be calculated by considering the path described by the arrows in the following diagram.

The enthalpy change, ΔH, can be calculated from the sum of the three steps.

$$\Delta H = \Delta H_1 + \Delta H_2 + \Delta H_3$$

$$\Delta H = \int_{283}^{273} C_{P(s)}dT + \Delta H_{fusion} + \int_{273}^{283} C_{P(l)}dT$$

Assuming the heat capacities are constant:

$$C_{P(s)} = 36.4 \text{ J/K-mol}$$
$$C_{P(l)} = 75.3 \text{ J/K-mol}$$

$$\Delta H = 36.4(273-283) + 6025 + 75.3(283-273) \text{ J/mol}$$
$$\Delta H = 6.40 \text{ kJ/mol}$$

Since this a condensed phase, $\Delta(PV)$ is quite small, and for all practical purposes,

$$\Delta U \approx \Delta H = 6.40 \text{ kJ/mol}$$

In this case, note that *ΔU and ΔH are positive (an increase in energy), yet the process is spontaneous.* (Ice obviously melts at 10°C.)

Problem 3-A-1. In a manner similar to that described in Example 3-A-2, show that

$$\Delta U \approx \Delta H = -5.65 \text{ kJ/mol}$$

for the process

$$H_2O(l, -10^oC) \rightarrow H_2O(s, -10^oC)$$

Here we note that the process, though associated with a negative value for ΔU and ΔH, is spontaneous. We have now encountered all three possibilities for ΔU and ΔH; namely, zero, negative, and positive, yet all three processes are spontaneous. Obviously then, *neither ΔU nor ΔH is a good criterion for spontaneity.*

B. Second Law of Thermodynamics

The answer to the establishment of a criterion for spontaneity can be found in the Second Law of Thermodynamics. There are several ways in which this law can be presented, *all of which can be shown to be equivalent.* The student should consult a more comprehensive physical chemistry text, where the equivalence between these statements of the Second Law is shown.

Of these various statements of the Second Law of Thermodynamics, some are easier to understand and accept because of our everyday experiences. Let us take, for example, the statement: "It is impossible for a body at a lower temperature to transfer heat to a body at a higher temperature, unless additional work is done in the process." Obviously, anyone who has had the experience of touching a hot frying pan with his cold hand has had the opportunity to agree with this statement of the Second Law. The statement does *not* say it is impossible to transfer heat from a cold to a hot body under all conditions—if this were true the air conditioning manufacturers would quickly go out of business. However, in the operation of an air conditioner, work must be added to the process, and for this reason, an air conditioner costs money to operate.

Although the preceding statement of the Second Law has the advantage that it is quickly understood, it has the disadvantage that it is not readily applicable to chemical problems. The statement of the Second Law which we will present is of the opposite nature, in that it is readily applicable to problems, but it will not be as easy to relate to our physical experiences. However, we will show that our statement of the Second Law is in agreement with several physical processes that occur spontaneously, and thus, its universal applicability should be more easily accepted.

A new term called entropy will be introduced in our statement of the Second Law, and, though it may seem foreign at first, it should not be any more puzzling than U, internal energy, in the First Law of Thermodynamics. U is probably readily acceptable because it has the word energy in it, which is familiar. But as the new term is used, it too will become acceptable to us.

Entropy

The function called **entropy**, **S**, is defined through its differential by the equation

$$dS = \frac{dQ \text{ reversible}}{T} \tag{3-B-1}$$

where $dQ_{reversible}$ refers to the differential of heat for a *reversible path* between the initial and final states, and T is equal to the absolute temperature. For a spontaneous process,

$$dS > \frac{dQ \text{ spontaneous}}{T} \qquad \textbf{Second Law of Thermodynamics} \qquad (3\text{-}B\text{-}2)$$

We take equations (3-B-1) and (3-B-2) to be our statement of the Second Law of Thermodynamics. dS is an exact differential, therefore, S is a state function. Consequently, ΔS is dependent only on the initial and final states of the system, and not on the particular path followed.

T in equations (3-B-1) and (3-B-2) is a variable called the thermodynamic temperature, which is an integrating factor that makes the inexact differential dQ an exact differential, dS; that is, S is a state function. We use the symbol T to represent the thermodynamic temperature scale, which is the same symbol used for the ideal gas temperature scale. Although the two scales are defined quite differently, they can be shown to be identical—a rather amazing result indeed. A comprehensive text on thermodynamics will show the equivalence between these two temperature scales. For this reason, we let T represent both temperature scales. The importance of the preceding statement of the Second Law of Thermodynamics is not readily apparent. However, after applying this criterion for spontaneity to numerous processes, you will gain a better appreciation for its general use and value. This will be especially true after we have defined a new function called free energy in terms of entropy, and have applied it to chemical transformations.

Calculation of Entropy Change

When calculating the entropy change for a given process, we must adhere to the exact definition given in equation (3-B-1). If the process is actually reversible, then this definition can be applied directly, when dQ is associated with the reversible process. On the other hand, if the process is not reversible, you might wonder how we can calculate ΔS for these processes to determine if the > sign of equation (3-B-2) applies. The answer to this question lies in the fact that S is a state function, and therefore, ΔS is independent of the path. Then, the calculation of the entropy change for the irreversible process, ΔS_{irrev}, can be accomplished by calculating the entropy change for a reversible path, ΔS_{rev}, which begins at the same initial state and ends at the same final state. We can depict this graphically, as in Figure 3-B-1. Since S is a state function,

$$\Delta S_{rev} = \Delta S_{irrev} = S_2 - S_1$$

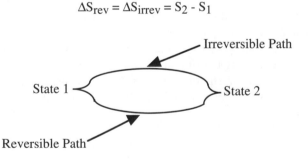

Figure 3-B-1. Calculation of Entropy Change Through the Reversible Path ΔS_{rev}

This is a most important concept to remember, since we are generally interested in the calculation of ΔS for irreversible or potentially spontaneous processes. This might seem contradictory at first, in view of equations (3-B-1) and (3-B-2); however, a simple example should convince you that it is consistent. For chemical reactions, ΔS can be calculated from the entropy of the products of the reaction minus the entropy of the reactants, and this

ΔS would apply, whether the process were carried out reversibly or irreversibly. However, for the reversible process,

$$\Delta S_{rev} = \int \frac{dQ_{rev}}{T}$$

where the subscript on Q specifies that this Q refers to the reversible process. For the irreversible process, providing it is spontaneous,

$$\Delta S_{irrev} > \int \frac{dQ_{irrev}}{T}$$

These results can be summarized in the following equation, again assuming the irreversible step is spontaneous.

$$\Delta S_{rev} = \int \frac{dQ_{rev}}{T} = \Delta S_{irrev} > \int \frac{dQ_{irrev}}{T}$$

If the process is not spontaneous, the > sign in equation (3-B-2) would be < and *the process would be spontaneous in the reverse direction.*

Calculation of ΔS for Reversible Processes

Since ΔS can only be calculated along a reversible path, we will first consider certain basic reversible processes and derive the formulas by which we can calculate ΔS. These can be applied to irreversible processes in order to obtain a reversible path between the same initial and final states.

Isothermal Expansion of an Ideal Gas: From Chapter 2 we know that the internal energy, U, of an ideal gas is a function of temperature only. Since the process is isothermal, the change in U must be zero and from the First Law

$$0 = \Delta U = Q + W$$

$$Q = -W$$

For a reversible process we know that the work is given by equation (1-E-8)

$$Q_{rev} = -W_{rev} = nRT \ln \frac{V_2}{V_1}$$

The entropy change is calculated from

$$dS = \frac{dQ_{rev}}{T}$$

Since T = constant for an isothermal process

$$\int dS = \int \frac{dQ_{rev}}{T} = \frac{1}{T} \int dQ_{rev}$$

$$\Delta S = \frac{Q_{rev}}{T} = \frac{nRT}{T} \ln \frac{V_2}{V_1}$$

$$\mathbf{\Delta S = nR \ln \frac{V_2}{V_1}} \tag{3-B-3}$$

Constant Pressure Process: If a substance is heated or cooled at constant pressure the process is reversible and the entropy change can be calculated from the heat change. For a constant pressure process the heat change dQ is equal to the enthalpy change dH as shown in equation (2-D-4). The enthalpy change for a constant pressure process is given by equation (2-D-12)

$$dQ_{rev} = dH_{rev} = nC_PdT \tag{2-D-12}$$

$$\int dS = \int \frac{dQ_{rev}}{T} = \int_{T_1}^{T_2} \frac{nC_PdT}{T}$$

$$\Delta S = n \int_{T_1}^{T_2} \frac{C_PdT}{T} \tag{3-B-4}$$

If C_P is known as a function of temperature, the entropy change can be calculated by evaluating the integral on the right side. If C_P is given in tabular form at different temperatures, a numerical integration of C_P/T versus T can be carried out as discussed in Appendix C. For small temperature changes, C_P may be assumed constant and the integral takes on a simple form

$$\Delta S = nC_P \ln \frac{T_2}{T_1}.$$

Constant Volume Process: Similarly, if a substance is heated or cooled at constant volume the process is reversible and the entropy change can be calculated from

$$dQ_{rev} = dU_{rev} = nC_VdT \tag{2-D-13}$$

$$\int dS = \int \frac{dQ_{rev}}{T} = n \int_{T_1}^{T_2} \frac{C_VdT}{T}$$

$$\Delta S = n \int_{T_1}^{T_2} \frac{C_VdT}{T} \tag{3-B-5}$$

Again C_V must be known as a function of temperature in order to evaluate the integral. If C_V is assumed constant

$$\Delta S = nC_V \ln \frac{T_2}{T_1}.$$

Reversible Phase Change: If a phase change is carried out where the two phases are in equilibrium, the process is reversible and the entropy change can be evaluated from the ΔH for the phase transition, ΔH_{trans}, and the temperature of the phase transition, T_{trans},

$$\Delta S = \frac{Q_{rev}}{T} = \frac{\Delta H_{trans}}{T_{trans}} \tag{3-B-6}$$

The values for ΔH_{trans} and T_{trans} must be known from experiment for each compound. This equation would apply to any phase transition solid --> liquid, liquid --> gas, or solid --> gas providing the two phases are in equilibrium. If the phase transition is carried out under irreversible conditions (phases not in equilibrium) then ΔS must be evaluated along some reversible path, as illustrated in Example 3-B-2.

Application of the Second Law of Thermodynamics

In order to illustrate the application of the Second Law to irreversible processes, we might investigate the examples we worked previously in Section 3-A to see if equation (3-B-2) can predict spontaneity of a process. As you recall, both ΔU and ΔH failed to do so.

Example 3-B-1. *Isothermal expansion of an ideal gas against a constant pressure.* Referring to Example 3-A-1 we note that $\Delta U = 0$. Thus from the First Law of Thermodynamics

$$dU = dQ_{irrev} + dW_{irrev} = 0$$

$$dQ_{irrev} = - dW_{irrev}$$

The right hand side of equation (3-B-2) is then

$$\int \frac{dQ_{irrev}}{T} = - \int \frac{dW_{irrev}}{T}$$

Since T is constant

$$\int \frac{dQ_{irrev}}{T} = - \frac{1}{T} \int dW_{irrev} = - \frac{W_{irrev}}{T}$$

W_{irrev} for an isothermal expansion against a constant external pressure equal to the final pressure is given by equation (1-E-6) in Chapter 1

$$- W_{irrev} = P_2(V_2 - V_1)$$

The integral of the right hand side of equation (3-B-2) then becomes

$$\int \frac{dQ_{irrev}}{T} = \frac{P_2(V_2 - V_1)}{T} \tag{3-B-7}$$

The ΔS can be calculated from the *reversible*, isothermal process which was derived in the previous section. The ΔS was given by equation (3-B-3).

$$\Delta S = \frac{nRT}{T} \; \ln \frac{V_2}{V_1} = nR\ln \frac{V_2}{V_1} \tag{3-B-3}$$

Now comparing

$$\Delta S \; ? \int \frac{dQ_{irrev}}{T}$$

and substituting in the results of equations (3-B-3) and (3-B-7),

$$nR\ln \frac{V_2}{V_1} \; ? \frac{P_2(V_2 - V_1)}{T}$$

It is not obvious if the left side is larger or smaller than the right side. However, from above one can see that the left side of the equation is $-W_{rev}/T$ and earlier we showed that the right side is $-W_{irrev}/T$.

$$-\frac{W_{rev}}{T} \quad ? \quad -\frac{W_{irrev}}{T}$$

In Chapter 1, Section E we showed that W_{rev} is the <u>maximum</u> work performed by the system and that

$$-W_{rev} > -W_{irrev}$$

Thus we conclude that

$$\Delta S = nR\ln\frac{V_2}{V_1} > \frac{P_2(V_2 - V_1)}{T} = \int\frac{dQ_{irrev}}{T}$$

and according to the Second Law of Thermodynamics the process is spontaneous, which is in agreement with our experience that the gas will expand from V_1 to V_2.

Problem 3-B-1. Suppose we have two ideal gases which are separated in a container by a partition, as shown in the following diagram. The two gases are at the same pressure,

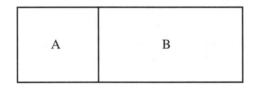

but occupy different volumes, V_A and V_B. If the partition is punctured, the two gases will mix until a homogeneous mixture in both compartments is attained.

(a) Show that the following expression for ΔS, commonly referred to as the *ideal entropy of mixing*, can be derived for this process.

$$\Delta S = n_A R \ln\frac{V_A + V_B}{V_A} + n_B R \ln\frac{V_A + V_B}{V_B}$$

where n_A, and n_B are the moles of A and B, respectively. (*Hint:* the entropy change of A can be calculated from an isothermal expansion from V_A to $V_A + V_B$.)

(b) Show that the above expressions for the ideal entropy of mixing can be expressed in terms of mole fractions X_A and X_B.

$$\Delta S = -n_A R \ln X_A - n_B R \ln X_B \tag{3-B-8}$$

(c) Show that the ideal entropy for one mole of mixture given by equation (3-B-9) can be derived from equation (3-B-8).

$$\overline{\Delta S} = -X_A R \ln X_A - X_B R \ln X_B = -R\sum X_i \ln X_i \tag{3-B-9}$$

where i is summed over all components in the mixture and $\overline{\Delta S}$ is the entropy change per mole of the mixture. Does the above process agree with the Second Law of Thermodynamics?

Example 3-B-2. *Melting of ice at 10ºC.* Referring to Example 3-A-2 we note that when ice melts at 10°C,

$$\Delta H = 6.40 \text{ kJ/mol}$$

Since the process was carried out at constant pressure,

$$\Delta H = Q_P = 6.40 \text{ kJ/mol}.$$

The process occurs spontaneously; hence, Q_P is the heat corresponding to an irreversible process carried out at T = 283K, and

$$\int \frac{dQ_{irrev}}{T} = \frac{Q_P}{T} = \frac{6.40 \text{ kJ/mole}}{283K} = 22.6 \text{ J/K-mol}$$

Now in order to calculate ΔS, we must construct a reversible path from the same initial state to the same final state. The path considered previously in example 3-A-2 can be constructed reversibly, so we will follow it and calculate $\Delta S_1, \Delta S_2, \Delta S_3$, corresponding to the three steps. Since S is a state function, ΔS for the irreversible process is

$$\Delta S = \Delta S_1 + \Delta S_2 + \Delta S_3$$

Since step 1 is at constant P,

$$\Delta S_1 = \int \frac{dQ_{rev}}{T} = \int_{283}^{273} \frac{C_{P(s)}dT}{T} = 36.4 \ln \frac{273}{283} \text{ J/K-mol}$$

Step 2 is at constant P and T:

$$\Delta S_2 = \int \frac{dQ_{rev}}{T} = \frac{Q_{rev}}{T} = \frac{\Delta H_{fus}}{T} = \frac{6025.5 J/mol}{273K}$$

Step 3 is also at constant P, so

$$\Delta S_3 = \int \frac{dQ_{rev}}{T} = \int_{273}^{283} \frac{C_{P(l)}dT}{T} = 75.3 \ln \frac{283}{273} \text{ J/K-mol}$$

Hence,

$$\Delta S = (-1.31 + 22.07 + 2.71) \text{ J/K-mol}$$

$$\Delta S = 23.5 \text{ J/K-mol}$$

Therefore, as the Second Law predicts for a spontaneous process,

$$\Delta S > \int \frac{dQ}{T}$$

$$23.5 \text{ J/K-mole} > 22.6 \text{ J/K-mol}.$$

Problem 3-B-2. Calculate ΔS for the process in Problem 3-A-1 and show that the Second Law predicts the spontaneity of the process.

Answer: $\Delta S = -20.6$ J/K-mol $> \int \frac{dQ_{irrev}}{T} = -21.5$ J/mol.

Problem 3-B-3. Calculate ΔS for the processes described in Problems 2-E-1, 2-E-2, 2-E-3, 2-E-4, 2-E-5, 2-E-6, 2-E-7. Answer: 2-E-1: -19.1 J/K-mol; 2-E-2: -19.1 J/K-mol; 2-E-3: 61.8 J/K; 2-E-4: -36.71 J/K; 2-E-5: -26.64 J/K; 2-E-6: $\Delta S_1 = \Delta S_2 = n \int_{T_1}^{T_2} \frac{C_P}{T} dT$

$+ nR\ln$; 2-E-7: $\Delta S_{ADC} = \Delta S_{ABC} = n \int_{T_1}^{T_2} \frac{C_V}{T} dT + nR\ln \frac{V_2}{V_1}$.

C. Third Law of Thermodynamics

In the previous discussions of the calculation of entropy change, one might note the following observations:

(1) Gas expansion $V_1 \longrightarrow V_2$, the entropy change is positive since $V_2 > V_1$.

(2) Heating a substance at constant pressure or constant volume $T_1 \longrightarrow T_2$, the entropy change is positive since $T_2 > T_1$ and C_p and C_v are always positive.

(3) Phase transition where ΔH_{trans} is positive (heat absorbed), the entropy change is positive.

In each of the processes there is an increase in the uncertainty of position or energy level for the atoms, molecules, or ions in the substances. This is most evident in phase transitions where ΔH_{trans} is positive and the final state of the transformation is more disordered than the initial state

$$\text{solid} \longrightarrow \text{liquid}$$
$$\text{liquid} \longrightarrow \text{gas}$$
$$\text{solid} \longrightarrow \text{gas}$$

For this reason we associate entropy with a measure of the disorder of the system. This "disorder" is measured by the uncertainty in the position of the atoms, molecules, or ions or their occupation of energy levels, which will be discussed in the chapters on Molecular Motions and Energies and Atomic and Molecular Structure. Based on this concept of disorder as a measure of entropy, we can expect the maximum order (minimum entropy) to be a pure solid at the lowest temperature possible, zero degrees Kelvin. On the basis of this rationale we state:

Third Law of Thermodynamics: The entropy of all perfect crystalline substances is zero at absolute zero; i.e., $S^o_{perfect\ crystal,\ 0K} = 0$.

The assignment of the entropy to be zero at absolute zero is actually arbitrary. The only real requirement is that all perfect substances have the <u>same entropy value</u> at absolute zero. The choice of zero, however, is convenient. By perfect crystalline substance is meant to be a solid of 100% purity and completely ordered array of atoms, molecules, or ions in the solid with no imperfections. Of course nothing can be absolutely perfect but we can attain a level of purity so that the error in S is less than the experimental error in the calorimetric measurements.

Absolute Entropy Values

With the Third Law of Thermodynamics we are able to determine entropy values at 298K by making only thermal measurements. This can be illustrated by the following sequence of reaction steps where we are assuming that the substance, A, has a single solid form and is a liquid at 298K and one atmosphere pressure. The solid undergoes a phase transition to the liquid at the melting point, T_{mp}, with a heat of fusion, ΔH_{fus}.

$$A(s, 0^o_K) \longrightarrow A(s, T<20K) \qquad \Delta S_1 = \int_0^{T<20K} \frac{C_{P(s)}}{T} dT$$

$$A(s, T<20K) \longrightarrow A(s, T_{mp}) \qquad \Delta S_2 = \int_{T<20K}^{T_{mp}} \frac{C_{P(s)}}{T} dT$$

$$A(s, T_{mp}) \longrightarrow A(l, T_{mp}) \qquad \Delta S_3 = \frac{\Delta H_{fus}}{T_{mp}}$$

$$A(l, T_{mp}) \longrightarrow A(l, 298K) \qquad \Delta S_4 = \int_{T_{mp}}^{298K} \frac{C_{P(l)}}{T} dT$$

Since we cannot attain absolute zero, we are limited to values of C_P in the vicinity of absolute zero. In order to calculate the entropy change from absolute zero, we must extrapolate the heat capacity curve to 0K. We do this using the Debye Equation which states that for most crystals C_P is directly proportional to T^3 for $T < 20K$.

$$C_P = kT^3$$

where k is a proportionality constant which varies from compound to compound. In order to evaluate k we need at least one C_P value at a temperature less that 20K. ΔS_1 can then be evaluated

$$\Delta S_1 = \int_0^{T<20K} \frac{kT^3}{T} dT = k \int_0^{T<20K} T^2 dT$$

$$\Delta S_1 = \frac{k}{3} [T^3]_0^{<20K}$$

Evaluation of ΔS_2, ΔS_3, ΔS_4 have been discussed previously in Section B under Constant Pressure Process and Reversible Phase Change. The Thermodynamic Tables in Appendix D have standard entropy values S^o at 298K. These values were obtained in a manner similar to that just described. In some cases the solid may exist in more than one form and it is necessary to consider the ΔH_{trans}, for each of these solid-solid transformations. The standard states for S are:

solid, liquid—pure material at one bar pressure and specified T;

ideal gas—pure material at one bar pressure and specified T;

real gas—pure material in a hypothetical ideal gas state at one bar pressure.

reference state—S = 0 for perfect crystalline material at 0K.

The standard state in the case of a real gas is defined so that it is consistent with the standard state for enthalpy. Since we are primarily interested in ideal gases, we will not be concerned with the calculation for real gases.

Change in Entropy for a Chemical Reaction

We can calculate the ΔS^o for a chemical reaction using the absolute standard entropies given in the Thermodynamic Tables. For the general chemical reaction

$$aA + bB = cC + dD$$

ΔS^o is calculated in a manner similar to the calculation of ΔH^o from $\Delta H^o_{f,i}$ values

$$\Delta S^o = cS^o_C + dS^o_D - aS^o_A - bS^o_B \qquad (3\text{-}B\text{-}10)$$

Note that the units of entropy are joules/K-mol and the resulting ΔS^o will have units of joules/K.

The value of ΔS^o represents the change in disorder in the reaction. A positive value of ΔS^o indicates an increase in disorder whereas a negative value indicates a decrease in disorder or an increase in order. One of the dominant factors influencing the ΔS^o of reaction is the change in moles of gaseous components. Gases are obviously more disordered than condensed phases since their molecules are free to move within the container with no restraint since there are no intermolecular interactions. The contribution of gases to the ΔS^o of reaction is revealed by observing the following reactions and the associated ΔS^o and Δn_{gas}.

With some exceptions one can see that ΔS^o per mole of change in gases is on the order of 140–180 J/K. This is a large entropy change and as we will see later, it plays an important role in the temperature dependence of reactions.

	ΔS^o(J/K)	Δn_{gas}	$\Delta S^o/\Delta n_{gas}$
$NH_4Cl_{(s)} \longrightarrow NH_{3(g)} + HCl_{(g)}$	284.6	2.0	142
$H_{2(g)} + Cl_{2(g)} \longrightarrow 2HCl_{(g)}$	19.8	0.0	—
$C_{(graphite)} + \frac{1}{2}O_{2(g)} \longrightarrow CO_{(g)}$	89.5	0.5	179
$CO_{(g)} + \frac{1}{2}O_{2(g)} \longrightarrow CO_{2(g)}$	-86.8	-0.5	173
$N_2O_{4(g)} \longrightarrow 2NO_{2(g)}$	176.6	1.0	177

Use of the Second and Third Laws of Thermodynamics to Predict Spontaneity of Chemical Reactions

In principle the ΔS^o of reaction could be used to predict spontaneity of a chemical reaction. Both sides of equation (3-B-2) could be evaluated

$$\int dS > \int \frac{dQ}{T}$$

and compared to see if the chemical reaction is spontaneous. The integral of the left side of equation (3-B-2) is simply the ΔS of the reaction which could be evaluated using the absolute entropies based upon the Third Law of Thermodynamics. However, evaluation of the right hand side of equation (3-B-2) could not be carried out unless we define the process by which we are carrying out the reaction. In Section D we will assume the reaction is being carried out at constant pressure and constant temperature. If only PV (mechanical) work is done then

$$dQ_P = dH \qquad\qquad (2\text{-}D\text{-}3)$$

and the integration can be performed since T = constant

$$\int dS > \int \frac{dQ_P}{T} = \frac{1}{T} \int dH$$

$$\Delta S > \frac{\Delta H}{T} \qquad (\text{T,P = constants})$$

The pressure and temperature would need to be specified in order to evaluate ΔH and ΔS for the reaction. Thus in principle we could use the Second and Third Laws to predict spontaneity of a chemical reaction. In the next section we will do this more succinctly by introducing a new thermodynamic function, called Gibbs Free Energy which includes both ΔH and ΔS of reaction.

D. Free Energy and Spontaneity

As mentioned at the outset of this chapter, one of the principal objectives of thermodynamics is the establishment of a criterion for spontaneity. The Second Law, presented in the previous section, can be used directly for this purpose. This criterion for spontaneity requires that *both* ΔS and $\int (dQ_{irrev}/T)$ be calculated or measured independently. If we place restrictions of constant P and T on the process, we can develop a standard for spontaneity that is simpler, because it involves the calculation of only *one* parameter. Since chemical reactions are generally carried out under constant T and P conditions, this new criterion for spontaneity can be readily applied.

We can develop this test for spontaneity under constant T and P conditions, starting with the Second Law of Thermodynamics for a spontaneous process,

$$dS > \frac{dQ}{T}$$

If the process is carried out at constant P and only PV work is done, then

$$dQ_P = dH_P$$

and the Second Law statement becomes

$$dS > \frac{dH_P}{T}$$

Rearranging gives

$$TdS > dH_P$$

$$0 > (dH_P - TdS)$$

$$(dH_P - TdS) < 0 \qquad\qquad (3\text{-}D\text{-}1)$$

This differential function is the basis for our new criterion for spontaneity.

We can now define a new thermodynamic function called **Gibbs Free Energy**, **G**, whose differential will give equation (3-D-1) when the temperature is held constant. By definition, this function is

$$\textbf{G = H - TS} \qquad\qquad (3\text{-}D\text{-}2)$$

The Gibbs Free Energy is obviously a state function, since H is a state function, as is also the TS product. The differential of G is

$$dG = dH - TdS - SdT$$

If T is held constant for the process, dT = 0, and

$$dG = dH - TdS \qquad\qquad (3\text{-}D\text{-}3)$$

The differential of G, given by equation (3-D-3), is identical to that in equation (3-D-1), if both T and P are held constant. Thus, our criterion for spontaneity becomes

$$\textbf{dG}_{\textbf{T,P}} \textbf{< 0} \qquad\qquad (3\text{-}D\text{-}4)$$

or, for a finite process,

$$\boldsymbol{\Delta}\textbf{G}_{\textbf{T,P}} \textbf{= (} \boldsymbol{\Delta}\textbf{H - T}\boldsymbol{\Delta}\textbf{S)}_{\textbf{T,P}} \textbf{< 0} \qquad\qquad (3\text{-}D\text{-}5)$$

where the subscripts T and P designate that the process is restricted to constant T and P conditions.

This criterion for spontaneity can be applied to any constant T and P process. Example 3-B-2 was a spontaneous phase transition at constant T and P, so the above criterion should apply to this process. ΔH, calculated in Example (3-B-2), is

$$\Delta H = 6.40 \text{ kJ/mol}$$

ΔS, calculated in Example 3-B-2, is

$$\Delta S = 23.3 \text{ J/K-mol} = 23.3 \times 10^{-3} \text{ kJ/K-mol.}$$

The temperature of the process is T = 283K, hence,

$$\Delta G = \Delta H - T\Delta S$$

$$\Delta G = 6.40 \text{ kJ/mol} - (283K)(23.3 \times 10^{-3} \text{ kJ/K-mol})$$

$$\Delta G = (6.40 - 6.57) \text{ kJ/mol} = -0.17 \text{kJ/mol}$$

$$\Delta G_{T,P} = -0.17 \text{ kJ/mol} < 0$$

Thus, this criterion does predict spontaneity for the process, which agrees with our experience that ice melts at $+10^{\circ}$C.

Problem 3-D-1. Use the change in Free Energy to predict the spontaneity of the process described in Problem 3-A-1. Recall that ΔS for this process was obtained in problem 3-B-2. Answer: $\Delta G = -0.23$ kJ/mole < 0.

The general criterion for equilibrium *at constant T and P* can be derived in a manner analogous to that for spontaneity. Since a reversible process is one which is in equilibrium at all times, we apply the definition of entropy which contains the equals sign, so that

$$dS = \frac{dQ}{T} \qquad \text{(equilibrium)}$$

Analogous to equation (3-D-1)

$$(dH_P - TdS) = 0 \qquad \text{(equilibrium)}$$

and in terms of the free energy equation, (3-D-5), our criterion for equilibrium becomes

$$\mathbf{\Delta G_{T,P} = (\Delta H - T\Delta S)_{T,P} = 0} \quad \text{(constant temperature and pressure)} \qquad \text{(3-D-6)}$$

Application to Chemical Reactions

The most important chemical application for this criterion of spontaneity is to chemical reactions. Suppose we consider the chemical system represented by the general, balanced equation given previously for the reaction in which all reactants and products are gases

$$aA(P_A) + bB(P_B) = cC(P_C) + dD(P_D)$$

where the P_i represent some partial pressure for the "ith" species in the mixture. Now, if the various reactants and products are present at these particular partial pressures P_A, P_B, P_C, and P_D, is it possible for the reaction to proceed in the forward direction? In other words, is further *net reaction* to produce C and D spontaneous? If the temperature and pressure are held constant, the decision will depend on the value of ΔG for the process. ΔG could be calculated from tabulated values of ΔH and ΔS at the appropriate T and P. This method is perfectly satisfactory, but of course requires that the necessary data be available. Since G is a state function, an alternate method would be to obtain ΔG by subtracting the G of the reactants from G of the products. However, since H is known only relatively (Chapter 2, Section F), and G is defined in terms of H, at best, we know only relative values of G. Also, with the use of only the Second Law of Thermodynamics, we know only relative values for entropy. However, with the Third Law of Thermodynamics, we can put entropy on an absolute basis, but even this does not help in the case of free energy, since enthalpy still remains on a relative basis.

Similar to our convention for enthalpy, we define **free energy of formation**, ΔG_f, as the free energy of formation of one mole of the material from its elements in their standard states, which will be defined shortly. Knowing ΔG_f for the products and reactants, ΔG for the reaction can be calculated from

$$\Delta G_{T,P} = c\Delta G_{f,C} + d\Delta G_{f,D} - a\Delta G_{f,A} - b\Delta G_{f,B} \qquad \text{(3-D-7)}$$

If $\Delta G_{T,P} < 0$, then the chemical reaction is spontaneous in the forward direction (left to right). If $\Delta G_{T,P} > 0$ for the reaction as written, it would obviously be less than zero for the reverse reaction, and in that case, it would be spontaneous in the reverse direction (right to left). If $\Delta G = 0$, the reaction is not spontaneous in either direction at the specific concentrations, that is, no net reaction is expected in either direction, and the reactants and products are said to be in equilibrium. There will be no change in the partial pressures with time, although the equilibrium is a dynamic one, in that A + B are continually being converted to C + D, and vice versa. The statement $\Delta G = 0$ simply asserts there will be *no net change* in concentration, but says nothing about the exchange, which may or may not be rapid.

Before proceeding, let us define the standard states for free energy and the zero reference. *The standard states for G are*:

solid and *liquid*—pure material at one bar pressure and specified temperature;
ideal gas—pure material at one bar pressure and specified temperature;
reference state—zero has been assigned to the elements in their most stable form in their standard states.

As mentioned in Chapter 2, one bar pressure is very close to one atmosphere and the standard state at one atmosphere can be considered to be equivalent to that at one bar pressure.

Calculation of ΔG for a Reaction

If the free energies of formation are known for both the products and reactants, then equation (3-D-7) can be used to calculate ΔG for the reaction. Also, if the ΔH_f and S are known for both reactants and products, then ΔH and ΔS for the reaction can be calculated from equation (2-F-2) and an analogous equation for entropy found in equation (3-B-10).

$$\Delta H = c\Delta H_{f,C} + d\Delta H_{f,D} - a\Delta H_{f,A} - b\Delta H_{f,B} \qquad \text{(2-F-2)}$$

$$\Delta S = cS_C + dS_D - aS_A - bS_B \qquad \text{(3-B-10)}$$

We can find $\Delta G_{T,P}$, that is, the change in free energy under constant T and P conditions, from

$$\Delta G_{T,P} = \Delta H - T\Delta S \qquad \text{(3-D-8)}$$

If $\Delta H_{f,i}$ and S_i are referred to standard states,

$$\Delta H^o = c\Delta H^o_{f,C} + d\Delta H^o_{f,D} - a\Delta H^o_{f,A} - b\Delta H^o_{f,B}$$

$$\Delta S^o = cS^o_C + dS^o_D - aS^o_A - bS^o_B$$

and

$$\Delta G^o_{T,P} = \Delta H^o - T\Delta S^o \qquad \text{(3-D-9)}$$

Problem 3-D-2. Consider the reaction

$$Br_{2(g)} + Cl_{2(g)} = 2BrCl_{(g)}$$

The following standard heats of formation and entropies have been determined for this reaction at T = 298K.

Compound	ΔH_f^o (kJ/mol)	S^o (J/K-mol)
$Br_{2(g)}$	30.9	245.5
$Cl_{2(g)}$	0	223.1
$BrCl_{(g)}$	14.6	240.1

(a) Calculate ΔG^o for this reaction at 298K.
(b) If you have a mixture of $Br_{2(g)}$, $Cl_{2(g)}$, $BrCl_{(g)}$, each in their respective standard states, is the reaction spontaneous at 298K as it is described by the above chemical equation? Answer: (a) - 5.16 kJ (b) yes.

ΔG and Net Work

An important relationship can be derived between the change in Gibbs Free Energy and the maximum net work for a process. We will define net work as all work in addition to mechanical work.

$$W = W_{mechanical} + W_{net}$$

or

$$dW = dW_{mechanical} + dW_{net}$$

We will consider the process to be carried out reversibly so the mechanical work will be PdV and the net work will be the maximum net work.

$$dW = - PdV + dW_{net,max}$$

Since G is given by

$$G = H - TS$$

the differential is

$$dG = dH - TdS - SdT$$

From the definition of enthalpy

$$H = U + PV$$

$$dH = dU + PdV + VdP$$

We can substitute the First Law for dU

$$dH = dQ + dW + PdV + VdP$$

Since the process is reversible we can substitute for dQ = TdS (Second Law) and for dW above

$$dH = TdS - PdV + dW_{net,max} + PdV + VdP$$

$$dG = TdS + dW_{net,max} + VdP - TdS - SdT = + dW_{net,max} + VdP - SdT$$

Now if T and P are held constant for the process,

$$dG_{T,P} = + dW_{net,max}$$

or

$$\Delta G_{T,P} = + W_{net,max} \qquad (3\text{-}D\text{-}10)$$

Equation (3-D-10) gives a rationale for the name "Free" Energy in that at constant T and P the ΔG is the energy change that is free to carry out work. Referring to equation (3-D-9) we see that $\Delta G_{T,P}$ is made up of two energy terms: ΔH and $T\Delta S$. The first term ΔH is the total energy change in the process but the $T\Delta S$ second term takes away that portion which is unavailable to do work. For a chemical reaction the $\Delta G_{T,P}$ is obtained from equation (3-D-9) and this gives the maximum net work that can be obtained according to equation (3-D-10).

E. Dependence of G on P and T

Quite often, we want to know if a reaction is spontaneous at a T and/or P other than that for which tabulated information is available. We may know, for example, from ΔG^o_f at 25°C, that a certain reaction is spontaneous under standard conditions, however, this does not imply that the reaction is spontaneous at some other set of conditions which are of interest. For this reason it is necessary to be able to calculate the change in G for a substance when the T and/or P is different from a value at which G is known; in other words, we must have expressions which relate a change in G to a change in T or P. Moreover, the expression for the change in G with pressure for an ideal gas is essential in establishing the expression for calculating the equilibrium yield.

In order to develop an equation relating the changes in G with T and P, we must utilize considerable differential calculus, in addition to fundamental laws and definitions encountered previously. The equations we are to develop will apply to a change in G for a single compound. Since G is a state function, it is independent of the path, so for convenience, we will take the reversible path. Furthermore, since it is simply a P and T change for a single substance, we will assume only PV work can be done. Since G is independent of the path, these conditions place no restrictions on the use of the resulting value of G. Starting with the basic definition of G and H

$$G = H - TS = U + PV - TS$$

and differentiating, we obtain

$$dG = dU + PdV + VdP - TdS - SdT \qquad (3\text{-}E\text{-}1)$$

The First Law of Thermodynamics, with the restriction of reversibility and PV work, is

$$dU = dQ + dW = dQ - PdV \qquad (3\text{-}E\text{-}2)$$

The Second Law, with the restriction of reversibility, is

$$dS = \frac{dQ}{T} \qquad (3\text{-}B\text{-}1)$$

or

$$TdS = dQ \qquad (3\text{-}E\text{-}3)$$

Substituting equation (3-E-3) into equation (3-E-2), we get

$$dU = TdS - PdV \qquad (3\text{-E-}4)$$

Substituting equation (3-E-4) into (3-E-1), and canceling terms, results in the fundamental relationship

$$\mathbf{dG = VdP - SdT} \qquad (3\text{-E-}5)$$

This, then, is the basic equation for the calculation of G with changing T and P. Note that VdP is not PV work, which is $P_{ex}dV$.

Dependence of ΔG on P for Reaction of Ideal Gases

In order to utilize equation (3-E-5) for our purpose, let us consider changes in T and P separately. First, consider a change in pressure at constant temperature. If T is constant, then obviously dT = 0, and equation (3-E-5) reduces to

$$dG = VdP \qquad \text{(constant T)} \qquad (3\text{-E-}6)$$

Integrating both sides, we obtain

$$\mathbf{G_2 - G_1 = \Delta G}_{P_1 \to P_2} = \int_{P_1}^{P_2} \mathbf{VdP} \qquad \text{(constant temperature)} \qquad (3\text{-E-}7)$$

This result is generally valid and applies to any single substance in any phase. It can be used to calculate ΔG for a change in P, provided a relationship between V and P for the particular substance is known; that is, an equation of state is necessary in order to carry out the integration.

The application of equation (3-E-7) is particularly easy for the special case of an ideal gas. In this situation, the equation of state is PV = nRT, or V = nRT/P, and when it is substituted into equation (3-E-7), we obtain

$$G_2 - G_1 = \int_{P_1}^{P_2} \frac{nRT}{P} dP = nRT \int_{P_1}^{P_2} \frac{dP}{P} = nRT \ln \frac{P_2}{P_1} \qquad \text{(at constant T)} \qquad (3\text{-E-}8)$$

Problem 3-E-1. Calculate the change in free energy for one mole of an ideal gas when the pressure is increased from one atmosphere to two atmospheres. Assume T = 298K.
Answer: 1.717 kJ.

Since the molar free energy of a gas is tabulated at standard conditions (P = 1 bar), we can calculate the free energy at any other pressure using equation (3-E-8) where $G_1 = G^o$ at $P_1 = 1$ bar.

$$G - G^o = RT \ln \frac{P}{1 \text{ bar}}$$

$$G = G^o + RT \ln P \qquad (3\text{-E-}9)$$

where it is understood that P is expressed in bars and the ratio P/1 bar is unitless. Note that the ln term is always unitless when P is expressed in bars. If we have a mixture of gases, as in a chemical reaction, we must designate each gas by a subscript. We will use the subscript "i" as a general designation of the ith compound.

$$G_i = G_i^o + RT \ln P_i$$

Since we do not know absolute values for G_i, we will use the free energy of formation scale defined earlier in Section D. In this expression we are assuming the elements remain in their standard state

$$\Delta G_{f,i} = \Delta G_{f,i}^o + RT \ln P_i \qquad (3\text{-}E\text{-}10)$$

where again the P_i must be expressed in bars and the logarithm term is unitless.

In order to derive an equation showing the pressure dependence of ΔG for a reaction, let us consider the general reaction used previously (equation 2-F-1) where all reactants and products are in the gas phase, each at its own partial pressure, P_i.

$$aA(g,P_A) + bB(g,P_B) = cC(g,P_C) + dD(g,P_D)$$

We will assume that each of the gases obeys the ideal gas equation of state even in this mixture of ideal gases. Equation (3-E-10) will then represent the molar free energy for each of these reactants and products in the reaction mixture.

Each of the gases in the reaction mixture is at its own partial pressure: P_A, P_B, P_C, P_D. These are simply partial pressures of the gases in the actual mixture, not necessarily the equilibrium mixture. The ΔG for this reaction is the change in free energy associated with the conversion of a moles of A and b moles of B into c moles of C and d moles of D where each of the species <u>remains</u> at the designated partial pressure. One may visualize this as a large reaction mixture such that the conversion of a moles of A etc. does not significantly change the partial pressures. ΔG is given by equation (3-D-7)

$$\Delta G = c\Delta G_{f,C} + d\Delta G_{f,D} - a\Delta G_{f,A} - b\Delta G_{f,B} \qquad (3\text{-}D\text{-}7)$$

We can express equation (3-E-10) for each of the reactants and products

$$\Delta G_{f,A} = \Delta G_{f,A}^o + RT \ln P_A$$

$$\Delta G_{f,B} = \Delta G_{f,B}^o + RT \ln P_B$$

$$\Delta G_{f,C} = \Delta G_{f,C}^o + RT \ln P_C$$

$$\Delta G_{f,D} = \Delta G_{f,D}^o + RT \ln P_D.$$

Substitution of these expressions into equation (3-D-7) gives

$$\Delta G = c(\Delta G_{f,C}^o + RT \ln P_C) + d(\Delta G_{f,D}^o + RT \ln P_D)$$

$$- a(\Delta G_{f,A}^o + RT \ln P_A) - b(\Delta G_{f,B}^o + RT \ln P_B)$$

Expanding and regrouping the $\Delta G_{f,i}^o$ terms and the $RT \ln P_i$ terms

$$\Delta G = (c\Delta G_{f,C}^o + d\Delta G_{f,D}^o - a\Delta G_{f,A}^o - b\Delta G_{f,B}^o)$$

$$+ (cRT\ln P_C + dRT\ln P_D - aRT\ln P_A - bRT\ln P_B)$$

Note that the first term contained in parentheses is simply ΔG^o for the reaction; i.e., change in free energy where reactants and products are in their standard state of one bar.

The second term can be simplified by factoring out the RT and noting that each logarithm term be simplified; e.g.,

$$c \ln P_C = \ln P_C^c$$

The expression then becomes

$$\Delta G = \Delta G^o + RT \, (\ln P_C^c + \ln P_D^d - \ln P_A^a - \ln P_B^b)$$

Since

$$\ln x + \ln y = \ln xy$$
$$\ln x - \ln z = \ln x/z$$

Our final equation becomes

$$\boldsymbol{\Delta G = \Delta G^o + RT \ln \frac{P_C^c P_D^d}{P_A^a P_B^b}} \qquad \text{(3-E-11)}$$

where again the P_i must be expressed in bar and the logarithm term is unitless. This expression can be used to calculate the ΔG for a reaction in which the reactants and products are not at standard conditions. Based on the sign of ΔG one can predict whether a reaction mixture under nonstandard conditions will be spontaneous or not. Note that the logarithm term contains the partial pressures of reactants and products and accounts for the difference between ΔG and the change in <u>standard</u> free energy, ΔG^o. The ratio of pressures in the logarithm term is of the same form as the expression for the equilibrium constant, which you may recall from a General Chemistry course. However, this ratio of pressures is not the equilibrium constant since the partial pressures are the values for the reaction mixture, which are <u>not</u> necessarily at equilibrium.

Problem 3-E-2.

(a) Calculate the change in free energy for the reaction given in Problem (3-D-2), if the reaction were carried out with each species at 2 atm pressure, rather than 1 atm. Assume T = constant = 298K.

(b) Would ΔG be the same as ΔG^o for the following reaction

$$N_{2(g)} + 3H_{2(g)} = 2\, NH_{3(g)}$$

if the partial pressure of each species was 2 atm?

Answer: (a) $\Delta G = - 5.16$ kJ (b) No, ΔG at 2 atm would be 3.43 kJ lower than ΔG^o.

Dependence of ΔG on T

The change of G with temperature at constant presure is obtained starting with equation (3-E-1), with the restriction of constant P, dP = 0, so that equation (3-E-5) becomes

$$dG = - SdT \qquad \text{(3-E-12)}$$

However, G and S have different scales: G is based on a relative scale where $\Delta G^o_{f,elements}$ = 0 at 298K and S is based on $S^o_{perfect\ crystal}$ = 0 at 0K. For this reason we wish to transform equation (3-E-12) into one which includes G and H. This can be performed if we consider the ratio (G/T) as a product (G)(1/T),

$$d\left(\frac{G}{T}\right) = (G)d\left(\frac{1}{T}\right) + \left(\frac{1}{T}\right)d(G)$$

$$d\left(\frac{G}{T}\right) = -\frac{G}{T^2}dT + \frac{dG}{T} \tag{3-E-13}$$

Substituting (3-E-12) into (3-E-13), we obtain

$$d\left(\frac{G}{T}\right) = -\frac{G}{T^2}dT - \frac{S}{T}dT = -\left(\frac{G}{T^2} + \frac{S}{T}\right)dT = -\left(\frac{G + TS}{T^2}\right)dT \tag{3-E-14}$$

From the basic definition of G

$$G = H - TS$$

or

$$H = G + TS$$

$$d\left(\frac{G}{T}\right) = -\frac{H}{T^2}dT \qquad \text{(constant P)} \tag{3-E-15}$$

This can also be written in the derivative form, recalling that P = constant.

$$\left(\frac{\partial \frac{G}{T}}{\partial T}\right)_P = \frac{-H}{T^2} \tag{3-E-16}$$

Since we do not have absolute values for ΔG, we will apply the general relationship (3-E-16) to the free energy of formation.

$$\left(\frac{\partial \frac{\Delta G_f}{T}}{\partial T}\right)_P = -\frac{\Delta H_f}{T^2} \tag{3-E-17}$$

From equation (3-D-7) applied to a reaction involving ideal gases,

$$\Delta G = c\Delta G_{f,C} + d\Delta G_{f,D} - a\Delta G_{f,A} - b\Delta G_{f,B} \tag{3-E-18}$$

If we divide (3-E-18) by T and take the partial derivative with respect to T, we obtain

$$\left(\frac{\partial \frac{\Delta G}{T}}{\partial T}\right)_P = c\left(\frac{\partial \frac{\Delta G_{f,C}}{T}}{\partial T}\right)_P + d\left(\frac{\partial \frac{\Delta G_{f,D}}{T}}{\partial T}\right)_P - a\left(\frac{\partial \frac{\Delta G_{f,A}}{T}}{\partial T}\right)_P - b\left(\frac{\partial \frac{\Delta(G_{f,B})}{T}}{\partial T}\right)_P$$

Substituting equation (3-E-17) for each of the terms on the right side of the equation, we obtain

$$\left(\frac{\partial \frac{\Delta G_f}{T}}{\partial T}\right)_P = -\frac{c\Delta H_{f,C}}{T^2} - \frac{d\Delta H_{f,D}}{T^2} + \frac{a\Delta H_{f,A}}{T^2} + \frac{\Delta H_{f,B}}{T^2}$$

Recalling equation (2-F-2)

$$\Delta H = c\Delta H_{f,C} + d\Delta H_{f,D} - a\Delta H_{f,A} - b\Delta H_{f,B}$$

then,

$$\left(\frac{\partial \frac{\Delta G}{T}}{\partial T}\right)_P = \frac{-\Delta H}{T^2} \tag{3-E-19}$$

From (3-E-19) we can calculate ΔG as a function of T. If standard conditions are employed,

$$\left(\frac{\partial \frac{\Delta G^o}{T}}{\partial T}\right)_P = \left(\frac{d \frac{\Delta G^o}{T}}{dT}\right)_P = \frac{-\Delta H^o}{T^2} \tag{3-E-20}$$

The designation of constant P in (3-E-20) can be eliminated, since the standard state condition infers P_i = constant = 1 bar; that is, ΔG^o is independent of pressure. By integrating equation (3-E-20) we could calculate ΔG^o at any temperature T providing we knew the temperature dependence of ΔH^o and some value of ΔG^o at a specified temperature, say 298K. This is a double integration if ΔCp is not equal to zero for the reaction and the resulting expression would include several terms.

　　An alternative procedure for the calculation of ΔG^o at temperature T is to calculate ΔH^o and ΔS^o at temperature T and to calculate ΔG^o from

$$\mathbf{\Delta G^o = \Delta H^o - T\Delta S^o}$$

The ΔH^o at temperature T is found from equation (2-F-3) when restricted to standard conditions.

$$\Delta H^o = \Delta H_1^o + \int_{T_1}^{T} \Delta C_P dT \tag{3-E-21}$$

The ΔS^o at temperature T is found from

$$\Delta S^o = \Delta S_1^o + \int_{T_1}^{T} \frac{\Delta C_P}{T} dT \tag{3-E-22}$$

Problem 3-E-3. Consider the process of sublimation

$$I_{2(s)} = I_{2(g)}$$

When this process is carried out at 25°C the $\Delta H^o_{298} = 62.4$ kJ/mol, $\Delta C^o_{P,298} = -17.5$ J/K-mol, and $\Delta S^o_{298} = 144.6$ J/K-mol. Calculate ΔG^o at 500°C assuming ΔC^o_P = constant. Answer: - 44.8 kJ.

Problem 3-E-4. Consider the following reaction which is carried out at nonstandard conditions

$$C_2H_{2(g,P\,=\,3\,atm)} + 2H_2O_{(g,P\,=\,0.3\,atm)} = 2CO_{(g,P\,=\,2\,atm)} + 3H_{2(g,P\,=\,4\,atm)}.$$

Calculate ΔG for this reaction at 900K, assuming the Cp's are constant. Hint: Calculate ΔG^o at 900K and then calculate ΔG at 900K; i.e. calculate the change in ΔG with temperature at standard conditions, ΔG^o, starting with ΔH^o_{298}, ΔS^o_{298}, ΔG^o_{298} and then calculate the change with pressure starting with ΔG^o_{900}. Answer: -114.6 kJ.

F. Chemical Equilibrium of Ideal Gases

The second of our objectives in thermodynamics was to be able to predict the equilibrium or maximum yield for a given chemical reaction. This can be accomplished through the free energy function introduced in the previous section. This function can be used to predict equilibrium yields for reactions in solutions, in addition to reactions in the gas phase. However, the development of the equation and the necessary calculations is greatly simplified when the reaction being considered is restricted to ideal gas reactants and products, as was done previously in deriving equation (3-E-11). This, of course, is an approximation to real gas mixtures. However, if the total gas pressure is low (less approximately 5-10 atm), then assumption of an ideal gas mixture is quite good.

In the previous section we derived in equation (3-E-11) an expression for calculating ΔG for a reaction in which the reactant and product gases are not at the standard state of one bar. This expression is applicable to any combination of ideal gas partial pressures. Now we want the reaction mixture to be at equilibrium, then the ratio of partial pressures is such that $\Delta G = 0$. In other words, we are placing a requirement of equilibrium on the partial pressures and we designate this by a subscript.

$$0 = \Delta G^o + RT\ln\left(\frac{P_C^c P_D^d}{P_A^a P_B^b}\right)_{equil.}$$

For a given reaction, ΔG^o is a constant at the designated temperature. Since R and T are constants, then the ratio of partial pressures at equilibrium must also be contant, a constant which we define as the equilibrium constant, K_P

$$K_P = \frac{P_C^c P_D^d}{P_A^a P_B^b} \qquad\qquad (3\text{-F-2})$$

where again the P_i must be expressed in bar and the K_P is unitless. Substitution for K_P into equation (3-F-1) gives

$$\mathbf{\Delta G^o = - RT\ln K_P} \qquad\qquad (3\text{-F-3})$$

This is the final equation which we have been seeking. Knowing ΔG^o for a reaction, K_P can be calculated and the equilibrium yield can be evaluated. Note that as ΔG^o becomes smaller, K_P becomes larger; a negative value for ΔG^o gives a value for K_P greater than unity, that is, the products are favored. You will recall that the concentration of a gas is proportional to its partial pressure, when the temperature is constant. Only the restriction of constant T is of significance, in regard to the use of equation (3-F-3). Some T is specified for the reaction, and ΔG^o and Kp in equation (3-F-3) must correspond to this temperature. ΔG^o is the change in free energy between the products and the reactants in their respective standard states, each at equal pressures of one atmosphere, and therefore, independent of pressure. This is not to infer, however, that the equilibrium yield cannot change with total pressure; in fact, it does change with pressure if the change in the number of moles in the reaction, Δn, is not zero.

$$\Delta n = c + d - a - b$$

The constancy of Kp with pressure simply infers that the ratio of equilibrium partial pressures raised to the appropriated powers, as defined by equation (3-F-2), is constant. An example will serve to clarify this point.

Example 3-F-1. Consider the Haber Process for the gas-phase production of ammonia at 500 K and a total pressure of P_T.

$$N_{2(g)} + 3H_{2(g)} = 2NH_{3(g)}$$

Treating each gas as if it were ideal, according to equation (3-F-3) we may write

$$\ln K_P = \frac{-\Delta G^o_{500K}}{R(500K)} \text{ or } K_P = \exp[-\Delta G^o_{500K}/R(500K)]$$

We see that Kp at 500K is independent of pressure; however, the actual maximum or equilibrium yield is the mole fraction of ammonia, X_{NH_3}. From equation (3-F-2) we can express Kp in terms of the equilibrium partial pressures.

$$K_P = \frac{P^2_{NH_3}}{P_{N_2} P^3_{H_2}}$$

Recalling that the partial pressure of an ideal gas is given by the product of the mole fraction, X, and the total pressure, and substituting into the above expression, we get

$$K_P = \frac{\left(X_{NH_3}P_T\right)^2}{\left(X_{N_2}P_T\right)\left(X_{H_2}P_T\right)^3} = \frac{X^2_{NH_3}}{X_{N_2}X^3_{H_2}}\frac{1}{P^2_T}$$

or,

$$K_X = \frac{X^2_{NH_3}}{X_{N_2}X^3_{H_2}} = P^2_T K_P$$

Thus, we see that while the ratio of the partial pressures is independent of the total pressure, the ratio of the mole fractions is a function of the total pressure. This is consistent with the conclusions in Problem (3-D-1). Even though ΔG for the reaction at any given T changes when the pressure is changed, ΔG^o (the standard free energy change) and consequently, Kp, remain constant.

Problem 3-F-1. Calculate the equilibrium constant K_P for the reaction given in Problem 3-D-2 assuming the temperature is 298K. If the total pressure were increased to 2 atm
  (a) would the equilibrium constant change?
  (b) would the relative concentrations at equilibrium change?
  Answer: $K_P = 8.02$ (a) No (b) No

Problem 3-F-2.
  (a) Calculate the equilibrium constant for the reaction at 25°C

$$CH_2=CH_{2(g)} + H_{2(g)} = CH_3\text{-}CH_{3(g)}$$

  (b) If the pressure of each component were increased to 2 atmospheres, calculate ΔG for the reaction.
  (c) Calculate K_P for the reaction, if the total pressure was 2 atmospheres.
  (d) Compared to the reaction at 1 atmosphere, would the amount of product $CH_3\text{-}CH_3$ increase, decrease, or remain the same at equilibrium if the total pressure were increased to 2 atmospheres? Explain.
  Answer: (a) $K_P = 3.8 \times 10^{17}$ (b) - 102.0 kJ (c) 3.8×10^{17} (d) The amount of product would increase if P = 2 atm, since $\Delta n_{gas} = -1$.

Temperature Dependence of K_P

It is now possible to evaluate the change in K_P with temperature. If equation (3-F-3) is divided by T, we obtain

$$\frac{\Delta G^o}{T} = - R \ln K_P$$

Substituting this expression into (3-E-20), we obtain

$$\frac{d(-R\ln K_P)}{dT} = - \frac{\Delta H^o}{T^2}$$

which gives, upon rearrangement

$$\frac{d\ln K_P}{dT} = \frac{\Delta H^o}{RT^2} \tag{3-F-4}$$

Again, you should be reminded that ΔG^o is defined at 1 bar (~1 atmosphere) pressure, and hence, K_P is independent of pressure.

During a reaction, the heat capacities of the products tend to cancel the heat capacities of the reactants (note the results of Problem 3-E-3 when the temperature is changed). Therefore, if we consider temperature changes over a restricted range (small), we can make the approximation that $\Delta C_P = 0$. From equation (2-F-3) then, at constant P, ΔH^o is approximately constant (independent of temperature). With this assumption, equation (3-F-4) can be readily integrated

$$\int d\ln K_P = \int \frac{\Delta H^o}{RT^2} dT = \frac{\Delta H^o}{R} \int \frac{dT}{T^2}$$

$$\ln K_P = - \frac{\Delta H^o}{R} \frac{1}{T} + C \tag{3-F-5}$$

where C is the constant of integration. The constant of integration can be evaluated by

$$\Delta G^o = - RT \ln K_P \qquad (3\text{-}F\text{-}3)$$

$$\ln K_P = - \frac{\Delta G^o}{RT}$$

Substitution into equation (3-F-5) gives

$$- \frac{\Delta G^o}{RT} = - \frac{\Delta H^o}{R}\frac{1}{T} + C$$

$$C = \frac{\Delta H^o - \Delta G^o}{RT}$$

Since

$$\Delta G^o = \Delta H^o - T\Delta S^o \qquad (3\text{-}D\text{-}9)$$

$$\Delta S^o = \frac{\Delta H^o - \Delta G^o}{T}$$

$$C = \frac{\Delta S^o}{R}$$

Substitution into equation (3-F-5) gives

$$\ln K_P = - \frac{\Delta H^o}{R}\frac{1}{T} + \frac{\Delta S^o}{R} \qquad (3\text{-}F\text{-}6)$$

which is a linear equation with $y = \ln K_P$ versus $x = 1/T$, slope $= -\Delta H^o/R$, intercept $= \Delta S^o$.

Problem 3-F-3. The ΔH^o, ΔS^o, and ΔG^o for the reaction

$$Br_{2(g)} + Cl_{2(g)} = 2\ BrCl_{(g)}$$

were calculated in Problem 3-D-2 with these results: $\Delta S^o_{298} = 11.6$ J/K, $\Delta H^o_{298} = -1.7$ kJ, $\Delta G^o_{298} = -5.16$ kJ. Using these results, calculate ΔG^o and K_P at 500°C. Assume $\Delta C_P = 0$. Answer: $\Delta G^o = -10.7$ kJ; $K_P = 5.26$

If ΔC_P for the reaction was assumed constant, then from equation (2-F-3) and integrating from T_1 to a variable temperature, T

$$\Delta H^o = \Delta H^o_1 + \int_{T_1}^{T} \Delta C_P dT$$

$$\Delta H^o = \Delta H^o_1 + \Delta C_P(T - T_1) \qquad (3\text{-}E\text{-}7)$$

Substitution of equation (3-F-7) into equation (3-F-4)

$$\frac{d \ln K_P}{dT} = \frac{\Delta H^o_1}{RT^2} + \frac{\Delta C_P(T - T_1)}{RT^2}$$

$$\frac{d \ln K_P}{dT} = \frac{(\Delta H_1^o - \Delta C_P T_1)}{RT^2} + \frac{\Delta C_P}{RT}$$

$$\int_{\ln K_{P,T_1}}^{\ln K_P} d\ln K_P = \left(\frac{\Delta H_1^o - \Delta C_P T_1}{R}\right)\int_{T_1}^{T}\frac{dT}{T^2} + \frac{\Delta C_P}{R}\int_{T_1}^{T}\frac{dT}{T}$$

$$\ln K_P = \ln K_{P,T_1} - \left(\frac{\Delta H_1^o - \Delta C_P T_1}{R}\right)\left(\frac{1}{T} - \frac{1}{T_1}\right) + \frac{\Delta C_P}{R}\ln\frac{T}{T_1} \qquad (3\text{-}F\text{-}8)$$

Problem 3-F-4. Calculate the equilibrium constant at 500K for the reaction

$$N_{2(g)} + 3H_{2(g)} = 2NH_{3(g)}$$

Assume ΔC_P = constant. Use data from Table 2 in Appendix D.
Answer: $K_{P,500K} = 0.095$.

"Turning" Temperature

It is apparent from equation (3-F-3) that the equilibrium constant is very small when ΔG^o is positive. For example, if ΔG^o were 83.7 kJ at 298K, the equilibrium constant would be 1.5×10^{-15}. This a very small equilibrium constant and most likely very little product would be formed at equilibrium. On the otherhand, if ΔG^o were - 83.7 kJ the equilibrium constant at 298K would be 4.7×10^{14}, a very large number, and one would expect a very large yield of products at equilibrium. In between these extremes we have $\Delta G^o = 0$ and $K_P = 1.00$.

Obviously we generally desire ΔG^o to be negative for a reaction where we wish to produce a high yield of product. If ΔG^o is positive, at what temperature can we make ΔG^o become negative? From equation (3-E-20) we see that when ΔH^o for the reaction is positive (endothermic reaction), ΔG^o will become smaller as the temperature is increased. The question one might ask is how high must the temperature be increased in order to make ΔG^o negative? An approximate answer to that question can be obtained by a very simple analysis.

If we knew the temperature at which the ΔG^o were reduced to zero, then a further increase in the temperarure would make ΔG^o become negative. In equation (3-D-7) we can solve for the temperature, T^*, which makes ΔG^o become zero.

$$0 = \Delta H^o - T^* \Delta S^o$$

We know that ΔH^o and ΔS^o are dependent on the temperature of the reaction, so this equation is not easily solved for T^*. However, as an approximation we will assume ΔH^o and ΔS^o are independent of temperature ($\Delta C_P = 0$) and an approximate value of T^* is simply

$$T^* = \frac{\Delta H^o}{\Delta S^o} \qquad (3\text{-}F\text{-}9)$$

This is a very simple way to estimate the temperature required to make an endothermic reaction proceed. T^* can be thought of as the "turning temperature" where the ΔG^o "turns" from a positive value to a negative value. At this point one would then want to calculate the ΔG^o precisely considering the temperature dependence of ΔH^o and ΔS^o from equations (3-E-20) and (3-E-21).

It should be noted that an important factor in making ΔG^o become negative is the value of ΔS^o. When ΔS^o is large, ΔG^o will be zero at a lower value of T^*. As you will recall, ΔS^o for gaseous reactions depends heavily upon Δn_{gas} for the reaction (a large increase in disorder), the reaction can be made to occur at a smaller T^*.

Problem 3-F-5. Calculate the turning temperature, T^*, for the following reaction, called the steam reforming reaction

$$CH_{4(g)} + H_2O_{(g)} \longrightarrow CO_{(g)} + 3\,H_{2(g)}$$

Answer: 960K.

G. Calculation of Equilibrium Yield

In Section D we described how ΔG^o and the equilibrium constant, K_P, can be calculated at any desired temperature from the known ΔH^o, ΔG^o, and ΔS^o. We now want to show how we can calculate the equilibrium yield from the known equilibrium constant, K_P. As you will recall from our discussion at the beginning of Chapter 2, one of the objectives of chemical thermodynamics is the calculation of equilibrium yield.

Reaction of Gaseous Reactants and Products

We will assume that the gaseous reaction is being carried out at <u>constant volume</u> and, as assumed previously in Section C, that all gases obey the ideal gas equation of state. Consequently, the partial pressure for a species is proportional to the number of moles for that species.

$$P_i V = n_i RT$$

$$P_i = \frac{RT}{V} n_i$$

Since R, T, and V are constant, this will allow us to use partial pressures to describe the molar changes in a chemical reaction.

For example, in the chemical reaction

$$2\,N_2O_{5(g)} = 4\,NO_{2(g)} + O_{2(g)}$$

2 moles of $N_2O_{5(g)}$ react to give 4 moles of $NO_{2(g)}$ and one mole of $O_{2(g)}$. If we had one mole of N_2O_5 initially before reaction occured, it could produce 2 moles of NO_2 and 0.5 mole of O_2 upon completion. Similarly, the reaction of one atmosphere of N_2O_5 would produce 2 atmospheres of NO_2 and 0.5 atmosphere of O_2.

Now when we have an equilibrium between reactants and products in a reaction, we obviously do not have complete reaction but rather we have a partial reaction (equilibrium yield) and this is what we want to calculate. If we start out with only the reactants, then we can calculate the amount of products produced at equilibrium. Using the previous reaction we would start out with an initial pressure of N_2O_5 which we designate by $P_{N_2O_5}^o$, the superscript zero designating the initial value. Let us represent the partial pressure of O_2 at equilibrium by x.

$$P_{O_2} = x$$

Since 4 moles of NO_2 are produced for every mole of O_2, then obviously

$$P_{NO_2} = 4x$$

In order to produce one mole of O_2 we must react 2 moles of N_2O_5, and the initial pressure of N_2O_5 must be decreased by 2 times the amount of O_2 produced.

$$P_{N_2O_5} = P^o_{N_2O_5} - 2x$$

We can now substitute the partial pressures into the equilibrium constant expression for this reaction.

$$K_P = \frac{P^4_{NO_2} P_{O_2}}{P^2_{N_2O_5}}$$

$$K_P = \frac{(4x)^4 (x)}{(P^o_{N_2O_5} - 2x)^2}$$

Since K_P is known, we can calculate x for some specified initial pressure of N_2O_5 and then we can determine the P_{O_2}, P_{NO_2} and $P_{N_2O_5}$ at equilibrium.

The exact numerical solution to the above equilibrium expression is impossible since it is a 5th degree polynomial. If K_P is small and the initial pressure is reasonably large, then one may simplify the equation by assuming that x must be small and

$$\left(P^o_{N_2O_5} - 2x \right) \sim P^o_{N_2O_5}$$

This then simplifies the equilibrium expression to a 5th power of x

$$K_P = \frac{(4x)^4 (x)}{(P^o_{N_2O_5})^2} = \frac{256 x^5}{(P^o_{N_2O_5})^2}$$

If K_P were 1×10^{-20} and $P^o_{N_2O_5} = 1$ atm, we find x to be 3.3×10^{-5}. The partial pressures would then be

$$P_{O_2} = x = 3.3 \times 10^{-5} \text{ atm}$$

$$P_{NO_2} = 4x = 1.32 \times 10^{-4} \text{ atm}$$

$$P_{N_2O_5} = P^o_{N_2O_5} - 2x = (1 - 6.6 \times 10^{-5}) \text{ atm}$$

Indeed one can see that the magnitude of x is small, $P^o_{N_2O_5} \sim 1$ atm, and the approximation made appears to be adequate. In order to test the adequacy of the approximation, one can substitute the calculated values into the equilibrium expression, calculate the K_P and compare this calculated K_P with the true value. For this example

$$K_P = \frac{P_{NO_2}^4 P_{O_2}}{P_{N_2O_5}^2} = \frac{(1.32 \times 10^{-4})^4 (3.3 \times 10^{-5})}{(0.99993)^2} = 1.002 \times 10^{-20}$$

As one can see, the calculated K_P deviates from the true K_P by only 0.2% and the approximation is satisfactory. Generally K_P values are not accurate to better than 1% and in many cases the error is much greater.

On the other hand, if K_P is very large and/or the initial pressure of N_2O_5 is small, then we obtain nearly complete reaction and the previous form of the equilibrium expression is difficult to solve for x. In this case it is best to assume complete reaction and then calculate the small amount of back reaction to establish equilibrium.

Let us assume in the previous example that $K_P = 1 \times 10^{20}$. Complete reaction would give initial values of $P_{O_2}^o = 0.5$ atm and $P_{NO_2}^o = 2.0$ atm. The equilibrium expression is now

$$K_P = 1 \times 10^{20} = \frac{\left(P_{NO_2}^o - 4x\right)^4 \left(P_{O_2}^o - x\right)}{(2x)^2}$$

Now since K_P is large, we will assume x will be small for the back reaction and

$$(P_{NO_2}^o - 4x) \sim P_{NO_2}^o$$

$$(P_{O_2}^o - x) \sim P_{O_2}^o$$

and the equilibrium expression is simplified to

$$K_P = 1 \times 10^{20} = \frac{\left(P_{NO_2}^o\right)^4 \left(P_{O_2}^o\right)}{(2x)^2} = \frac{(2.0)^4 (0.5)}{(2x)^2}$$

Solving for $x = 1.4 \times 10^{-10}$ and the equilibrium pressures are now

$$P_{NO_2} = P_{NO_2}^o - 4x = 2.0 - 4(1.4 \times 10^{-10}) = 1.99999999936 \text{ atm}$$

$$P_{O_2} = P_{O_2}^o - x = 0.5 - 1.4 \times 10^{-10} = 0.49999999986 \text{ atm}$$

$$P_{N_2O_5} = 2x = 2(1.4 \times 10^{-10}) = 2.8 \times 10^{-10} \text{ atm}$$

Again we must check our approximation by substituting into the equilibrium expression to calculate K_P.

$$K_P = \frac{(1.99999999936)^4 \ (0.49999999986)}{(2.8 \times 10^{-10})^2} = 1.0003 \times 10^{20}$$

This calculated K_P is 0.03% greater than the true value of 1.0×10^{20} and the approximation of a small x for the back reaction is justified.

These examples are extreme cases where the K_P is small and large. As one gets closer to an intermediate K_P nearer unity, the approximation of a small x is in more error. For example, if K_P were assumed to be 1×10^{-5} and $P(N_2O_5)^o = 1.0$ atm, then the approximation leads to $x = 0.033$ atm. The calculated K_P is

$$K_P = \frac{[(4)(0.033)]^4 \, (0.033)}{[1 - 2(0.033)]^2} = 1.15 \times 10^{-5}$$

In this case the calculated K_P is in error by 15% from the true value of 1.0×10^{-5}. If K_P is assumed to be 1×10^{-3} the x is calculated to be 0.083 and the calculated K_P is 1.437×10^{-3}, a deviation of 43.7%. In these cases a more precise method of numerical calculation is required.

A very simple method of calculating the equilibrium partial pressures is to calculate K_P for various assumed values of x. A graph is then made of the calculated K_P's versus x and the true x value estimated from the graph that gives the true K_P. For example, in the above case we calculated $K_P = 1.437 \times 10^{-3}$ for a value of x = 0.083. Since the true value of K_P is 1.0×10^{-3} we will lower x by a small amount so that the calculated K_P will be reduced. An x value of 0.080 gives a calculated K_P of 1.189×10^{-3}. Tabulating these results gives:

x	Calculated K_P
0.083	1.437×10^{-3}
0.080	1.189×10^{-3}
0.078	1.038×10^{-3}
0.077	0.968×10^{-3}
0.07746	0.9996×10^{-3}

Since we have not reduced K_P sufficiently, we must reduce x further, this time to 0.078. This value of x = 0.078 gives a K_P value of 1.038×10^{-3}. Again K_P is larger than the true value so x is reduced to 0.077 which gives a calculated K_P of 0.968×10^{-3}. Obviously the true x value falls between 0.078 and 0.077. A graph of the calculated K_P's versus x will show that a value of x = 0.0775 will give the desired K_P of 1.0×10^{-3}. An alternative of graphing the calculated K_P versus x is to linearly interpolate between the two data points straddling the true K_P value. For example the values of x of 0.078 and 0.077 would entail the calculation

$$\frac{1.038 \times 10^3 - 0.968 \times 10^{-3}}{0.078 - 0.077} = \frac{1.0 \times 10^{-3} - 0.968 \times 10^{-3}}{x - 0.077}$$

which gives x = 0.07746.

Problem 3-G-1. For the following reaction

$$2\, N_2O_{(g)} = 2N_2(g) + O_{2(g)}$$

the equilibrium constant is 1×10^{-10}. If the initial pressure of N_2O is 0.1 atm, calculate the pressure of N_2, O_2, and N_2O. Answer: $P_{O_2} = 6.3 \times 10^{-5}$ atm, $P_{N_2} = 1.26 \times 10^{-4}$ atm, $P_{N_2O} = 0.09987$ atm.

Problem 3-G-2. If the reaction described in problem 3-G-1 had an equilibrium constant of 1×10^{10}, calculate the equilibrium partial pressures of N_2, O_2 and N_2O. Answer: $P_{O_2} = 0.049999888$ atm, $P_{N_2} = 0.099999776$ atm; $P_{N_2O} = 2.24 \times 10^{-7}$.

Problem 3-G-3. If the reaction described in problem 3-G-1 had an equilibrium constant of 1×10^{-2}, calculate the equilibrium partial pressures of N_2, O_2, and N_2O. Answer: $P_{N_2} = 0.0411$ atm, $P_{O_2} = 0.02055$, $P_{N_2O} = 0.0589$ atm.

Heterogeneous Equilibria

Heterogeneous equilibrium is one in which a solid and/or liquid plus one or more gases are in equilibrium. The equilibrium expression for such a reaction is simplified greatly since the free energies of the solid and/or liquid do not change during the course of the reaction to attain equilibrium. It is assumed that the solid and/or liquid are pure phases and consequently their standard free energies of formation would remain constant. The free energy of condensed phases (solids and liquids) are essentially independent of pressure and we can therefore assume the free energy is the standard free energy (1 atm).

Since the free energy of the condensed phases are independent of pressure, there will <u>not</u> be an RT lnP term in the free energy expression.

$$\Delta G_{f,gas} = \Delta G_{f, gas}^{o} + RT \ln P$$

$$\Delta G_{f,solid} = \Delta G_{f, solid}^{o}$$

Consequently, the condensed phases will not have any pressure terms in K_P. As an example let us consider the following hetergeneous reaction carried out at 25°C

$$Hg_2Cl_{2(s)} + H_{2(g)} = 2Hg_{(l)} + 2HCl_{(g)}$$

$$\Delta G^{o} = 2\Delta G_{f,Hg(l)}^{o} + 2\Delta G_{f,HCl(g)}^{o} - \Delta G_{f,Hg_2Cl_2(s)}^{o} - \Delta G_{f,H_2(g)}^{o}$$

$$\Delta G^{o} = [2(0) + 2(-95.3) - (-210.7) - 0] \text{ kJ}$$

$$\Delta G^{o} = 20.1 \text{ kJ}$$

$$\Delta G^{o} = -RT \ln K_P$$

$$K_P = \frac{(P_{HCl})^2}{P_{H_2}} = 3.0 \times 10^{-4}$$

In order to calculate the equilibrium pressures you must have $Hg_2Cl_{2(s)}$ and $Hg_{(l)}$ present. The amounts of $Hg_2Cl_{2(s)}$ and $Hg_{(l)}$ do not influence the equilibrium since they are pure materials and their nature does not change during the course of the rection.

Problem 3-G-4. For the following reaction

$$CaCO_{3(s)} + H_{2(g)} = CaO_{(s)} + CO_{(g)} + H_2O_{(g)}$$

where $CaCO_{3(s)}$ is in the more stable calcite form.
a) Calculate the equilibrium constant at 700°C. Assume $\Delta C_P = 0$.
b) If the initial pressure of H_2 was 4.0 torr with no products present, calculate the equilibrium pressures for the gaseous species. Assume both solids are present, when the reaction is complete.
Answers: (a) $K_P = 0.051$, (b) $P_{CO} = P_{H_2O} = 0.004805$atm, $P_{H_2} = 0.000458$ atm.

H. Calculation of Thermodynamic Parameters from Experimental Measurements of Equilibrium

The equations developed in the previous section also serve a useful purpose in the evaluation of thermodynamic parameters for a reaction from experimental equilibrium measurements. In fact, without the Third Law of Thermodynamics, it is necessary that at least one equilibrium measurement be made with a compound before the free energy of formation can be established. ΔH can be evaluated from thermochemical measurements (Chapter 2, Section F), but we have no way of measuring ΔS directly.

Suppose we have made experimental measurements of the equilibrium mixture for a given reaction at various temperatures. The equilibrium constants can then be evaluated from the known equilibrium concentrations. From equation (3-F-4)

$$\frac{d\ln K_P}{dT} = \frac{\Delta H^O}{RT^2} \qquad (3\text{-}F\text{-}4)$$

it can be shown that

$$\frac{d\ln K_P}{dT} = \frac{d\ln K_P}{d(1/T)}\frac{d(1/T)}{dT} = \frac{d\ln K_P}{d(1/T)}\left(-\frac{1}{T^2}\right) = \frac{\Delta H^O}{RT^2}$$

Therefore,

$$\frac{d\ln K_P}{d(1/T)} = -\frac{\Delta H^O}{R} \qquad (3\text{-}H\text{-}1)$$

The value of $\ln K_P$ at each temperature can be plotted versus the reciprocal of the temperature, as in Figure 3-H-1.

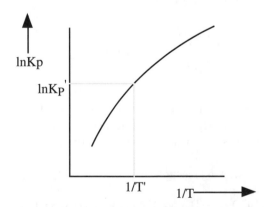

Figure 3-H-1. Relationship Between Equilibrium Constant and AbsoluteTemperature

The slope of the tangent to the curve in Figure 3-H-1 at any point, T', for example, can be evaluated graphically. Since the derivative, $d\ln K_p/d(1/T)$, at any point is equal to the slope of the tangent at any point,

$$(\text{slope})_{T'} = -\frac{\Delta H^O}{R}$$

$$\Delta H^O = -R\,(\text{slope})_{T'}$$

ΔG^o can be evaluated from the relation,

$$\Delta G^o = - RT' \ln K_P{}'$$

Finally, employing equation (3-D-9) at $T = T'$, we can obtain ΔS^o.

$$\Delta G^o = \Delta H^o - T' \Delta S^o$$

or,

$$\Delta S^o = \frac{\Delta H^o - \Delta G^o}{T'}$$

Thus, we have evaluated ΔH^o, ΔG^o, and ΔS^o for the reaction at the temperature T'.

Problem 3-H-1. The dimerization of RbBr

$$2 \, RbBr_{(g)} = Rb_2Br_{2(g)}$$

was measured by the deviation from ideal gas behavior (K. Hagemark and D. Hengstenberg, *J. Phys Chem.,* **1967**, *71*, 3337]. The following gas pressures were measured as a function of temperature at two different gas concentrations.

Concentration = 3.680×10^{-4} mol/L   Concentration = 5.049×10^{-4} mol/L

T(K)	Pressure(mm)	$K_P(atm^{-1})$	T(K)	Pressure (mm)	$K_P(atm^{-1})$
1310	26.74	4.51	1307	34.97	5.74
1339	27.48	4.10	1336	36.74	3.66
1366	28.60	----	1352	37.43	3.88
1376	28.91	3.05	1375	39.10	2.60
1397	29.74	2.30	1392	40.15	2.13
1410	30.49	1.77	1421	41.55	1.70

(a) For ideal gas behaviour, show that the concentration (c) is related to the ideal gas pressure (P_{id})

$$P_{id} = cRT$$

(b) The ideal pressure (P_{id}) corresponds to the pressure, as if no dimers existed. Since every dimer that is formed requires two monomers, the partial pressure of dimer (P_2) must be multiplied by 2 to give this contribution to the P_{id}. If we represent the partial pressure of monomer in the mixtures as P_1,

$$P_{id} = P_1 + 2P_2$$

The total pressure of the mixture of monomer and dimer must be the sum of the partial pressures,

$$P_{total} = P_1 + P_2$$

Solve these two equations simultaneously to give

$$P_1 = 2P_{total} - P_{id}$$

$$P_2 = P_{id} - P_{total}$$

(c) Using these expressions for P_1 and P_2, calculate K_P at $T = 1366K$ and

$$c = 3.680 \times 10^{-4} \text{ mol/liter}$$

(d) Graphically evaluate ΔH^o for this reaction and calculate ΔG^o and ΔS^o at 1350K. Answer: (c) 3.1 (d) ΔH^o = - 167 kJ, ΔG^o = - 14.0 kJ, ΔS^o = - 113 J/K. (Your answers may differ since it is a graphical solution.)

If ΔCp = 0, the graph of $\ln Kp$ versus $1/T$ would be linear, corresponding to constant ΔH^o and ΔS^o, independent of temperature. This can be seen from equation (3-F-6) given previously:

$$\ln Kp = - \frac{\Delta H^o}{R}\left(\frac{1}{T}\right) + \frac{\Delta S^o}{R} \qquad (P = \text{constant}, \Delta Cp = 0) \qquad (3\text{-}F\text{-}5)$$

where $\Delta H^o = \Delta H_1^o$ = constant and $\Delta S^o = \Delta S_1^o$ = constant. Since ΔH^o and ΔS^o are constant when ΔCp = 0, the slope ($-\Delta H^o/R$) and intercept ($\Delta S^o/R$) are thus constant. Hence, equation (3-F-5) defines a straight line.

Problem 3-H-2. Observe the graph of $\ln Kp$ versus $1/T$ in Problem 3-H-1. Would you conclude that ΔH^o and ΔS^o are approximately constant over that temperature range (within experimental error)? Answer: Yes. When you consider data sets at both concentrations, the graph appears to be linear.

I. Summary

The following is a summary of the essential concepts from this chapter:

1. Second Law of Thermodynamics: $dS \geq dQ/T$ where S = entropy is a state function, > refers to a spontaneous process, = refers to a reversible process.

2. Third Law of Thermodynamics: <u>All</u> substances in the form of a perfect crystal at 0^oK have an entropy of zero.

3. Based on the Third Law, absolute entropies can be calculated using the formulas

$$\Delta S = \int \frac{Cp}{T} \, dT$$

$$\Delta S_{trans} = \frac{\Delta H_{trans}}{T_{trans}}$$

4. The standard free energy change for a reaction, ΔG^o, can be calculated from

$$\Delta G^o = \sum \Delta G_{f,products}^o - \sum \Delta G_{f,reactants}^o$$

$$\Delta G^o = \Delta H^o - T\Delta S^o$$

where ΔH^o and ΔS^o are the changes in standard enthalpy and entropy for the reaction.

5. The standard free energy change for a reaction at nonstandard conditions can be calculated from

$$\Delta G = \Delta G^o + RT \ln \frac{P_C^c P_D^d}{P_A^a P_B^b}$$

6. The equilibrium constant K_P, is related to the change in <u>standard</u> free energy for the reaction by

$$\Delta G^o = - RT \ln K_P$$

Exercises

1. Consider the change from state A to state C as described in Excercise 1 at the end of Chapter 2.
   (a) How would ΔS and ΔG for the path AC compare with the path ABC?
   (b) Derive an expression for $\Delta S = S_C - S_A$. Use the same symbols for temperature and heat capacity as described in Exercise 1 at the end of Chapter 2.
   Answer: (a) $\Delta S_{AC} = \Delta S_{ABC}$, $\Delta G_{AC} = \Delta G_{ABC}$

$$(b)\ \Delta S = n \int_{T_A}^{T_B} \frac{C_P}{T} dT + n \int_{T_B}^{T_C} \frac{C_V}{T} dT = n \int_{T_A}^{T_C} \frac{C_V}{T} dT + n R \ln \frac{T_B}{T_A}$$

2. Calculate ΔS and ΔG for the vaporization process described in Exercise 4 at the end of Chapter 2. Answer: $\Delta S = 83.8$ J/K, $\Delta G = 0$.

3. Derive an expression for ΔS and ΔG for the isothermal, nonideal gas described in Exercise 5 at the end of Chapter 2.
   Answer: $\Delta S = R \ln \dfrac{V_2 - b}{V_1 - b}$, $\Delta G = b(P_2 - P_1) - RT \ln \dfrac{V_2 - b}{V_1 - b}$

4. Suppose we go from state A to state B by two different paths: path No. 1 and path No. 2.

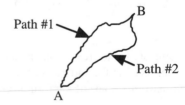

If this is the only data specified, which of the following equations are definitely applicable *without further restrictions*? In other words, answer "yes" if the equation rigorously applies to the above described processes, or "no" if you cannot be absolutely certain that the equation applies.
(a) $Q_1 = Q_2$
(b) $\Delta H_1 = \Delta H_2$
(c) $\Delta H_2 = \Delta U_1 + \Delta(PV)_1 = \Delta U_1 + (P_B V_B - P_A V_A)$
(d) $\Delta G_1 = \Delta H_1 - T \Delta S_1$
(e) $W_2 = - P_2 (V_B - V_A)$
(f) $\Delta S_1 = \int \dfrac{dQ_{rev}}{T}$
(g) $\Delta S_1 = \int \dfrac{dQ_1}{T}$
(h) $\Delta G_1 = < 0$
Answers: (a) No (b) Yes (c) Yes (d) No (e) No (f) Yes (g) No (h) No.

5. Suppose we go from state A to state B by two different paths, path No.1 and path No. 2. In this case, however, the paths are further specified, thusly:

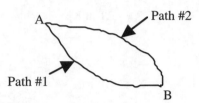

path No. 1--constant pressure, irreversible, only PV work
path No. 2--constant pressure, reversible, PV work and W_{net}

W_{net} is all the work other than PV work, that is $dW = PdV + dW_{net}$.
$W_{net,max}$ is the maximum net work which one obtains from the reversible path.
Which of the following equations are applicable to these processes without making any further restrictions?
(a) $Q_1 = Q_2$
(b) $\Delta H_1 = \Delta H_2$
(c) $\Delta H_1 = Q_1$
(d) $\Delta H_2 = Q_2$
(e) $\Delta G_1 = \Delta G_2$
(f) $\Delta G_1 = W_{net,max}$
(g) $\Delta U_1 = Q_2 - P_A(V_B - V_A) + W_{net,max}$
(h) $\Delta U_1 = \Delta U_2$
(i) $\Delta U_1 = Q_1 + W_1$
(j) $\Delta U_1 = Q_1 - P_1(V_B - V_A)$
(k) $\Delta S_1 = \int \dfrac{dQ_2}{T}$
(l) $\left(\dfrac{\partial U}{\partial V}\right)_T = 0$

Answer: (a) No (b) Yes (c) Yes (d) No (e) Yes (f) No (g) Yes (h) Yes (i) Yes (j) Yes (k)Yes (l) No.

6. Derive expressions for ΔS and ΔG for the reversible, isothermal expansion of an ideal rubber described in Exercise 3 at the end of Chapter 1 and Exercise 7 at the end of Chapter 2. The force was assumed to be of the form $f = -kT(L - L_o)$.

Answer: $\Delta S = -\dfrac{k\left(L_2^2 - L_1^2\right)}{2} + kL_o(L_2 - L_1)$, $\Delta G = -\dfrac{1}{2}kT\left(L_2^2 - L_1^2\right)$.

7. Derive expressions for ΔS and ΔG for the isothermal expansion of an ideal rubber at constant f described in Exercise 4 at the end of Chapter 1 and Exercise 8 at the end of Chapter 2. Compare with Exercise 6. Explain. Answer: Same as in Exercise 6 above since the process involves the same initial and final states.

8. (a) Using thermodynamic data from tables in Appendix D, predict whether or not the following reaction is spontaneous at 500K at Standard Conditions. Assume $\Delta C_p = 0$. Show your calculations.

$$HCN_{(g)} + 3H_{2(g)} = CH_{4(g)} + NH_{3(g)}$$

(b) For the reaction given above, calculate ΔG at 500K when the pressures of the gases are $P_{HCN(g)} = 3 \times 10^{-8}$ atm, $P_{H2(g)} = 2.0$ atm, $P_{CH4(g)} = 0.5$ atm, $P_{NH3(g)} = 0.5$ atm. As before, assume $\Delta C_p = 0$.
Answer: (a) Reaction is spontaneous (b) $\Delta G = -90.4$ kJ

9. Using the Second Law of Thermodynamics, prove that upon touching a hot frying pan the heat will be transferred from the hot frying pan to your cold hand, and not vice versa. Let Q_p represent the heat change of the pan and Q_h the heat change of your hand. Note that

$$Q_p = -Q_h$$

and that the surroundings undergo no heat change. Therefore, according to the Second Law,

$$\Delta S = (\Delta S_p + \Delta S_h) > 0$$

in order for the process to be spontaneous. Assume that Q heat is transferred, but it is small enough that the temperature of the pan, T_p, and the temperature of the hand, T_h, do not change significantly. Show that

$$\Delta S = Q_h \left(\frac{1}{T_h} - \frac{1}{T_p} \right)$$

and that heat must flow from the pan at a higher temperature, T_p, to the hand at a lower temperature, T_h, in order to satisfy the Second Law of Thermodynamics.

10. Calculate the equilibrium constant at 25°C for the reaction

$$2H_{2(g)} + O_{2(g)} = 2H_2O_{(g)}$$

Use thermodynamic values found in the Appendix, Section D. Answer: $K_P \sim 10^{80}$.

11. Consider the process of evaporation of water at 85°C.

$$H_2O_{(l)} \longrightarrow H_2O_{(g)} \quad (t = 85°C)$$

In the following questions assume the vapor is an ideal gas.
(a) If the pressure of the liquid and the vapor were one atmosphere, what sign do you expect for ΔG°? Is this in accord with what you expect to happen physically? Explain.
(b) If the pressure of the liquid and the vapor were the vapor pressure of the H_2O at 85°C, what do you expect ΔG to be?
(c) If the pressure of the liquid and the vapor were one atmosphere would you expect

$$\Delta S > \int \frac{dQ}{T}$$

Explain. Answer: (a) $\Delta G^\circ = (+)$; Yes, since the process of $H_2O_{(l)} \longrightarrow H_2O_{(g)}$ at 85°C is not spontaneous. (b) $\Delta G = 0$ (c) No, since the process is not spontaneous.

12. The standard heats of formation, absolute entropies at 298K (based on the Third Law of Thermodynamics), and densities for diamond and graphite are given in the following table.

	ΔH°_f(kJ/mol)	S°(J/K-mol)	density (g/ml)	C_P(J/K-mol)
diamond	1.9	2.4	3.51	6.1
graphite	0.0	5.7	2.25	8.5

Consider the transformation

$$\text{diamond} \longrightarrow \text{graphite}$$

(a) Can this process occur spontaneously at 298K and P = 1 atm?

(b) Considering the results of (a), why shouldn't women owning expensive diamond rings worry about their diamonds turning into graphite overnight?

(c) Can diamonds be prepared from graphite by changing the temperature? Assume $\Delta C_P = 0$. (In Exercise 13, the temperature dependence of C_P will be taken into account, but the same conclusion is reached when the C_P's are considered constant and equal.)

(d) The molar volume, $\overline{V}$, (liters/mole), can be calculated from the reciprocal of the density times the atomic weight of carbon (g/gram-atom) times the factor (1 liter/1000 mL). Calculate the molar volumes for graphite and diamond.

(e) Recalling that

$$\left(\frac{\partial \Delta G}{\partial P} \right)_T = \Delta \overline{V}$$

and assuming the $\overline{V}$'s are constant, calculate the pressure at which graphite and diamond are in equilibrium. In what pressure range can graphite be converted to diamond? Answer: (a) Yes (b) The process is slow (c) No (d) $\overline{V}_{graphite} = 5.33 \times 10^{-3}$ L/mol, $\overline{V}_{diamond} = 3.43 \times 10^{-3}$ L/mol (e) $P > P_{eq} \sim 15{,}000$ atm.

13. The heat capacities as function of T for graphite and diamond are:

T(K)	C_P graphite	C_P diamond	T(K)	$C_{P,}$ graphite	$C_{P,}$ diamond
25	0.025	0.0001	350	2.520	2.012
50	0.120	0.0025	400	2.960	2.500
100	0.400	0.060	450	3.262	2.970
150	0.780	0.240	500	3.500	3.325
200	1.200	0.600	600	3.900	3.90
250	1.625	1.000	900	4.95	4.95
300	2.085	1.500	1200	5.52	5.52

Show how ΔG° can be calculated as a function of T from the above data. A graphical technique would be the simplest. Assume thermodynamic data in Exercise 12. The units for C_P are cal/K-mol.

14. Calculate the equilibrium constant for the reaction (at 25°C)

$$A = B$$

if ΔG° is (a) 0, (b) - 4.184 kJ/mol, (c) - 41.84 kJ/mol, (d) + 4.184 kJ/mol, (e) + 41.84 kJ/mol. Note the relationship between the sign of ΔG° and the relationship with K_P.

(f) Sketch out a graph of K_P versus ΔG^o. Calculate additional K_P values for other ΔG^o. (g) The fraction of A converted to B is given by

$$\alpha = \frac{[B]}{[A]+[B]}$$

Calculate α for the values of ΔG^o in (a) to (e), then sketch α versus ΔG^o, calculating additional α values in order to sketch the graph. Note that the reaction is essentially complete when $\Delta G^o \leq -41.84$ kJ/mol and the extent of reaction is negligible when $\Delta G^o \geq 41.84$ kJ/mol. Answer: (a) 1.0 (b) 5.4 (c) 2.1 x 10^7 (d) 0.185 (e) 4.7 x 10^{-8} (g) $\alpha = 0.50, 0.842, 0.99999995, 0.156, 4.7$ x 10^{-8}.

15. Consider the following reaction:

$$CH_3CN_{(g)} + 2H_2O_{(g)} = 2CO_{(g)} + \frac{1}{2} N_{2(g)} + \frac{7}{2} H_{2(g)}$$

(a) Using the thermodynamic data in Appendix D, determine if this reaction is spontaneous at standard conditions at 298K. Briefly explain what is meant by standard conditions for this reaction. Assume the C_Ps are constant.
(b) Would this reaction be spontaneous at 298K if the pressures of the reactants and products were:

$$P_{CH_3CN_{(g)}} = 100 \text{ atm}, P_{H_2O_{(g)}} = 200 \text{ atm}, P_{CO_{(g)}} = 1 \text{ x } 10^{-7} \text{atm}$$

$$P_{N_{2(g)}} = 0.5 \text{ x} 10^{-7} \text{atm}, P_{H_{2(g)}} = 2 \text{ x } 10^{-7} \text{ atm? Show your calculations.}$$

(c) Calculate the equilibrium constant for this reaction at 298K.
(d) Calculate the partial pressures at equilibrium if the initial pressures are:

$$P^o_{CH_3CN} = 2.5 \text{ atm}, P^o_{H_2O} = 10 \text{ atm}, P^o_{CO} = P^o_{N_2} = P^o_{H_2} = 0$$

(e) Do you expect the yield to increase or decrease if we were to raise the temperature of the reaction to greater than 298K? Briefly explain and show all calculations.
(f) Calculate the change in entropy for this reaction and account for its magnitude in terms of the reaction.
(g) If this reaction were carried out at constant pressure, would the equilibrium yield increase, decrease, or show no change if the pressure were increased? Show all equations that you based your answer upon.
Answer: (a) Reaction is not spontaneous at standard conditions. Standard conditions means that all reactants and products are at the standard pressure of one bar~one atmosphere. (b) Yes (c) 1.9 x 10^{-18} (d) $P_{CH_3CN} = 2.499$ atm, $P_{H_2O} = 9.998$ atm, $P_{CO} = 0.00226$ atm, $P_{N_2} = 0.000565$, $P_{H_2} = 0.00396$ (e) Increase since K_P will increase ($\Delta H^o = +$) (f) $\Delta S^o = 325.0$ J/K since $\Delta n_{gas} = 3$ (g) Decrease.

16. Consider the reaction

$$PCl_{5(s)} = PCl_{3(g)} + Cl_{2(g)}$$

Note that PCl_5 is a solid whereas PCl_3 and Cl_2 are gases. The thermodynamic data for these compounds is as follows:

	state	$\Delta H_f^o(kJ)$	$\Delta G_f^o(kJ)$	$S^o(J/K)$
Cl_2	g	0.0	0.0	223.0
PCl_3	g	- 306.4	- 286.3	311.7
PCl_5	s	- 398.9	- 324.6	352.7

(a) Calculate the equilibrium constant for this reaction at 50°C assuming $\Delta C_p = 0$.

(b) Calculate the partial pressures of PCl_3 and Cl_2 if 20 mg of PCl_5 is added to a one liter flask at 50°C. M.W. PCl_5 = 208.24.

(c) Calculate the partial pressures of PCl_3 and Cl_2 if 1 mg of PCl_5 is added to a one liter flask at 50°C. Answer: (a) 3.5×10^{-6} (b) 1.9×10^{-3} atm (c) 1.3×10^{-4} atm.

17. The equilibrium constant for the reaction

$$CH_3CHO + I_2 = CH_3COI + HI$$

was measured experimentally at 481K and found to be K = 0.488.

(a) Calculate the ΔG^o in kcal at 481K for this reaction.

(b) The following table gives values for S_{298}^o; calculate ΔS_{298}^o for this reaction.

Species	S^o (gibbs/mol)	$\Delta H_{f,298K}^o$ (kcal/mol)	$C_{P,298K}^o$ (gibbs/mol)	$C_{P,500K}^o$ (gibbs/mol)
CH_3CHO	63.15	- 39.67	13.06	18.27
I_2	62.28	14.92	8.81	8.95
CH_3COI	76.0	?	17.49	20.90
HI	49.35	6.30	6.97	7.11

1 gibb = 1 cal/K

(c) Assume $\Delta S_{298}^o = \Delta S_{481}^o$ and calculate ΔH_{481}^o using the results of (a) and (b).

(d) Calculate $\Delta C_{P,298K}^o$ and $\Delta C_{P,500K}^o$ for the reaction and calculate the average

$$\overline{\Delta C_P^o} = \frac{1}{2}(C_{P,298K}^o + C_{P,500K}^o).$$ Then using the average ΔC_P (assume constant), calculate ΔH_{298K}^o.

(e) Using the heats of formation at 298K along with the result of (d), calculate the $\Delta H_{f,CH_3COI_{(g)}}^o$ (1 gibb = 1 cal/K). Answer: (a) 0.69 kcal (b) - 0.08 gibb (c) 0.65 kcal (d) 0.34 kcal (e) - 30.71 kcal.

18. The temperature dependence for the dimerization of acetic acid in the gas phase has been found to obey the equation

$$\log K = - 25.732 + \frac{4120}{T} + 4.910 \log T$$

where the units of K are in mm^{-1} (J.O. Halford, *J. Chem. Phys.*, **1941**, 9, 859) and log is base 10. Derive expressions for ΔG^o, ΔH^o, and ΔS^o for this dimerization. Note that ΔG^o must be calculated when K has its units expressed in terms of bar rather than mm. Answer: ΔG^o = - 9488.4R + 52.627RT - 4.910RT lnT, ΔH^o = - 9488.4R + 4.910RT, ΔS^o = - 47.717R + 4.910R lnT.

19. Consider the reaction

$$N_2O_{4(g)} = 2NO_{2(g)}$$

The thermodynamic data can be found in the Appendix, Section D.

(a) Calculate the equilibrium constant, K_P, for this reaction at 25°C , assuming ideal gas behavior.

(b) In a 500 mL bulb containing an equilibrium mixture of these two substances, the partial pressure of the NO_2 is found to be 500mm. What is the partial pressure of the N_2O_4?

(c) If the equilibrium mixture in (b) was produced by introducing NO_2 into the flask, what would be the quantity of this substance initially present?

(d) If the gaseous mixture described in (b) were allowed to expand into a total volume of 2 liters, would you expect the relative amount of NO_2 to N_2O_4 to increase or decrease? Calculate the partial pressures of each of the gases.

(e) If the temperature is increased to 50°C, would the relative amount of NO_2 to N_2O_4 increase or decrease? Explain how you would do the calculations.

(f) If K_P were measured at 50°C , how would you calculate ΔS^o for the reaction? Answer: (a) 0.15 (b) 2,200 mm (c) 4,900 mm (d) increase; P_{NO_2} = 237 mm, $P_{N_2O_4}$ = 494 mm (e) ΔH^o_{298} = 57.2 kJ. Since ΔH^o is positive, then K_P would increase with an increase in T and the relative amount of NO_2 to N_2O_4 would increase. (f)ΔS^o_{323} = (ΔH^o_{323} - ΔG^o_{323})/323K where ΔG^o_{323} = - R(323K) lnK_P and
$$\Delta H^o_{323} = \Delta H^o_{298} + \int_{298}^{323} \Delta C_P dT .$$

20. Calculate the equilibrium constant for the reaction at 25°C.

$$2 C_2H_{6(g)} + 7 O_{2(g)} = 4 CO_{2(g)} + 6H_2O_{(l)}$$

Use data from the Appendix, Section D. In making heat of combustion measurements, it is convenient to assume the reaction has gone to completion. Do you think this is justified in this case for the combustion of C_2H_6? Answer: $K_P \sim 1 \times 10^{514}$, Yes.

21. The equilibrium constant, K_P, for the following reaction at T= 298K is 2.8 x 10^6.

$$HCN_{(g)} + H_2O_{(l)} = CO_{(g)} + NH_{3(g)}$$

Note that H_2O is a liquid in this reaction.

(a) Write the equilibrium constant expression for this reaction.

(b) If the initial pressure of HCN is 0.035 atm, calculate the equilibrium pressures of all gases at T = 298K. Assume an excess of $H_2O_{(l)}$. Answer: $P_{HCN} = 4.4 \times 10^{-10}$ atm, $P_{NH_3} = P_{CO} = 0.035$ atm.

22. Consider the following reaction in answering the following questions.

$$C_2H_5OH_{(g)} \longrightarrow CO_{(g)} + C_{(graphite)} + 3H_{2(g)}$$

(a) Using thermodynamic data from tables in Appendix D, determine if this reaction is spontaneous at T = 298K at standard conditions.

(b) Calculate the equilibrium constant at 298K.

(c) Assume the initial pressure of C_2H_5OH is 1.0 X 10^{-10} atm. Set up the equilibrium expression required to calculate the equilibrium partial pressure for CO at 298K. Let x = P_{CO} and assume the volume is constant.

(d) Determine if this reaction is spontaneous at 298K when the partial pressures in atmospheres are: $P_{C_2H_5OH} = 1 \times 10^{-10}$, $P_{CO} = 1 \times 10^{-8}$, $P_{H_2} = 3 \times 10^{-8}$.

(e) If the reaction were carried out at a constant pressure, show what effect increasing the pressure would have on the equilibrium yield.

(f) What effect would increasing the temperature have on the equilibrium yield? Briefly explain.

(g) Assuming ΔC_P^o is a constant, calculate the equilibrium constant for this reaction at 400°C. Answer: (a) Reaction is not spontaneous (b) 3.26×10^{-6} (c) K = $\dfrac{(x)(3x)^3}{(1 \times 10^{-10} - x)}$ (d)Reaction is spontaneous (e) Decrease yield (f) $\Delta H^o = (+)$; Increase equilibrium yield (g) 2.7×10^7.

23. Consider the following reaction for the following questions.

$$CH_3Cl_{(g)} + H_{2(g)} \longrightarrow CH_{4(g)} + HCl_{(g)}$$

(a) Calculate ΔG^o and ΔS^o for this reaction, using thermodynamic values in Appendix D.

(b) Is the reaction spontaneous at standard conditions and 298K? Explain.

(c) Calculate the equilibrium constant at 298 K for the reaction.

(d) Calculate the equilibrium partial pressures for each of the species for the reaction. Assume the initial pressure of H_2 is 1×10^{-4} atm and the initial pressure of CH_3Cl is 1×10^{-6} atm. Hint: You do not need to use the quadratic equation.

(e) Do you expect the equilibrium constant to be larger or smaller if the temperature is increased? Explain.

(f) Do you expect the equilibrium constant to change if you were to change the initial pressure of the individual species? Explain.

(g) Calculate the equilibrium constant for the reaction at 1000°C = 1273 K assuming ΔC_p = constant. Answer: (a) $\Delta G^o = -87.2$ kJ, $\Delta S^o = 7.9$ J /K (b) Yes, since $\Delta G_{298}^o < 0$ (c) 1.9×10^{15} (d) $P_{CH_3Cl} = 5.3 \times 10^{-24}$ atm, $P_{H_2} = 0.99 \times 10^{-4}$ atm , $P_{CH_4} = P_{HCl} = 1 \times 10^{-6}$ atm (e) Since $\Delta H^o < 0$, the equilibrium constant will be smaller (f) No, the equilibrium constant is independent of pressure (g) 8.9×10^3.

24. At 80°C the vapor pressure of water is 355 torr and the heat of vaporization is 41.50 J/mol.

(a) Calculate ΔS and ΔG for the vaporization at P = 355 torr and 80°C.

(b) Is ΔG^o for this vaporization at 80°C positive, negative, or zero? Briefly explain. Answer: $\Delta S = 0.118$ J/K-mol, $\Delta G = 0$ (b) $\Delta G^o > 0$ since the process is not spontaneous at standard conditions.

25. Nitrogen dioxide dimerizes to dinitrogen tetroxide and an equilibrium is established.

$$2NO_{2(g)} \rightleftarrows N_2O_{4(g)}$$

2.82 g of this gas mixture was put into a 500 mL container. What is the total pressure of this gas at 298 K? Hint: thermodynamic data are available in the Appendices for NO_2 and N_2O_4. Answer: 1.70 atm.

26. The following is a graph of $\ln Kp$ versus $1/T$. Predict the signs of ΔH^o and ΔCp.

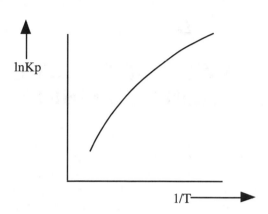

Answer: $\Delta H^o = (-)$, $\Delta C_P = (-)$.

27. Consider the following reaction which is being run at the pressures specified.

$$CH_3OH_{(g,\ P\ =\ 0.10\ atm)} + C_{(graphite)} \longrightarrow CO_{(g,\ P\ =\ 0.01\ atm)} + CH_{4(g,\ P\ =\ 0.01\ atm)}$$

   (a) Is this reaction spontaneous at standard conditions at 298K? Show calculations.
   (b) Is this reaction spontaneous at the pressures specified in the above equation at 298K? Show calculations.
   (c) Calculate the equilibrium constant at 298K.
   (d) Would the equilibrium constant increase or decrease with increasing temperature? Explain.
   (e) Calculate the equilibrium partial pressures at 298K if the initial pressures were those specified in the above equation.
   (f) Calculate the equilibrium constant at 150°C, assuming $\Delta C_P = 0$.
   Answer: (a) Yes, since $\Delta G^o_{298} < 0$ (b) Yes, since $\Delta G_{298} < 0$ (c) $K_P = 2.3 \times 10^4$ (d) Increase, since $\Delta H^o_{298} > 0$ (e) $P_{CH_3OH} = 5.3 \times 10^{-7}$ atm, $P_{CO} = P_{CH_4} = 0.11$ atm (f) 1.45×10^5.

4
THERMODYNAMICS OF SOLUTIONS AND SOLUTION EQUILIBRIA

When a reaction is carried out in the gas phase, the intermolecular interactions are small and possibly negligible at reasonably low pressures, say a few atmospheres at most (Ideal Gas Mixture). However, if a reaction is carried out in solution, there are significant solute-solvent interactions that are present even if the solute is at low concentration. The interactions can be significant enough that the change in standard free energy for the reaction can be altered sufficiently to affect the reaction yield and possibly the course of the reaction. Understanding the details of these interactions is difficult to carry out experimentally. Alternatively, theoretical calculations give a more accurate account of the nature of the interactions and the effect on the molecular structure. This is especially true for water interactions with large biochemical molecules, such as Bovine Pancreatic Trypsin Inhibitor (BPTI). The polar nature of BPTI interacts with water molecules principally through hydrogen bonds whereas the non-polar portions interact through van der Waals' forces. Compared to other biological species, BPTI is relatively small, yet the calculations are very complex and quite time-consuming on the computer. The results of the theoretical calculations showed that the structure of BPTI was highly influenced by the presence of the H_2O molecules compared to a previous calculation on the structure of BPTI "in vacuo"; i.e., without water present. For a review of this subject you should read an article by M. Gerstein and M. Levitt in Scientific American, *November, 1998. With the rapid advancement in the computer field, both with respect to speed and memory, these calculations will become more feasible in the future.*

A. Partial Molar Quantities

In Chapters 2 and 3 the principles of thermodynamics were applied to single substances and ideal gas mixtures. Several useful equations relating thermodynamic variables were developed, and these were applied to pure substances or ideal gas mixtures. With these equations we were able to reach the goals we set out to accomplish; namely, the prediction of spontaneity, and the calculation of the equilibrium yield. However, before we can apply these principles to solutions, we must make some slight but important modifications. These modifications, the principle subject matter of this chapter, will enable us to apply thermodynamic principles to solutions in order to predict spontaneity and calculate equilibrium yield for reactions in solution.

The word solution refers to a homogeneous mixture of two or more components in a single phase. There are several different types of solutions, with regard to the natural phase of each component when it exists in its pure state at the same temperature and pressure. For example, one could consider gases, liquids, and solids dissolved in solids, as well as liquids. In addition to the general properties of solutions, there are special properties related to each particular type of solution. In this chapter, however, we will limit our treatment to certain properties and considerations concerning solutions in the liquid phase; that is, solutions of gases, liquids, or solids in liquids.

Generally, the discussion will be limited to two-component solutions. The term solvent will refer, as usual, to the component in excess, while solute will refer to the other components. This distinction is not fundamental, and in some cases it completely loses its significance, as for example, when the components are both of comparable concentration in the same phase, such as two liquids.

In Chapter 3 we were able to apply a thermodynamic analysis to pure substances and ideal gas reactions. The reason we could treat both of these with the same principles is that in an ideal gas mixture, each of the components behaves as if it were in the pure state; that is, there is no interaction between the different gases. However, in a solution, this is generally not the case. The presence of one of the species in the solution, say the solvent, affects the thermodynamic properties of the solute. A simple example of this is the volume that results from a mixture of ethyl alcohol and water. 1.00 liter of ethyl alcohol when mixed with 1.00 liter of water gives a total volume of 1.93 liters, not 2.00 as one might expect. This result is due to a strong attractive force between water and ethyl alcohol. In this case, the variable, volume, is a function of the relative number of moles of each component in the solution, in addition to the total number of moles, temperature, and pressure. Mathematically, this can be expressed by

$$V = V(X_2, n, T, P) \tag{4-A-1}$$

where X_2 is the mole fraction of solute. (X_1, the mole fraction of the solvent, is equal to $1 - X_2$.) For an ideal gas mixture V would not be dependent on X_2.

If T and P are held constant, then the volume of a solution is dependent only on the composition and the total number of moles, and equation (4-A-1) becomes

$$V = V(X_2, n) \tag{4-A-2}$$

Since the moles of solute (n_2) is given by $n_2 = X_2 n$, and the moles of solvent (n_1) by $n_1 = X_1 n = (1 - X_2)n$, we can replace the variables (X_2, n) by (n_1, n_2) and

$$V = V(n_1, n_2) \tag{4-A-3}$$

Similar statements can be written for other extensive thermodynamic variables such as H, S, and G.

In the thermodynamic analysis of solutions, it is convenient to express a given extensive variable, such as total volume (V), in terms of the contributions of its components. In the case of volume,

$$V = V_1 + V_2 \tag{4-A-4}$$

where V_1 and V_2 are the contributions of components 1 and 2 to the total volume. In solution, V_1 and V_2 are not the volumes of the pure components before mixing. This should be obvious from the example of the mixing of ethyl alcohol and water, in which the total volume (1.93 liters) was not equal to the sum of the volume of water (1 liter) and ethyl alcohol (1 liter). In the case of ideal gas mixtures, V_1 and V_2 are the volumes of the pure components. This will also be true for what we shall later define as an ideal solution (not to be confused with an ideal gas mixture).

The volume of a solution can be written in terms of its components by the expression

$$V = n_1 \overline{V}_1 + n_2 \overline{V}_2 \tag{4-A-5}$$

$\overline{V}_1$ and $\overline{V}_2$ are new variables which must be introduced in order to consider the thermodynamic properties of the different components in solution. They are called *partial molar quantities*, and for the particular variable, volume, are called *partial molar volumes*. Mathematically, they are given by

$$\overline{V}_1 = \left(\frac{\partial V}{\partial n_1}\right)_{n_2, T, P} \qquad \text{(Partial Molar Volume of Component 1)}$$

$$\overline{V}_2 = \left(\frac{\partial V}{\partial n_2}\right)_{n_1, T, P} \qquad \text{(Partial Molar Volume of Component 2)}$$

For $\overline{V}_1$, the n_2, T,P are held constant as indicated in the partial derivative; similarly for $\overline{V}_2$, n_1, T,P are held constant. The T and P are generally held constant, and for this reason, constancy is not usually indicated in the partial derivative. Thus,

$$\overline{V}_1 = \left(\frac{\partial V}{\partial n_1}\right)_{n_2} \qquad \overline{V}_2 = \left(\frac{\partial V}{\partial n_2}\right)_{n_1}$$

and it is understood that T and P are constant.

Comparing equation (4-A-5) with (4-A-4), the contribution of component 1 can be expressed through the partial molar volume:

$$\overline{V}_1 = n_1 \overline{V}_1$$

This relationship and the interpretation of $\overline{V}_1$ can be better understood when we consider that $\overline{V}_1$ is a partial derivative. Being a derivative, $\overline{V}_1$ is the rate at which the volume changes with number of moles of component 1, the composition or concentration of the solution being held constant. $\overline{V}_1$ thus has the units of volume/mole, and when multiplied by n_1 (moles), gives the volume contributed by component 1 to the solution. A similar interpretation can be made of $\overline{V}_2$ and V_2.

For ideal gas mixtures and ideal solutions, $\overline{V}_1$ and $\overline{V}_2$ are the same as for pure components. Thus, the sum of the volumes of the pure components gives the total volume of the solution. For the pure phases, the molar volume is indicated by a solid dot as a superscript; for example, $\overline{V}_2^{\bullet}$ is the molar volume for pure component 2, that is, the volume of one mole of pure component 2. For an ideal gas mixture or an ideal solution,

$$\overline{V}_1 = \overline{V}_1^{\bullet} \qquad \text{(for ideal gas mixture or ideal solution)} \qquad (4\text{-}A\text{-}7)$$
$$\overline{V}_2 = \overline{V}_2^{\bullet}$$

The discussion we have had concerning total volume and partial molar volumes applies to other extensive variables in thermodynamics. Equations similar to equation (4-A-5) can be written for H, S, and G, namely,

$$H = n_1 \overline{H}_1 + n_2 \overline{H}_2$$
$$S = n_1 \overline{S}_1 + n_2 \overline{S}_2$$
$$G = n_1 \overline{G}_1 + n_2 \overline{G}_2$$

Furthermore, equations which were developed in Chapters 2 and 3 for pure components can also be used for the components in solution, providing the partial molar quantities are used. For example,

$$\overline{H} = \overline{U} + P\overline{V}$$
$$\overline{G} = \overline{H} - T\overline{S}$$

$$\left(\frac{\partial \dfrac{\overline{G}}{T}}{\partial T} \right)_P = - \frac{\overline{H}}{RT^2}$$

$$\left(\frac{\partial \overline{G}}{\partial P} \right)_T = \overline{V}$$

Before looking at the partial molar free energies of components in solution, it would seem proper to treat the subject of molar free energies of pure substances in the condensed phase. Since we already know how to deal with the free energy of a substance in the gas phase, $G_{i(vap)}$, we will relate it to the molar free energy of the substance in the liquid or solid phase, $\overline{G}_{i(cond)}$. We can accomplish this task by utilizing the principle that the change in free energy corresponding to a change between two states in equilibrium is zero. In other words, the molar free energy is the same for a substance in either of the two states in equilibrium. Consider a liquid in equilibrium with its vapor at a pressure equal to the equilibrium vapor pressure at that temperature. We can write

$$\overline{G}_{i(liquid)}^{\cdot} = \overline{G}_{i(vapor\ at\ P_i^{\cdot})}^{\cdot} \qquad (4\text{-}A\text{-}8)$$

where $P^{\cdot}$ is the equilibrium vapor pressure of the pure liquid at the temperature being considered.

Equation (3-E-6) can be written in terms of molar quantities as

$$d\,\overline{G}_i = \overline{V}_i\,dP$$

Integrating as we did in Chapter 3 from a standard state of 1 bar, which we designate by a superscript zero

$$\int_{\overline{G}_i^o}^{\overline{G}_i} d\overline{G}_i = \int_1^P \overline{V}_i\,dP \qquad (4\text{-}A\text{-}9)$$

Assuming the vapor obeys the ideal gas equation of state,

$$\overline{G}_i - \overline{G}_i^o = \int_1^P \frac{RT}{P}\,dP = RT\,\ln P \qquad (4\text{-}A\text{-}10)$$

If the vapor is at the vapor presure of the liquid ($P_i^{\cdot}$),

$$\overline{G}_{i(vapor\ at\ P_i^{\cdot})}^{\cdot} = \overline{G}_{i(vapor)}^o + RT\,\ln P_i^{\cdot} \qquad (4\text{-}A\text{-}11)$$

From equation (4-A-8), and the relationship between the free energy of a gas and the pressure given in equation (4-A-11), we can obtain the expression

$$\overline{G}_{i(liquid)}^{\cdot} = \overline{G}_{i(vapor)}^o + RT\,\ln P_i^{\cdot} \quad \textbf{Molar Free Energy of a Pure Liquid} \qquad (4\text{-}A\text{-}12)$$

Similarly, in the case of a solid substance,

$$\overline{G}^{\cdot}_{i \text{ (solid)}} = \overline{G}^{\cdot}_{i(\text{vapor at } P_i)}$$

and

$$\overline{G}^{\cdot}_{i(\text{solid})} = \overline{G}^{o}_{i(\text{vapor})} + RT \ln P^{\cdot}_i \quad \textbf{Molar Free Energy of a Pure Solid} \qquad \text{(4-A-13)}$$

where $P^{\cdot}_i$ is the equilibrium vapor pressure of the solid at temperature T.

The next step is to extend this method of representation of the partial molar free energy to the components of a solution. Consider a solution of composition X_1 and X_2, the mole fractions of components 1 and 2, respectively. The vapor above the solution will also be a mixture of components 1 and 2; however, its composition, Y_1 and Y_2, will not be the same as the composition of the liquid phase. The vapor will be richer in the more volatile component. The partial pressure of each component in the vapor phase is equal to its equilibrium vapor pressure, P_i, above the solution at the temperature. The vapor pressure of a component in solution is generally less than the vapor pressure exerted by the same substance in the pure state at the same temperature.

Analogous to the case of pure condensed substances, the partial molar free energy of a component in solution is equal to the molar free energy of the component in the vapor phase above the solution at a partial pressure equal to the equilibrium vapor pressure of the particular component. This can be expressed mathematically as

$$\overline{G}_{i(\text{solution})} = \overline{G}_{i(\text{vapor at } P_i)} \qquad \text{(4-A-14)}$$

or from equation (4-A-11),

$$\overline{G}_{i(\text{solution})} = \overline{G}^{o}_{i(\text{vapor})} + RT \ln P_i \qquad \textbf{Partial Molar Free Energy} \qquad \text{(4-A-15)}$$
$$\textbf{of a Component in Solution}$$

Using equation (4-A-15), the partial molar free energies of components in solution can be evaluated from experimentally measured vapor pressures, P_i. Thus, we must determine how the vapor pressure of a component in solution varies as a function of its concentration or mole fraction in the solution. This information will also be valuable in finding the total vapor pressure above the solution and the composition of the vapor phase in equilibrium with the solution.

B. Vapor Pressure of Liquid Solutions

The vapor pressure of a component in solution depends upon the concentration of the component and the nature of the other components. In particular the nature of the intermolecular interactions between the different components plays an important role upon the shapes of the vapor pressure curves. We will start out with the simpler situation, an ideal solution, and consider the more complex solutions later.

Ideal Solution

An *ideal solution* is one which arises from a solution of similiar components. For example, two saturated hydrocarbons, say hexane and heptane, would be expected to form an ideal solution. The interaction between hexane molecules would be expected to be similar to that between heptane molecules, and consequently, the interaction between hexane and heptane would be similar but probably intermediate. In any event when a solution contains similar type compounds, the vapor pressure of any of the components is simply reduced by its mole fraction, which is a measure of the fraction of the molecules of this component near the surface. For an ideal solution the vapor pressure of any component in the solution is given by Raoult's Law

$$P_i = X_i P_i^{\cdot} \qquad \text{Raoult's Law} \qquad (4\text{-}B\text{-}1)$$

For a two component solution the vapor pressures of the components can be displayed on a two-dimensional graph as shown in Figure 4-B-1. The mole fraction of component #2, X_2, runs from left to right along the X-axis. The vapor pressure for component #2 at $X_2 = 0$ is obviously zero since there is no component #2 present and then increases linearly, according to Raoult's Law, to $X_2 = 1$ where the vapor pressure is the vapor pressure of pure liquid, $P_2^{\cdot}$. Obviously for a mole fraction $X_2 = 0.5$ the vapor pressure of component #2 will be 1/2 that of pure component #2 and the surface would contain about 1/2 component #2 and of course 1/2 component #1.

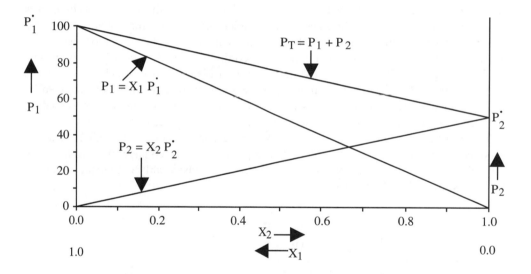

Figure 4-B-1. Vapor Pressures for Ideal Solution

The vapor pressure curve for component #1 is a fraction X_1, which runs from right to left along the X-axis in Figure 4-B-1. At $X_1 = 0$ the vapor pressure of component #1 is zero and increases linearly, according to Raoult's Law, to $P_1^{\cdot}$ at $X_1 = 1.0$. The total pressure, P_T, is the sum of the vapor pressures for the components and is the staight line connecting the points $X_1 = 1.0$, $P = P_1^{\cdot}$ and $X_2 = 1.0$, $P = P_2^{\cdot}$. We can show this mathematically by expressing the total pressure in terms of the Raoult's Law expressions for the vapor pressures of the components

$$P_T = P_1 + P_2 \qquad (4\text{-}B\text{-}2)$$

$$P_T = X_1 P_1^{\cdot} + X_2 P_2^{\cdot}$$

Substituting for $X_1 = 1 - X_2$

$$P_T = (1 - X_2) P_1^{'} + X_2 P_2^{'}$$

$$P_T = (P_2^{'} - P_1^{'}) X_2 + P_1^{'} \qquad (4\text{-B-}3)$$

Equation (4-B-3) is a linear function between the variables P_T and X_2 with a slope of $(P_2^{'} - P_1^{'})$ and an intercept of $P_1^{'}$.

Problem 4-B-1. If the vapor pressure of a certain liquid, A, is 79.1 torr above a solution which contains 79 moles A per mole B, what would be the vapor pressure of pure A, according to Raoult's Law? Answer: 80.1 torr.

Problem 4-B-2. At 35°C the vapor pressures of ethanol and water are 100 torr and 41.9 torr, respectively. What is the total vapor pressure of a solution which contains 2 moles of ethanol and 6 moles of water, if the solution were assumed to be ideal? Answer: 56.1 torr.

Real Solutions

When the two components in a solution are dissimilar in nature, the vapor pressure curves deviate from Raoult's Law. If the interaction of molecules between components #1 and #2 is less than that between molecules of the same component #1 or #2, the vapor pressure of either component is greater than that expected according to Raoult's Law. We refer to this as a phase diagram with positive deviations from an ideal solution. Such a situation is shown in Figure 4-B-2 where the actual vapor pressures of the components is depicted by the solid lines that are curved. Raoult's Law for each of the components is represented by the dashed straight lines similar to those shown in Figure 4-B-1. As one can see, the curved lines representing the actual vapor pressures deviate considerably from Raoult's Law at low concentrations. This is true for both components: P_2 is considerably higher than Raoult's Law: $P_2 = X_2 P_2^{'}$ at low X_2 values and P_1 is considerably higher than Raoult's Law: $P_1 = X_1 P_1^{'}$ at low X_1 values.

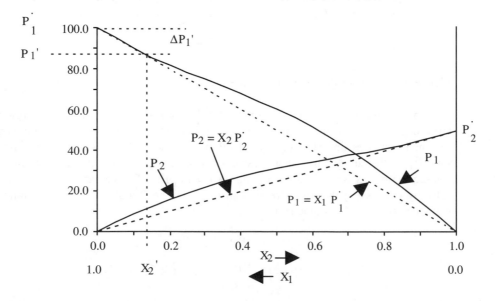

Figure 4-B-2. Vapor Pressures for Real Solution Showing Positive Deviations from Raoult's Law

On the other hand, at high concentrations of either component the actual vapor pressures are very similar to those represented by Raoult's Law. Note as $X_1 \rightarrow 1.0$ the curved vapor pressure curve P_1 coincides with the dashed straight line representing Raoult's Law ($P_1 = X_2 P_1^*$) for component #1. Similarly as $X_2 \rightarrow 1.0$ the curved vapor pressure curve, P_2, coincides with the dashed straight line representing Raoult's Law: $P_2 = X_2 P_2^*$ for component #2. In a solution the major component is referred to as the solvent and for this reason we conclude that the solvent obeys Raoult's Law closely.

If the two components making up the solution have intermolecular attractive forces which are significantly greater than those between the individual components, one will observe negative deviations from Raoult's Law. This is shown in Figure 4-B-3 where the solid curves representing the true vapor pressures fall <u>below</u> the dashed straight lines representing Raoult's Law for components #1 and #2. Again note the large deviations from Raoult's Law at low concentrations but very close agreement with Raoult's Law at high concentration.

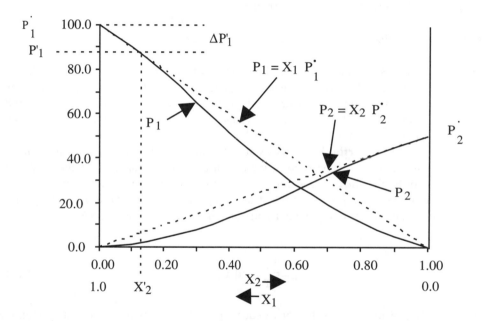

Figure 4-B-3. Vapor Pressure Curves for Real Solution Showing Negative Deviations from Raoult's Law

Decrease in Vapor Pressure Due to a Non-volatile Solute

If a solution consists of a volatile solvent and a non-volatile solute, the vapor pressure of the solution will diminish as the non-volatile solute concentration is increased. This decrease in vapor pressure–non-volatile solute concentration relationship can be derived since the solvent will obey Raoult's Law

$$P_1 = X_1 P_1^*$$

Substituting in $X_1 = (1 - X_2)$

$$P_1 = (1 - X_2) P_1^* = P_1^* - X_2 P_1^*$$

Solving for the decrease in vapor pressure, ΔP_1,

$$P_1^{\cdot} - P_1 = P_1^{\cdot} X_2$$

$$\Delta P_1 = P_1^{\cdot} X_2 \qquad (4\text{-}B\text{-}4)$$

This decrease in vapor pressure is shown in Figure 4-B-2 at a specific solute concentration, X_2'. The ΔP_1 corresponding to this specific solute concentration is shown as $\Delta P_1'$ in Figure 4-B-2. The same $\Delta P_1'$ would be observed for the solution with negative deviations shown in Figure 4-B-3.

Problem 4-B-3. If the vapor pressure of benzene at 26°C is decreased by 10 torr when 12.8 g of naphthalene are dissolved in 70.2 g benzene, calculate the value of the equilibrium vapor pressure of pure benzene at 26°C. Assume the vapor pressure of benzene obeys Raoult's Law. Answer: 100 torr.

Vapor Composition

The vapor above a solution of two volatile components also contains a mixture of the two components. The partial pressure of each component is equal to its vapor pressure above the solution. If the solution is considered to be ideal, then the partial pressures are given by Raoult's Law, and the total pressure in the vapor phase, being the sum of the partial pressures, is given by equation (4-B-3).

Dalton's Law of Partial Pressures states that the mole fraction of a component in a mixture of gases is equal to the ratio of the partial pressure of that component to the total pressure. Accordingly, the composition of the vapor phase above a solution can be determined, if the individual vapor pressures of the components are known. If Y_1 and Y_2 represent the mole fractions of component 1 and component 2 in the vapor phase, respectively, then

$$Y_1 = \frac{P_1}{P_T} \quad \text{and} \quad Y_2 = \frac{P_2}{P_T} \qquad (4\text{-}B\text{-}5)$$

We need only consider one of the above equations, since in a two-component system, the value of one mole fraction determines the value of the other. For convenience, let us choose component 1. In light of Raoult's Law, $P_i = X_i P_i^{\cdot}$, equation (4-B-5) can be written in the form

$$Y_1 = \frac{X_1 P_1^{\cdot}}{P_T} = X_1 \frac{P_1^{\cdot}}{P_T}$$

or for the ith component,

$$Y_i = X_i \frac{P_i^{\cdot}}{P_T} \qquad (4\text{-}B\text{-}6)$$

Equation (4-B-6) states that the mole fraction of a component in the vapor phase is equal to the mole fraction of the component in the solution multiplied by the ratio of the vapor pressure of the pure component to the total vapor pressure of the solution.

Equation (4-B-6) illustrates the fact that, in general, the composition of the vapor in equilibrium with a liquid solution is not the same as the composition of the solution. This result is in obvious agreement with experimental observations—for example, distillation phenomena. Since the total vapor pressure of an ideal solution, P_T, is always less than the vapor pressure of the more volatile component, P_1^*, the ratio of the vapor pressure of the more volatile component to the total vapor pressure is always greater than unity. When this situation occurs, theory predicts that the mole fraction of the component in the equilibrium vapor is greater than its mole fraction in the solution, as readily seen by equation (4-B-6). In other words, the vapor is richer in the more volatile component than is the liquid, while the liquid is richer in the less volatile component than is the vapor. This result is also consistent with experimental information.

As the difference between the vapor pressures of the two pure components becomes smaller and smaller, the ratio of the vapor pressure of either pure component to the total vapor pressure approaches unity. The closer the pressure ratio value is to unity, the nearer the composition of the vapor, Y_i, is to the composition of the liquid, X_i. In the rare event that the two pure components have the same vapor pressures, the ratio of the vapor pressure of the pure component to the total vapor pressure would be unity. In such an instance, the equilibrium vapor composition would be the same as the composition of the liquid.

Problem 4-B-4. Find the composition of the vapor which would be in equilibrium with the liquid solution in Problem 4-B-2. Answer: $Y_{C_2H_5OH} = 0.443$, $Y_{H_2O} = 0.557$.

Problem 4-B-5. Find the expression for the vapor composition, Y_1, in terms of X_1, P_1^* and P_2^*. Show that as P_1^* and P_2^* approach the same value, the composition of the vapor approaches the composition of the solution. Answer: $Y_1 = X_1 \dfrac{P_1^*}{X_1 P_1^* + X_2 P_2^*}$,

$$Y_1 = X_1 \frac{P_1^*}{X_1 P_1^* + X_2 P_1^*} = X_1 \frac{P_1^*}{(X_1 + X_2)P_1^*} = X_1$$

Problem 4-B-6. What characteristic enhances the ease of separation of two miscible liquids by distillation? What condition would make it impossible to separate two miscible liquids by distillation? Answer: Large difference in vapor pressure or boiling point of the two components. If the two components had the same boiling point.

Henry's Law

Recalling Figure 4-B-2, as the mole fraction of a component in solution becomes smaller and smaller, the vapor pressure curve deviates significantly from the linear Raoult's Law. What this means is that Raoult's Law is not a good approximation of the vapor pressure of a component as its mole fraction becomes small and approaches zero. Thus, Raoult's Law does not work very well in the case of the solute. However, the actual vapor pressure curve of a component does approach another line as the mole fraction approaches zero, namely, the tangent to the actual vapor pressure curve at the point $X_2 = 0$, $P_2 = 0$, as illustrated in Figure 4-B-4. Figure 4-B-4 has the same vapor pressure curves as in Figure 4-B-2 but in Figure 4-B-4 we focus on the vapor pressure for the solute, component #2, at <u>low</u> concentration of solute.

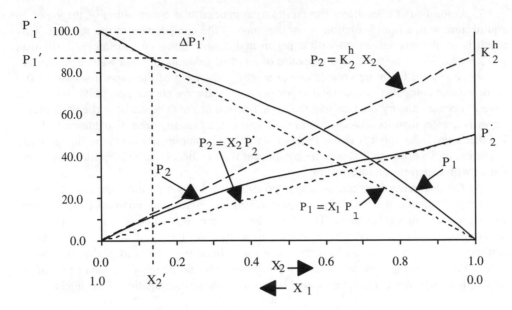

Figure 4-B-4. Vapor Pressure Curves for Real Solution Illustrating Use of Henry's Law for the Solute

The actual vapor pressure curve approaches the tangent line as X_2 approaches zero. Thus, at low solute concentrations the vapor pressure of the solute in a solution is given by the equation of the line tangent to the actual vapor pressure curve at zero, known as Henry's Law.

$$P_2 = K_2^h \; X_2 \qquad \textbf{Henry's Law} \qquad (4\text{-}B\text{-}7)$$

The constant K_2^h is known as the Henry's Law constant and is the value of P_2 calculated by equation (4-B-7) when the mole fraction X_2 is set equal to unity. In other words, K_2^h is the value which Henry's Law would predict for the vapor pressure of the pure solute, pure component 2. The Henry's Law constant for a solute may be greater than or less than the vapor pressure of the pure solute. It will be greater than $P_2^\cdot$ when the actual vapor pressure curve lies above the Raoult's Law line, positive deviations, as shown in Figure 4-B-4, and less than $P_2^\cdot$ in the case when the actual vapor pressure curve dips below the Raoult's Law line, Figure 4-B-3.

Henry's Law is similar to Raoult's Law, in the sense that both laws insist that the vapor pressure of a component in solution is directly proportional to its mole fraction. However, the two laws differ in their designation of the proportionality constant. According to Henry's Law, the vapor pressure, P_2' corresponding to the composition X_2' is $K_2^h \; X_2'$ while Raoult's Law predicts it is equal to $P_2^\cdot \; X_2'$. Figure 4-B-5, an expanded version of the lower left hand portion of Figure 4-B-4, illustrates that the value predicted by Henry's Law is much closer to the actual vapor pressure than is the value calculated with Raoult's Law in the instance where the mole fraction is less than 0.15.

The value of the Henry's Law constant is not a property unique to the solute, but is dependent upon the nature of the solvent as well. The values of K_2^h for a particular solute in several solutions can be similar, however, if the solvents are similar in regard to their

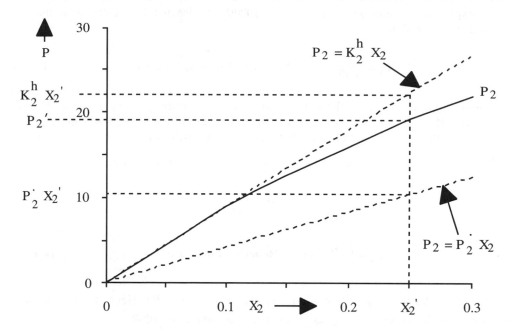

Figure 4-B-5. Expanded View of Figure 4-B-4 at Low Solute Concentrations—Agreement Between Vapor Pressure of Solute and Henry's Law

chemical and physical properties. Needless to say, the value of the Henry's Law constant is extremely sensitive to any specific interaction between the solute and the solvent. As would be expected, Henry's Law constants also vary with the temperature of the solution.

The value of the Henry's Law constant is not a property unique to the solute, but is dependent upon the nature of the solvent as well. The values of K_2^h for a particular solute In several solutions can be similar, however, if the solvents are similar in regard to their chemical and physical properties. Needless to say, the value of the Henry's Law constant is extremely sensitive to any specific interaction between the solute and the solvent. As would be expected, Henry's Law constants also vary with the temperature of the solution.

Henry's Law is usually associated with the solute, since it more accurately represents the vapor pressures of a component whose mole fraction lies in the range usually labeled solute. It is a limiting law, being completely valid only in the limiting vapor pressure law for the solute. It is convenient to express equation (4-B-7) in terms of molality of the solute rather than the mole fraction of the solute. Before we can accomplish this task, we must first find the relationship between the molality, m_2, and the mole fraction. By definition, the mole fraction of the solute is given by the expression

$$X_2 = \frac{\text{moles solute}}{\text{moles solute} + \text{moles solvent}}$$

If M represents the molar mass (gram molecular weight) of the solvent, every 1000 grams of solvent will contain 1000/M moles of solvent and will be associated in solution with m_2 moles of solute (molality). Thus, the above expression can be rewritten in terms of M and m_2 as

$$X_2 = \frac{m_2}{m_2 + \dfrac{1000}{M}}$$

When the solution is dilute and m_2 is small compared to $1000/M$, the denominator in the above expression is approximately equal to $1000/M$. Consequently, we can write the approximate relationship:

$$X_2 \approx \frac{M}{1000} m_2 \qquad \textbf{Mole Fraction and Molality} \qquad (4\text{-}B\text{-}8)$$

The more dilute the solution, the more valid is the relationship given in equation (4-B-8). Thus, both Henry's Law and equation (4-B-8) are useful in the same, low concentration range. Substituting the relationship given in equation (4-B-8) into Henry's Law, ($P_2 = K_2^h X_2$), the expression of Henry's Law in terms of the molality is obtained.

$$P_2 = K_2^h \frac{M}{1000} m_2$$

or,

$$P_2 = \left(K_2^h\right)' m_2 \qquad \textbf{Henry's Law (modified)} \qquad (4\text{-}B\text{-}9)$$

where P_2 is the vapor pressure of the solute and $\left(K_2^h\right)'$ is the Henry's Law constant multiplied by the molecular weight of the solvent and divided by 1000.

Problem 4-B-7. If the vapor pressure of SO_2 above a 2.6 mole percent solution of SO_2 in water is 1 atm at 25°C, what is the value of Henry's Law constant for SO_2 in water? What is the vapor pressure of SO_2 for a one mole percent solution? Assume Henry's Law is valid in the concentration range < 2.6 mole percent SO_2. Answer: K_2^h = 38.5 atm, $P_2 = 0.385$ atm at 1 mole percent.

Problem 4-B-8. If 32 g of SO_2 are disolved in 700 g water, what portion of the vapor pressure of the solution would be due to SO_2? The vapor pressure of pure water at 25°C is 23.8 torr. Assume SO_2 obeys Henry's Law and water obeys Raoult's Law. Answer: $Y_2 = 0.941$.

Solubility of Gases

Experimentally, Henry's Law applies quite well to the solubility of gases in many solvents. In such cases, if the amount of dissolved gas is kept well below its maximum solubility, that is, as long as the liquid phase does not approach the saturation point, the equilibrium molality of the dissolved gas appears to be directly proportional to the applied partial pressure of the particular gas. The proportionality constant in this instance is comparable to the reciprocal of the modified Henry's Law constant. To a first approximation, the solubility of any one gaseous solute is dependent only on the applied pressure of that particular gas and is independent of the concentrations of other dissolved gases.

Problem 4-B-9. If the ratio of the modified Henry's Law constant for gas A to that for gas B is 0.25, what is the ratio of partial pressures of the two gases required for the molality of A to be 10 times as great as the molality of B in the solution? Answer: 2.5.

Problem 4-B-10. Suppose the right side of equation (4-B-9) is multiplied by a factor, γ, the values of which will be chosen is such a way that Henry's Law (modified) will be valid over the entire concentration range. Is the factor, γ, a constant, or must its value change? How might we determine the value or values of γ experimentally? Answer: γ must change. We must measure P_2 at various m_2 and calculate γ from $\gamma = P_2/(K_2^h)'m_2$.

C. Colligative Properties

Vapor Pressure Lowering of Solvent (ΔP_1)

In the previous section we derived in equation (4-B-4) an expression for the vapor pressure lowering of the solvent in dilute solutions where Raoult's Law is obeyed.

$$\Delta P_1 = \left(P_1^{\cdot}\right)X_2 \tag{4-B-4}$$

Note in equation (4-B-4) that the vapor pressure lowering depends only on the concentration of the solute, X_2, and <u>not</u> on the nature of the solute. On the other hand, the vapor pressure lowering is dependent on the nature of the solvent, which in equation (4-B-4) is the vapor pressure of the solvent $P_1^{\cdot}$.

We will derive in this section similar expressions for boiling point elevation, freezing point lowering, and osmotic pressure for dilute solutions. These expressions will be similar to the vapor pressure lowering in that they will depend only on the concentration of the solute and <u>not</u> the nature of the solute. Since these properties have this common character, we refer to them as <u>colligative</u> properties. The term colligative refers to a "collection" of properties.

Boiling Point Elevation (ΔT_b)

In order to derive the expression for boiling point elevation, ΔT_b, we should recall the expression for the partial molar free energy of a component in solution, equation (4-A-15).

$$\overline{G}_{i(solution)} = \overline{G}^o_{i(vapor)} + RT \ln P_i \tag{4-A-15}$$

Since we are concerned with the boiling point elevation of the solvent, we will use equation (4-A-15) for the solvent where i =1. For dilute solutions Raoult's Law applies to the solvent, $P_1 = X_1 P_1^{\cdot}$, which gives

$$G_{1(solution)} = \overline{G}^o_{1(vapor)} + RT \ln X_1 + RT \ln P_1^{\cdot} \tag{4-C-2}$$

However, from equation (4-A-12) we note that

$$\overline{G}^{\cdot}_{1(liquid)} = \overline{G}^o_{1(vapor)} + RT \ln P_1^{\cdot} \tag{4-A-12}$$

and equation (4-C-2) becomes

$$\overline{G}_{1(solution)} = \overline{G}^{\cdot}_{1(liquid)} + RT \ln X_1 \tag{4-C-3}$$

For the process of boiling, the solvent in solution is in equilibrium with the solvent in the vapor state, and equation (4-C-3) becomes

$$\overline{G}_{1(vapor)} = \overline{G}^{\cdot}_{1(liquid)} + RT \ln X_1 \tag{4-C-4}$$

If we specify that the boiling point elevation be at the normal boiling point (P = 1 atm), then the vapor is at one atmosphere and (4-C-4) becomes

$$\overline{G}^o_{1(vapor)} = \overline{G}^{\cdot}_{1(liquid)} + RT \ln X_1 \tag{4-C-5}$$

The difference in the $\overline{G}$ values is the change in standard free energy of vaporization

$$\Delta G^{\circ}_{vaporization} = \overline{G}^{\circ}_{1(vapor)} - \overline{G}^{\circ}_{1(liquid)} = RT \ln X_1 \qquad (4\text{-}C\text{-}6)$$

Since $\Delta G = \Delta H - T\Delta S$,

then $\Delta G^{\circ}_{vap} = \Delta H^{\circ}_{vap} - T\Delta S^{\circ}_{vap} \qquad (4\text{-}C\text{-}7)$

where T = boiling point for this mixture with a solvent concentration of X_1. Dividing by RT

$$\frac{\Delta H^{\circ}_{vap}}{RT} - \frac{\Delta S^{\circ}_{vap}}{R} = \ln X_1 \qquad (4\text{-}C\text{-}8)$$

When $X_1 = 1$, we have pure solvent and the temperature is the normal boiling point of the solvent, which we will designate as Tb

$$\text{EMBED Equation.DSMT36} \quad = \quad \text{EMBED Equation.DSMT36} \qquad (4\text{-}C\text{-}9)$$

Substituting (4-C-9) into equation (4-C-8) and expressing the solution composition in terms of X2

$$\text{EMBED Equation.DSMT36} \quad - \quad \text{EMBED Equation.DSMT36} \quad = \ln X1 = \ln (1 - X2).$$

If X_2 is small, it can be shown that $\ln (1 - X_2) \sim - X_2$ and

$$- \frac{\Delta H^{\circ}_{vap}}{R} \left(\frac{T - T_b}{TT_b} \right) = - X_2 \qquad (4\text{-}C\text{-}10)$$

Solving for the normal boiling point elevation = ΔT_b

$$\Delta T_b = (T - T_b) = \frac{RTT_b}{\Delta H^{\circ}_{vap}} X_2 \qquad (4\text{-}C\text{-}11)$$

Since the boiling point elevation is small, $T \approx T_b$

$$\boldsymbol{\Delta T_b} = \left(\frac{RT_b^2}{\boldsymbol{\Delta H^{\circ}_{vap}}} \right) \boldsymbol{X_2} \qquad (4\text{-}C\text{-}12)$$

We note in equation (4-C-12) that the boiling point elevation depends only on the concentration of solute, X_2, and <u>not</u> on the properties of the solute. The boiling point elevation, however, does depend upon the properties of the solvent; namely, T_b and $\Delta H^{\bullet}_{vap}$.

Freezing Point Depression (ΔT_f)

This derivation of the freezing point depression assumes that the solid phase consists of only the solvent and that the solute is dissolved in the liquid solution phase. At the freezing point for the solution, the solution and the pure solid phase are in equilibrium

$$\overline{G}^{\cdot}_{1(\text{solid})} = \overline{G}_{1(\text{solution})} \qquad (4\text{-}C\text{-}13)$$

The $\overline{G}_{1(\text{solution})}$ depends upon the concentration of the solvent, as given in equation (4-C-3), and equation (4-C-13) becomes

$$\overline{G}^{\cdot}_{1(\text{solid})} = \overline{G}^{\cdot}_{1(\text{liquid})} + RT \ln X_1$$

where T is the freezing point of the solution. And as before, we can relate the ΔG_f for the process to the freezing point depression ΔT_f by

$$\Delta \overline{G}_f = \overline{G}^{\cdot}_{1(\text{solid})} - \overline{G}^{\cdot}_{(\text{liquid})} = RT \ln X_1$$

$$\Delta H^{\cdot}_f - T\Delta S^{\cdot}_f = RT \ln X_1$$

$$\frac{\Delta H^{\cdot}_f}{RT} - \frac{\Delta S^{\cdot}_f}{R} = \ln X_1 \qquad (4\text{-}C\text{-}14)$$

When $X_1 = 1$ we have pure solvent–solid equilibrium at $T = T_f$

$$\frac{\Delta H^{\cdot}_f}{RT_f} - \frac{\Delta S^{\cdot}_f}{R} = 0 \qquad (4\text{-}C\text{-}15)$$

Substituting for $\Delta S^{\cdot}_f / R$ from equation (4-C-15) into equation (4-C-14)

$$\frac{\Delta H^{\cdot}_f}{RT} - \frac{\Delta H^{\cdot}_f}{RT_f} = \ln X_1 \qquad (4\text{-}C\text{-}16)$$

Proceeding as before, equation (4-C-16) can be modified

$$\frac{\Delta H^{\cdot}_f}{R} \left(\frac{1}{T} - \frac{1}{T_f} \right) = \ln (1 - X_2) \sim - X_2$$

$$\frac{\Delta H^{\cdot}_f}{R} \left(\frac{T_f - T}{TT_f} \right) = - X_2$$

Since $T \approx T_f$ and $\Delta T_f = T_f - T =$ freezing point depression

$$\Delta T_f = T_f - T = - \frac{RT_f^2}{\Delta H_f^{\cdot}} X_2 \qquad (4\text{-}C\text{-}17)$$

The $\Delta H_f^{\cdot}$ in equation (4-C-17) is the heat change of freezing and must be a negative value since heat is <u>released</u> upon freezing. The heat of fusion, $\Delta H_{fus}^{\cdot}$ is the same magnitude as the heat of freezing but with the opposite sign.

$$\Delta H_{fus}^{\cdot} = - \Delta H_f^{\cdot}$$

Equation (4-C-17) then becomes

$$\Delta T_f = \left(\frac{RT_f^2}{\Delta H_{fus}^{\cdot}} \right) X_2 \qquad (4\text{-}C\text{-}18)$$

Again we see that for dilute solutions, ΔT_f, as well as ΔT_b and ΔP, depends only upon the concentration of solute, X_2, but depends upon the properties of the solvent, namely, T_f and $\Delta H_{fus}^{\cdot}$. In summary we note that equations (4-C-1), (4-C-12), and (4-C-18) have the same general form involving a product, the first term of which depends upon the solvent, and the second term which depends only upon X_2.

Phase Diagram Illustrating Colligative Properties

These three colligative properties can be visualized by observing the effect that a non-volatile solute has on the phase diagram of the solvent. The phase diagram for the pure solvent is given in Figure 4-C-1 by the solid lines. The solid lines represent the P and T at which the two phases shown on each side of the line are in equilibrium. For example, at point A the liquid and vapor of the pure solvent are at equilibrium. The pressure at point A is the vapor pressure of the solvent at this temperature. Similarly point B represents an equilibrium between the solid and liquid phases and point C an equilibrium between the solid and vapor. At one atmosphere pressure the normal boiling point, T_b, is shown as the temperature at which liquid and vapor are in equilibrium. Similarly T_f is the freezing point at which solid and liquid are in equilibrium at one atmosphere pressure. The equations of the vapor pressure of the liquid and solid can be derived from equations (4-A-12) and (4-A-13) in a manner similar to that used in the derivation of the colligative property ΔT_b.

When a non-volatile solute is added to the liquid phase of the solvent to make a solution, the vapor pressure of the solvent is lowered according to Raoult's Law, equation (4-B-1). This is shown by the dashed line in Figure 4-C-1 separating the vapor from the liquid solution in this case. The equilibrium along this dashed line is between pure solvent vapor and solvent in solution. For example, point A, representing equilibrium between pure liquid and pure vapor, would be lowered to point A' representing equilibrium between solvent in solution and pure vapor. At P = 1 atm one can see that the boiling point of the solution must be increased by ΔT_b in order to make the solvent in solution and vapor be at equilibrium, both at one atmosphere pressure. Similarly, the freezing point must be lowered by ΔT_f in order to establish equilibrium between the solvent in solution and pure solid solvent. The vapor pressure is decreased by ΔP_1 due to the non-volatile solute, as also shown in Figure 4-C-1.

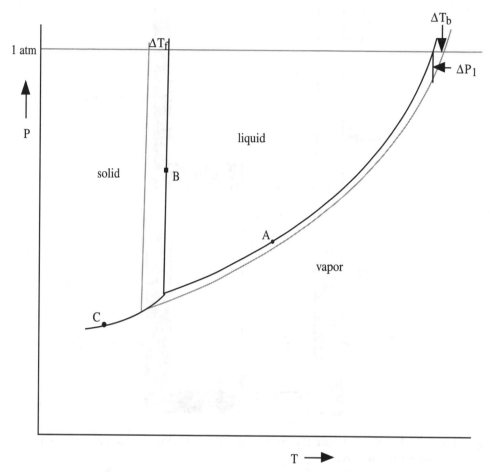

Figure 4-C-1. Phase Diagram of Pure Solvent (solid lines) and Solution Containing Non-volatile Solute (dashed lines)

Osmotic Pressure

The fourth colligative property is called Osmotic Pressure, commonly represented by the symbol Π. The process of *osmosis* is the passage of pure solvent through a membrane permeable only by the solvent into a solution containing a non-volatile solute. This is illustrated in Figure 4-C-2(a) where the passage of solvent increases the volume of the solution and decreases the volume of the solvent. The solvent would continue to pass through the semi-permeable membrane until the height of the solution causes a sufficient hydrostatic pressure at the membrane. This hydrostatic pressure is the osmotic pressure.

A more appropriate way to measure the osmotic pressure is illustrated in Figure 4-C-2(b). In this case pressure is applied to a piston above the solution until the solvent is prevented from flowing into the solution. Since the liquid levels are equal, the hydrostatic pressure (P) from the columns of liquids would be equal. The added pressure to the piston is the osmotic pressure (Π). In this situation there is no flow of solvent and the concentration of the solution remains constant.

The derivation of the relationship between osmotic pressure and solute concentration can be carried out similar to that used previously for ΔT_b and ΔT_f except that we will now use a change in pressure to establish the equilibrium rather than change in temperature. We

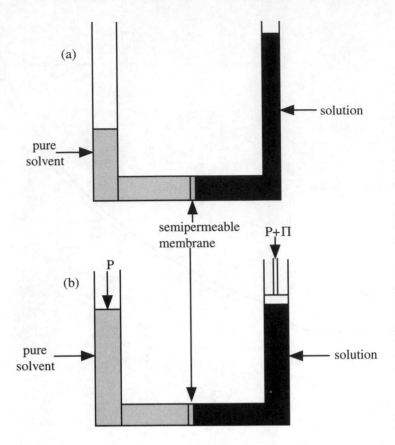

Figure 4-C-2. Osmotic Pressure (Π) Required to Establish Equilibrium Between Pure Solvent and Solvent in Solution

start with the equilibrium between pure solvent at pressure P and solvent in solution at pressure $(P + \Pi)$

$$\text{solvent (P)} \rightleftharpoons \text{solvent } (X_1, P+\Pi).$$

At equilibrium the partial molar free energies must be equal

$$\overline{G}_1^{\cdot}(P) = \overline{G}_1(X_1, P+\Pi) \tag{4-C-19}$$

The partial molar free energy of the solvent in solution is given by equation (4-C-3)

$$\overline{G}_1^{\cdot}(P) = \overline{G}_1^{\cdot}(P+\Pi) + RT \ln X_1 \tag{4-C-20}$$

where we have left off the subscripts for solution and liquid. Substituting in for the solute concentration X_2 and rearranging

$$- RT \ln X_1 = - RT \ln (1 - X_2) = RT\, X_2 = \overline{G}_1^{\cdot}(P+\Pi) - \overline{G}_1^{\cdot}(P) \tag{4-C-21}$$

The partial molar free energy is related to partial molar volume by

$$\left(\frac{\partial \overline{G}}{\partial P}\right)_T = \overline{V} \tag{4-C-22}$$

In order to find the change in going from pressure P to $(P + \Pi)$ we integrate

$$\int_{\overline{G}_i(P)}^{\overline{G}_i(P+\Pi)} d\overline{G} = \int_P^{P+\Pi} \left(\frac{\partial \overline{G}}{\partial P}\right)_T dP = \int_P^{P+\Pi} \overline{V}_1 dP$$

Assuming $\overline{V}_1$ is independent of P, which is reasonable since liquids are very incompressible,

$$\overline{G}_1(P + \Pi) - \overline{G}_1(P) = \overline{V}_1(P + \Pi - P) = \overline{V}_1 \Pi \qquad (4\text{-C-23})$$

Substituting into equation (4-C-21)

$$RT X_2 = \overline{V}_1 \Pi.$$

Solving for Π

$$\Pi = \frac{RT}{\overline{V}_1} X_2 \qquad (4\text{-C-24})$$

Again we see that the osmotic pressure is simply a function of the concentration of solute X_2, but dependent on the nature of the solvent in terms of $\overline{V}_1$.

The collective nature of the four colligative properties can be understood more clearly by examining the four equations at the same time. The four expressions are

$$\Delta P_1 = \left(P_1\right) X_2 \qquad (4\text{-B-4})$$

$$\Delta T_b = \left(\frac{RT_b^2}{\Delta H_{vap}^0}\right) X_2 \qquad (4\text{-C-12})$$

$$\Delta T_f = \left(\frac{RT_f^2}{\Delta H_{fus}}\right) X_2 \qquad (4\text{-C-18})$$

$$\Pi = \left(\frac{RT}{\overline{V}_1}\right) X_2 \qquad (4\text{-C-24})$$

In all four expressions the solute appears only as the mole fraction, X_2, and the terms in parentheses are a function of the solvent.

Problem 4-C-1. Consider solutions using toluene as the solvent. Evaluate the constants relating the magnitude of the colligative property with X_2 as shown in the above four expressions. Assume T = 298K for osmotic pressure. Use the following data for toluene which was obtained from the *Handbook of Chemistry and Physics*: Molecular Weight = 92.15, density = 0.8669 g/mL, Normal Boiling Point = 110.6 °C, Melting Point = - 95 °C, ΔH_{vap} = 49.85 kJ/mol, ΔH_{fus} = 36.62 kJ/mol. Calculate the vapor pressure depression at the Normal Boiling Point of toluene. Answer: ΔP_1 = (1 atm) X_2, ΔT_b = (24.6K) X_2, ΔT_f = (7.2K) X_2, Π = (230.1 atm) X_2.

D. Molar Free Energies of Components in Solution–Activities

After the preceding discussion of Raoult's Law and Henry's Law, we can define what is called the infinitely dilute solution. As its name indicates, it is a solution in which the concentration of the solutes are extremely small and approach infinite dilution. It can also be defined quantitatively as a solution for which Raoult's Law applies to the solvent and Henry's Law applies to the solute, differing from the ideal solution, in which both solvent and solute obey Raoult's Law. The infinitely dilute solution is generally applicable to all solutions; however, the concentration range over which the infinitely dilute approximaation is valid will vary for different types of solutes. This discussion will be restricted to nonelectrolytes, while electrolytic solutions will be included in a later section.

As we will see shortly, it is convenient to use the infinitely dilute solution as the standard state. Many applications of chemical equilibrium involve solutes at low concentrations and Henry's Law is closely obeyed at these low concentrations. At low solute concentrations the solvent must necessarily be at high concentrations and Raoult's Law is closely obeyed by the solvent. Deviations from Henry's and Raoult's Laws will be taken into account by activity coefficients, which will be defined shortly. In the infinitely dilute solution there will be no deviation from Henry's and Raoult's Laws and the activity coefficient will be unity.

Partial Molar Free Energy of the Solvent, $\overline{G}_1$, in Dilute Solutions

The partial molar free energy of a component in solution was expressed in equation (4-A-15) as

$$\overline{G}_{i(solution)} = \overline{G}^o_{i(vapor)} + RT \ln P_i \qquad (4\text{-}A\text{-}15)$$

in terms of the vapor pressure of the ith component, P_i. In the case of the solvent,

$$\overline{G}_{1(solution)} = \overline{G}^o_{1(vapor)} + RT \ln P_1$$

In the limiting case of a dilute solution, or when the solution can be considered to be approximately ideal with regard to the solvent, the vapor pressure of the solvent can be expressed by Raoult's Law. Substituting $X_1 P_1^{\bullet}$ for P_1 in the above expression, we obtain

$$\overline{G}_{1(solution)} = \overline{G}^o_{1(vapor)} + RT \ln X_1 P_1^{\bullet}$$

or,

$$\overline{G}_{1(solution)} = \overline{G}^o_{1(vapor)} + RT \ln P_1^{\bullet} + RT \ln X_1 \qquad (4\text{-}D\text{-}1)$$

According to equation (4-A-12)

$$\overline{G}^{\bullet}_{1(liquid)} = \overline{G}^o_{1(vapor)} + RT \ln P_1^{\bullet} \qquad (4\text{-}A\text{-}12)$$

the sum of the first two terms on the right side of equation (4-D-1) is equal to the molar free energy of pure component 1, $G^{\bullet}_{1(liquid)}$. The standard state for the free energy of the solvent is defined as the pure liquid under a total pressure of one atmosphere. Thus, $G^{\bullet}_{1(liquid)}$ can be set equal to G^o_1. Incorporating these relations into equation (4-D-1), we obtain

$$\overline{G}_{1(solution)} = \overline{G}^o_1 + RT \ln X_1 \qquad (4\text{-}D\text{-}2)$$

Equation (4-D-2) gives the partial molar free energy of the solvent in terms of its mole fraction and the standard molar free energy of the solvent $\overline{G}_1^o$. The term $RT \ln X_1$, in equation (4-D-2) is negative, since X_1 is a fraction greater than zero, but less than one. Consequently, the partial molar free energy of the solvent is less than the molar free energy of the pure liquid, by an amount which is dependent on the composition. Thus, the equation illustrates that the presence of a solute lowers the partial molar free energy of the solvent.

Activity of Solvent

The relationship between partial molar free energy and concentration of the solvent given in equation (4-D-2) is restricted to the region in which solutes are at low concentrations and the solvent concentration is high. This condition will not be met if the solute concentrations are higher and we must take this into account in order to have a general relationship between partial molar free energy and concentration of both solute and solvent.

We define the activity of the solvent, a_1, through its vapor pressure using a mathematical form similar to Raoult's Law

$$P_1 = P_1^{\bullet} a_1 \qquad (4\text{-}D\text{-}3)$$

The activity, a_1, replaces the mole fraction, X_1, in Raoult's Law and is the quantity that describes the vapor pressure of the solvent over all concentration ranges. The activity is related to the concentration by a factor, γ_1, defined as the activity coefficient

$$a_1 = \gamma_1 X_1 \qquad (4\text{-}D\text{-}4)$$

If we substitute equation (4-D-4) into (4-D-3) we obtain

$$P_1 = P_1^{\bullet} \gamma_1 X_1 \qquad (4\text{-}D\text{-}5)$$

At low solute concentrations we expect equation (4-D-5) to become Raoult's Law, representing the vapor pressure of the solvent with the mole fraction of the solvent. Obviously in order for this to occur, γ_1 must become unity. Consequently, we have the following limiting properties

$$\lim_{X_1 \to 1} \gamma_1 = 1 \qquad (4\text{-}D\text{-}6)$$

$$\lim_{X_1 \to 1} a_1 = X_1 \qquad (4\text{-}D\text{-}7)$$

As we will see shortly, this is very convenient when working with chemical equilibria in which the solute concentrations are low and there are only minor deviations from Henry's and Raoult's Laws.

Partial Molar Free Energy of the Solvent

Previously in the derivation of equation (4-D-1), Raoult's Law was substituted for the vapor pressure of solvent. In light of equation (4-D-3), which applies at all concentrations, we obtain a general expression for the partial molar free energy of the solvent in terms of the activity.

$$\overline{G}_{1(\text{solution})} = \overline{G}_1^o + RT \ln a_1 \tag{4-D-8}$$

Substituting in equation (4-D-4) to obtain the expression in terms of activity coefficient

$$\overline{G}_{1(\text{solutions})} = \overline{G}_1^o + RT \ln \gamma_1 X_1 \tag{4-D-9}$$

It is quite obvious that in the limit of dilute solutions, γ_1 becomes 1, and equation (4-D-9) becomes equation (4-D-2).

Partial Molar Free Energy of the Solute, $\overline{G}_2$, in Dilute Solutions

Following the same course as in the previous case of the solvent, the partial molar free energy of the solute (nonelectrolyte) can be expressed in accord with equation (4-A-15) as follows:

$$\overline{G}_{2(\text{solution})} = \overline{G}_{2(\text{vapor})}^o + RT \ln P_2$$

Again, in the case of a dilute solution, the vapor pressure can be expressed in terms of the appropriate limiting law, in this instance, Henry's Law. Choosing the modified form of Henry's Law, equation (4-B-9), and substituting it into the above relationship for P_2, leads to the equation

$$\overline{G}_{2(\text{solution})} = \overline{G}_{2(\text{vapor})}^o + RT \ln \left(K_2^h \right)' m_2$$

and,

$$\overline{G}_{2(\text{solution})} = \overline{G}_{2(\text{vapor})}^o + RT \ln \left(K_2^h \right)' + RT \ln m_2 \tag{4-D-10}$$

Since both $\overline{G}_{2(\text{vapor})}^o$ and $RT \ln \left(K_2^h \right)'$ are constants independent of concentration, they can be combined under the symbol $\overline{G}_2^o$, and equation (4-D-10) becomes

$$\overline{G}_{2(\text{solution})} = \overline{G}_2^o + RT \ln m_2 \tag{4-D-11}$$

where

$$\overline{G}_2^o = \overline{G}_{2(\text{vapor})}^o + RT \ln \left(K_2^h \right)'.$$

Equation (4-D-11) expresses the partial molar free energy of the solute, $\overline{G}_{2(\text{solution})}$, in terms of a constant independent of concentration, G_2^o, and a term containing the molality of the solution with respect to the solute. *The standard state for free energy of the solute is defined as a solution of unit molality under a total pressure of one atmosphere, existing in a hypothetical Henry's Law state, that has the properties of an infinitely dilute solution.*

   When using equation (4-D-11) for the calculation of the molar free energy of the solute, the solute is assumed to have the properties of an infinitely dilute solution, arising from the use of Henry's Law. Thus, using equation (4-D-11), we automatically obtain the molar free energy for a solution having the properties of infinite dilution. Hence, the standard molar free energy of the solute is obtained by setting m equal to unity in equation (4-D-11).

$$\overline{G}^o_{2(solution)} = \overline{G}^o_2 = \overline{G}^o_{2(vapor)} + RT \ln \left(K^h_2 \right) \qquad \begin{array}{c} \text{Standard Molar} \\ \text{Free Energy of} \\ \text{Solute} \end{array} \qquad \text{(4-D-12)}$$

Just as in the previous case of the solvent, it should be pointed out that equation (4-D-11) is a valid representation of the partial molar free energy of the solute only in the extreme of infinite dilution. The important aspects concerning two-component, infinitely dilute solutions are summarized in the table that follows.

Table 4-D-1 Two-component, Infinitely Dilute Solutions

Component	Limiting vapor pressure law	Standard state	Partial molar free energy
Solute	Henry's Law $P_2 = \left(K^{\cdot}_h \right)' m_2$	One molal solution under one atm pressure having the properties of infinite dilution (Henry's Law).	$\overline{G}_{2(solution)} = \overline{G}^o_2 + RT \ln m_2$
Solvent	Raoult's Law $P_1 = P^{\cdot}_1 X_1$	Pure liquid under a pressure of one atm obeying Raoult's Law.	$\overline{G}_{1(solution)} = \overline{G}^o_1 + RT \ln X_1$

Problem 4-D-1. What increase in partial molar free energy of the solvent would be expected when a solution is diluted from a solute concentration of 1 mole percent to 0.1 mole percent at 25°C? Answer: - 5.71 kJ.

Problem 4-D-2. What overall change in free energy, if any, would be expected for the transformation of one mole of solute from an infinite quantity of 0.1 molal solution to an infinite quantity of 0.01 molal solution at 25°C? If the solutions were separated by a membrane through which only solute could pass, what would be expected to occur? Answer: - 5.71 kJ. Solute will pass from the solution at 0.1 molal to the solution at 0.01 molal.

Problem 4-D-3. Consider two immiscible liquids in contact, containing a common solute. At equilibrium the concentrations of the solute in the two liquids, m_1 and m_2, will be such that no further net transfer of solute between the two liquids will occur. Recalling the free energy criterion for equilibrium (no net change), and the equation for the partial molar free energy of the solute, derive a relationship between m_1 and m_2 at constant temperature.

Answer: $\Delta G^o = RT \ln \dfrac{m_1}{m_2} = RT \ln \dfrac{\left(K^{h'}_2 \right)^\alpha}{\left(K^{h'}_2 \right)^\beta}$ where α refers to the solution at m_1 molality

and β at m_2 molality.

Problem 4-D-4. Reconsidering Problem 4-B-10, derive a more accurate representation of the partial molar free energy of the solute than equation (4-D-11).

Answer: $\overline{G}_2 = \overline{G}^o_2 + RT \ln a_2$ where $a_2 = \gamma_2 m_2$.

Activity of the Solute

We define the activity of the solute, a_2, through its vapor pressure using a mathematical form similar to Henry's Law

$$P_2 = \left(K_2^h\right)' a_2 \qquad \text{Activity of the Solute} \qquad (4\text{-}D\text{-}13)$$

The activity replaces the solute molality, m_2, in Henry's Law and is the quantity that describes the vapor pressure of the solute over all concentration ranges. The activity, a_2, is related to the concentration of the solute by a factor, γ_2, defined as the activity coefficient.

$$a_2 = \gamma_2 m_2 \qquad \text{Activity Coefficient of the Solute} \qquad (4\text{-}D\text{-}14)$$

γ_2 is equal to the activity at any particular molality divided by the molality. γ_2 is usually not constant, but changes as the composition changes. Consequently, the activity, a_2, is not a linear function of the molality, m_2. As the solution becomes more and more dilute, Henry's Law becomes more accurate and the activity approaches the molality. In the limit of infinite dilution, the activity becomes identical in value to the molality and, consequently, the activity coefficient attains a value of unity. These results can be summed up in the following form:

$$
\begin{aligned}
&\text{as } m_2 \to 0 && \text{Limiting Conditions} \\
&a_2 \to m_2 && \text{for Activity of the} \\
&\gamma_2 \to 1 && \text{Solute} && (4\text{-}D\text{-}15)
\end{aligned}
$$

Using equation (4-D-13), rather than Henry's Limiting Law, to describe the vapor pressure of the solute leads to an expression for the partial molar free energy of the solute, analogous in form to equation (4-D-11).

$$\overline{G}_{2(\text{solution})} = \overline{G}_2^o + RT \ln a_2 \qquad \text{Molar Free Energy of the Solute} \qquad (4\text{-}D\text{-}16)$$

Equation (4-D-16) has the advantage, however, that it is valid over the entire concentration range. *The standard state for the molar free energy of the solute* ($\overline{G}_2^o$) *is redefined to be the solution under a total pressure of one atmosphere, and at unit activity of the solute.* Since a_2 is relative to the standard state of $a_2 = 1$, the a_2 in the logarithm term is actually the ratio $a_2/a_2 = 1$ molal and is unitless. This is similar to the dimensionless P in the lnP term in equation (3-E-9) if P is expressed in bar. This standard state corresponds to a one molal solution with the properties of the infinitely dilute solution. In an infinitely dilute solution, $\gamma_2 = 1$ and the activity becomes equal to molality.

$$
\begin{aligned}
&\gamma_2 = 1 \\
&a_2 = m_2 = 1 \text{ (standard state)}
\end{aligned}
$$

If we have a solution containing more than one solute, we can use equations (4-D-14) and (4-D-16) to represent each solute. Let us use i to designate any of the individual solutes. Equation (4-D-16) then becomes

$$\overline{G}_{i\,(\text{solution})} = \overline{G}_i^o + RT \ln a_i \qquad (4\text{-}D\text{-}17)$$

and equation (4-D-14) becomes

$$a_i = \gamma_i\, m_i \qquad (4\text{-}D\text{-}18)$$

Problem 4-D-5. If the activity coefficient of acetic acid for a 1.03 molal solution of acetic acid in toluene is 0.759, and if the Henry's Law constant, $(K_2^h)'$, is equal to 38.7 torr/mol HAc/kg toluene, what is the vapor pressure of the acetic acid above the solution? Answer: 30.3 torr.

Problem 4-D-6. Information concerning activity is usually tabulated in terms of the activity coefficient, γ_2. Derive an expression relating γ_2 to the actual measured vapor pressure, P_2, and the vapor pressure given by Henry's Law (modified), $P_2 = \left(K_2^h\right)' m_2$.

Answer: $\gamma_2 = \dfrac{P_2}{\left(K_2^h\right)' m_2}$.

Problem 4-D-7. What is the difference in free energy per mole for the solute between solutions in which its activity is a_2 and a_2' at the same temperature?

Answer: $\Delta \overline{G}_2 = RT \ln \dfrac{a_2'}{a_2}$.

Use the following vapor pressure data for toluene (P_1) and acetic acid (P_2) to work Problem 4-D-8:

Mole Fraction		Vapor Pressure (torr)	
toluene	acetic acid	toluene	acetic acid
0.0000	1.0000	0.0	136.0
0.1250	0.8750	54.8	120.5
0.2310	0.7690	84.8	110.8
0.3121	0.6879	101.9	103.0
0.4019	0.5981	117.8	95.7
0.4860	0.5140	130.7	88.2
0.5349	0.4651	137.6	83.7
0.5912	0.4088	145.2	78.2
0.6620	0.3380	155.7	69.3
0.7597	0.2403	167.3	57.8
0.8289	0.1711	176.2	46.5
0.9058	0.0942	186.1	30.5
0.9565	0.0435	193.5	17.2
1.0000	0.0000	202.0	0.0

Problem 4-D-8. Graph the vapor pressure of toluene and acetic acid versus mole fraction, analogous to the graphs shown in Figures 4-B-2 and 4-B-3.
(a) Does the graph show positive or negative deviations from Raoult's Law?
(b) Show on the graph Raoult's Law for both toluene and acetic acid. Estimate the range of concentrations over which Raoult's Law fits the experimental vapor pressure data.
(c) Show on the graph Henry's Law for both toluene and acetic acid. Estimate the range of concentrations over which Henry's Law fits the experimental vapor pressure data.
(d) Calculate the modified Henry's Law constant $\left(K_2^h\right)'$ for acetic acid. Hint: Graph vapor pressure of acetic acid versus molality of acetic acid.
(e) Calculate the activity coefficients for acetic acid in the range 0–3 molal, assuming acetic acid is the solute.
Answers: (a) positive deviations (b) acetic acid: $X_2 = 0.9 - 1.0$, Toluene: $X_1 = 0.95 - 1.0$
(c) acetic acid: $X_2 = 0.0 - 0.3$, toluene: $X_1 = 0.0 - 0.07$ (d) $\left(K_2^h\right)' = 45$ torr/mol/kg,
(e) $m_2 = 0.494, 1.130, 2.244, 3.438$; $\gamma_2 = 0.774, 0.600, 0.460, 0.374$.

The Chemical Potential

According to the thermodynamic criterion for spontaneity, under the conditions of constant temperature and pressure, spontaneous change will occur when accompanied by a net decrease in the free energy. Moreover, the criterion further states that under the same conditions, equilibrium prevails when a postulated change involves no increase or decrease in the free energy. This leads immediately to the concept that the partial molar free energy of a substance is an indication of whether or not any change in the system can be expected to occur. In other words, a physical change such as a change in the composition of a system or a change in phase must lead to a decrease in the free energy of a system. For these reasons the partial molar free energy has come to be thought of as the measure of some fundamental driving force and, consequently, has been given the name chemical potential, symbolized by μ.

$$\mu_i = \text{chemical potential of} = \overline{G}_i \qquad (4\text{-D-}19)$$
$$\text{the ith component}$$

The use of μ_i rather than $\overline{G}_i$ is simpler since the molar quantity designated by the bar is left off and it is assumed that μ_i pertains to a molar quantity. The relationship between chemical potential of the solute and the activity of the solute would then be given by

$$\mu_i = \mu_i^o + RT \ln a_i \qquad (4\text{-D-}20)$$

and as before a_i is given by

$$a_i = \gamma_i m_i \qquad (4D\text{-}21)$$

E. Chemical Equilibrium for Reactions in Solution

Chemical reactions in solution and solution equilibria are of principal concern to the chemist. In our discussion on thermodynamics in Chapter 3, we treated the case of chemical equilibrium in the gas phase on the basis of ideal gas behavior. Though not the most common case of chemical equilibrium, in a practical sense, the gas phase situation was immediately amenable to thermodynamic analysis. Recalling the mathematical relationships governing chemical equilibrium in the vapor phase, the derived relationships depended inherently on the equations expressing the molar free energies of the gaseous components in terms of the partial pressures of the components and on the free energy criterion of equilibrium. Through equations such as (4-D-20), we are now able to express the partial molar free energies of components in solution in terms of concentration and are consequently prepared to treat the problem of chemical equilibrium in solution.

Consider the chemical reaction between various solutes in a dilute liquid solution at constant temperature and pressure. This reaction can be represented in the following manner

$$aA(a_A) + bB(a_B) = cC(a_C) + dD(a_D) \qquad \Delta\mu = \Delta G \qquad (4\text{-E-}1)$$

where a_A, a_B, a_C, and a_D represent the activities of the components A, B, C, and D in the mixture and $\Delta\mu$ is the change in chemical potential (free energy) for the conversion of a moles of A and b moles of B to products at these activities and conditions of temperature and pressure.

$$\Delta\mu = c\mu_C + d\mu_D - a\mu_A - b\mu_B \qquad (4\text{-E-}2)$$

According to the thermodynamic criterion of equilibrium, $\Delta\mu$ will be zero when a_A, a_B, a_C, and a_D are equilibrium activities.

Since the partial molar free energy of a component in solution can be expressed in terms of its activity, we are now in a position to derive a relationship between $\Delta\mu$ and activities of the reactants and products. From equation (4-D-20),

$$\mu_i = \mu_i^o + RT \ln a_i \tag{4-D-20}$$

we obtain the following expressions:

$$\mu_A = \mu_A^o + RT \ln a_A;$$

$$\mu_B = \mu_B^o + RT \ln a_B;$$

$$\mu_C = \mu_C^o + RT \ln a_C;$$

$$\mu_D = \mu_D^o + RT \ln a_D.$$

Substituting the above equations into equation (4-E-2), and collecting the logarithm terms, we obtain

$$\Delta\mu = c\,\mu_C^o + d\mu_D^o - a\mu_A^o - b\,\mu_B^o + RT \ln \frac{a_C^c\, a_D^d}{a_A^a\, a_B^b} \tag{4-E-3}$$

The four terms to the right of the equals sign in (4-E-3) are the difference in the standard free energies of the products and reactants and, therefore, represents the change in free energy for the reaction when the reactants and products are present in their standard states, $\Delta\mu^o$. Equation (4-E-3) can thus be written:

$$\Delta\mu = \Delta\mu^o + RT \ln \frac{a_C^c\, a_D^d}{a_A^a\, a_B^b} \tag{4-E-4}$$

Since $\Delta\mu$ is equal to zero when the activities are at equilibrium, equation (4-E-4) reduces to

$$\Delta\mu^o = -RT \ln K_{eq} \qquad \text{Equilibrium in Solution} \tag{4-E-5}$$

K_{eq} is the familiar equilibrium constant which is equal to the ratio of the equilibrium activities of the products to the reactants, each raised to a power equal to its coefficient in the stoichiometric equation. K_{eq} is defined as

$$K_{eq} = \frac{a_C^c\, a_D^d}{a_A^a\, a_B^b} \tag{4-E-6}$$

Equation (4-E-5) relates the equilibrium constant for chemical equilibrium in solution to thermodynamic properties, namely, the standard free energy change for the process. Since at a given temperature $\Delta\mu^o$ is a constant, according to equation (4-E-5), K_{eq} is truly a constant dependent only on the absolute temperature. In the development of equation (4-E-5) and previously related expressions, the solute is presumed to be a nonelectrolyte. The same equilibrium expression results for electrolytes, but as will be seen in the next section, the relationship between ion activity coefficients and activities must be treated differently.

Since the activity of a solute is related to the molality of the solute by equation (4-D-21), equation (4-E-6) becomes

$$K_{eq} = \frac{(\gamma_C m_C)^c (\gamma_D m_D)^d}{(\gamma_A m_A)^a (\gamma_B m_B)^b} = \frac{\gamma_C^c \gamma_D^d}{\gamma_A^a \gamma_B^b} \frac{m_C^c m_D^d}{m_A^a m_B^b}$$

$$K_{eq} = K_\gamma K_m \qquad\qquad\qquad (4\text{-E-7})$$

where K_γ is the equilibrium constant expression in terms of activity coefficients and K_m is the equilibrium constant expression in terms of molalities. From the manner in which we defined the standard state of the solute as the infinitely dilute solution and the fact that the solute obeys Henry's Law at low concentrations

$$\lim_{m_i \to 0} K_\gamma = 1$$

and

$$\lim_{m_i \to 0} K_{eq} = K_m$$

This is very convenient since we frequently work at low solute concentrations, and the equilibrium concentrations can be calculated using K_m.

Problem 4-E-1. If the equilibrium constant K_a for the ionization of acetic acid in water solution at 25°C is 1.8×10^{-5}, what is the value of the standard free energy change for the process at 25°C? Assume that the equations developed in this section apply to the ionic equilibrium. Does the answer seem surprising in any way? Answer: 27.1 kJ/mol, $\Delta G^\circ > 0$ so the reaction is not spontaneous at standard conditions where <u>all</u> $a_i = 1$. If one starts with only acetic acid in solution some dissociation will occur, but these are not standard conditions where <u>all</u> species are at standard conditions including Ac^- and H^+.

Problem 4-E-2. Write an equation similar to equation (4-E-4) for the specific case of the ionization of acetic acid in water solution. If the ratio of the concentrations of Ac^- ions to HAc molecules is unity in a neutral solution at 25°C, what spontaneous change would be expected to occur? (Hint: $\Delta \mu^\circ$ for the ionization of acetic acid at 25°C was calculated in Problem 4-E-1). Answer: Dissociation of HAc into Ac^- and H^+ would occur.

Problem 4-E-3. What significance does the sign $\Delta \mu^\circ$ have for a process

$$A + B = C + D?$$

Answer: Evaluates if the reaction is spontaneous at standard conditions where <u>all</u> $a_i = 1.0$.

F. Activity of Ionic Solutes

Thus far, there has been no elaboration on the methods for determining activities, although the general method employed in the experimental determination of activities and activity coefficients using vapor pressures was indicated in the previous section. The several methods which do exist all have one thing in common—they involve the measurement of a property, or the change in some property of the system as a whole in relation to the composition of the system. Thus, in the case of a solute which ionizes in solution, the combined effect of both the positive and negative ions is observed. It is impossible to distinguish that portion of the observed effect due to the positive ions from the other portion due to the negative ions. We are unable, therefore, to obtain experimentally the activities of the individual positive and negative ions. However, we can define ionic activities and relate them through a mean ionic activity to the measured activity of the solute. This will provide considerable facility to any discussion of the properties of electrolytic solutions.

Consider the salt, A_xB_y, where A is the positive ion and B the negative ion. A_xB_y ionizes in solution according to the stoichiometric equation

$$A_xB_y \rightarrow xA + yB$$

The mean ionic activity, $a_\pm$, is defined in terms of the measured activity of A_xB_y, a_i, by

$$(a_\pm)^{(x+y)} = a_i \qquad \text{Mean Ionic Activity} \qquad (4\text{-}F\text{-}1)$$

$$(a_\pm)^{x+y} = (a_+)^x(a_-)^y \qquad \text{Geometric Mean} \qquad (4\text{-}F\text{-}2)$$

The reason for the geometric mean should be understood after the following discussion. An equation relating partial molar free energy and molality of solute was derived in equation (4-D-20)

$$\mu_i = \mu_i^o + RT \ln a_i \text{ (nonelectrolyte)} \qquad (4\text{-}D\text{-}20)$$

It is convenient to retain this same relationship between μ_i and a_i for each of the ions in an electrolytic solution. For the salt, A_xB_y, dissociation, as indicated above, gives x cations and y anions for each molecule of A_xB_y. The partial molar free energies for each of the ions would then be related to the activities by

$$x\mu_+ = x\mu_+^o + xRT \ln a_+$$
$$y\mu_- = y\mu_-^o + yRT \ln a_-$$

The free energy of the solute is the sum of the ionic free energies.

$$\mu_i = x\mu_+ + y\mu_-$$

Substituting for $x\mu_+$ and $y\mu_-$ yields

$$\mu_i = x\mu_+^o + xRT \ln a_+ + y\mu_-^o + yRT \ln a_-$$

Grouping the logarithm terms,

$$\mu_i = x\mu_+^o + y\mu_-^o + RT \ln a_+^x a_-^y \qquad (4\text{-}F\text{-}3)$$

In this equation the sum of the first two terms on the right are defined in terms of $\mu_\pm^o$, the algebraic mean of the ionic partial molar free energies.

$$\mu_i^o = x\mu_+^o + y\mu_-^o = (x + y) \mu_\pm^o \qquad (4\text{-}F\text{-}4)$$

The activities in the logarithm term must then be defined in terms of the geometric mean of the ionic activities.

$$a_i = a_\pm^{(x+y)} = a_+^x a_-^y \qquad (4\text{-}F\text{-}1, 4\text{-}F\text{-}2)$$

The difference between these means, algebraic versus geometric, arises simply from the logarithmic relationship between μ_i and a_i.

a_+ and a_- in (4-F-2) represent the activities of the positive and the negative ions, respectively, and, as previously discussed, are not experimentally determinable. The individual ionic activities have the same relevance and meaning as the activity of any solute. Thus, by preimposed definition,

$$a_+ = \gamma_+ m_+ \text{ and } a_- = \gamma_- m_- \tag{4-F-5}$$

where γ_+ and γ_- are the activity coefficients, and m_+ and m_- are the molalities of the positive and the negative ions, respectively.

$$m_+ = xm \text{ and } m_- = ym \tag{4-F-6}$$

where m is the molality of the solution with respect to the electrolyte, $A_x B_y$.

A mean ionic activity coefficient, $\gamma_\pm$, can be defined according to the equation

$$(\gamma_\pm)^{x+y} = (\gamma_+)^x (\gamma_-)^y \tag{4-F-7}$$

By replacing γ_+ and γ_- in this equation with a_+/m_+ and a_-/m_- from (4-F-5), and subsequently substituting for m_+ and m_- according to (4-E-6), we obtain

$$(\gamma_\pm)^{x+y} = \frac{(a_+)^x (a_-)^y}{(xm)^x (ym)^y}$$

or

$$\gamma_\pm^{x+y} = \frac{a_i}{m^{x+y} (x^x y^y)} \qquad \text{Mean Ionic Activity Coefficient} \tag{4-F-8}$$

Equation (4-F-8) expresses the mean ionic activity coefficient in terms of the numbers of cations and anions per molecule of solute, x and y, the measured activity of the solute, a_i, and the molality of the solution, m.

The relationship between a_i and m^{x+y} in (4-F-8) dictates the manner in which the standard state is defined. Henry's Law for electrolytes takes the form $P_i = (K_2^h)'(x^x y^y) m^{(x+y)}$. The standard partial molar free energy of the solute can then be evaluated with this value of $(K_2^h)'$.

Example 4-F-1. Consider the electrolyte, $Al_2(SO_4)_3$. For this case, x has the value 2 and y has the value 3. Therefore, the mean ionic activity is

$$a_\pm = (a_{\text{measured}})^{1/(2+3)} = a^{1/5}$$

according to its definition, equation (4-F-1). From equation (4-F-8), the mean ionic activity coefficient of $Al_2(SO_4)_3$ is given by

$$(\gamma_\pm)^{2+3} = \frac{a}{m^{2+3}(2^2 \cdot 3^3)}$$

or

$$\gamma_\pm = \frac{a^{1/5}}{m(108)^{1/5}} = \frac{a_\pm}{m(108)^{1/5}}$$

From the above results, we can readily obtain values of both $a_\pm$ and $\gamma_\pm$ for a solution of $Al_2(SO_4)_3$ of any molality, provided the activity has been measured.

Problem 4-F-1. Calculate the mean ionic activity and the mean ionic activity coefficient for a one molal solution of $CaCl_2$ in water at 25°C, if the measured value of the activity of the solute is 1.372. Answer: $a_\pm = 1.111$, $\gamma_\pm = 0.2778$.

G. Bioenergetics

A living biological cell is vigorously engaged in energy transformations. When a cell dies, its substance undergoes chemical reaction, largely hydrolysis, the macroscopic structures deteriorate, and the composition of the cell and its surrounding fluid becomes uniform. Since a biological cell and its surrounding nutrient solution, if thermostatted and open to the atmosphere, is a system at constant temperature, the direction of these changes is dictated by the need for the system to attain a minimum free energy.

Chemical Process in Cells

A living cell, with its evident, highly ordered structure, is obviously not in equilibrium with its surroundings, but although the macroscopic structures appear unchanged, the chemical substances making up these structures are being continually degraded and resynthesized. In resynthesizing its structure, the living cell exists in a dynamic state. In the process, the living cell is continually creating order; it is constantly involved in processes which apparently decrease the entropy and increase the free energy. This creation of order and free energy is even more evident in a growing cell, which increases in size, and finally, divides to produce two cells. The cell does not violate the laws of thermodynamics. Clearly, we have been looking at only part of the process; there must be concomitant parts to the processes which are less evident, and which involve a decrease in free energy, sufficiently large to more than compensate for the increase in free energy attending the creation of another living cell. This habit of looking at only a part of a process arises from the experimental nature of biochemistry, in which the investigator is mostly concerned with a physiological process, and it leads to a characteristic manner of communication. Thus, one says that the growth of the cell is coupled to *exergonic* processes ($\Delta G < 0$). This will be made clearer in the following illustration. If acetate and choline are placed in a buffered medium containing slices of brain tissue, or a homogenate of brain tissue, the synthesis of acetylcholine can be observed.

$$CH_3COOH + (CH_3)_3N^+C_2H_4OH \longrightarrow CH_3\overset{\overset{\displaystyle O}{\|}}{C}\text{-O-}C_2H_4N^+(CH_3)_3 + H_2O$$

$$\Delta G^{0'} = +27.6 \text{ kJ} \qquad\qquad (4\text{-G-}1)$$

This reaction can be brought about by a catalyst, but then, only the very small amount of acetylcholine that can exist in equilibrium with acetate and choline would be formed. Very much more acetylcholine is in fact synthesized. Considering only the synthesis of acetylcholine, free energy has been created. Since the creation of free energy in an isothermal, isobaric process would clearly be a violation of the Second Law of Thermodynamics, we must assume that the situation is more complicated than it appears and cannot be described by only the single reaction (4-G-1).

The solution of this apparent ambiguity is that the process (4-G-1) which produces acetylcholine is only a part of a larger, more complex process which, taken as a whole, results in a net decrease in the free energy of the system. Nevertheless, it is convenient to discuss portions of such biochemical processes. Obviously, there must be another portion of the biochemical process occurring simultaneously and involving a large enough decrease in the free energy to compensate for the increase in the free energy accompanying the production of acetylcholine. We say that the process producing acetylcholine is coupled to another process involving a decrease in free energy. This other process is the hydrolysis of adenosine triphosphate (ATP) to form adenosine monophosphate (AMP) and pyrophosphate (PP_i).

$$\text{ATP} + H_2O \xleftrightarrow{\hspace{1cm}} \text{AMP} + PP_i \quad \Delta G^{o'} = -31.8 \text{ kJ} \qquad (4\text{-G-}2)$$

The standard free energy change for the reaction (4-G-1) is + 27.61 kJ, and for (4-G-2), - 31.8 kJ. Therefore, the total standard free energy change is - 4.19 kJ, a value which is independent of the actual manner in which the two processes are carried out, since energy changes depend only upon the initial and final states. However, it is clear that if ATP were to hydrolyze in a manner indicated by (4-G-2), the decrease in free energy could not be used to drive another reaction; it would be lost. What must occur is some more complicated reaction, which is stoichiometrically and, therefore, energetically, the equivalent of the sum of the two separate reactions:

$$\text{acetate} + \text{choline} + \text{ATP} \xleftrightarrow{\hspace{1cm}} \text{AMP} + PP_i + \text{acetylcholine} \quad \Delta G^{o'} = -4.19 \text{ kJ} \quad (4\text{-G-}3)$$

Even so, biochemists speak of the free energy of hydrolysis of ATP being coupled to the synthesis of acetylcholine.

The reader will note that even if all reactants (except water) are at the same concentration, the actual free energy change of reactions (4-G-1) and (4-G-2) will still depend upon what that concentration is, because there are different numbers of reactant molecules and product molecules. (We should speak of the activities of the reactants, but since these can seldom be measured, in actual practice they are approximated by the concentrations.) The complete reaction (4-G-3), however, as a first approximation, is independent of the actual concentrations, provided they are all equal. This important conclusion should be remembered, because it is common for some biochemists to speak loosely in terms of standard free energy changes (corresponding to 1 M concentrations), when discussing biological reactions that occur at much lower concentrations. Thus, the free energy change for reaction (4-G-3), with all reactants (except water) at 0.001 M, is still - 4.19 kJ. However, it is unlikely that the actual concentrations will be equal, and also, not all reactions have equal numbers of reactant and product molecules. The equilibrium constant for reaction (4-G-3) can be calculated from the relationship

$$\Delta G^{o'} = -RT \, 2.3 \log K = 5.69 \log K \text{ (kJ) at } 25^{\circ}C$$

The actual biosynthesis of acetylcholine is rather complicated and involves the following steps:

$$\text{acetate} + \text{ATP} \xleftrightarrow{\hspace{1cm}} \text{acetyl AMP} + PP_i \qquad (4\text{-G-}4)$$

$$\text{acetyl AMP} + \text{coenzyme A} \xleftrightarrow{\hspace{1cm}} \text{acetyl coenzyme A} + \text{AMP} \qquad (4\text{-G-}5)$$

$$\text{acetyl coenzyme A} + \text{choline} \xleftrightarrow{\hspace{1cm}} \text{acetylcholine} + \text{coenzyme A} \qquad (4\text{-G-}6)$$

Coenzyme A has a sulfhydryl group that is acetylated in the above scheme. Reaction (4-G-3), then, is also the sum of reactions (4-G-4), (4-G-5), and (4-G-6). It is customary to say that the free energy of hydrolysis of ATP is used to synthesize acetylcholine, a statement that corresponds to reactions (4-G-1) and (4-G-2), whereas, the actual mechanism which is said to couple these two reactions consists of reactions (4-G-4), (4-G-5), and (4-G-6). We can say that reactions (4-G-1) and (4-G-2) are scrambled in the latter three reactions.

Biochemists do not use thermodynamic equilibrium constants. A thermodynamic equilibrium constant corresponds to a balanced equation. Thus, the equilibrium constant for reaction (4-G-1), written as a hydrolysis, is

$$\frac{[CH_3COOH][\text{choline}]}{[\text{acetylcholine}]} = e^{-\Delta G^O/RT} \tag{4-G-7}$$

This expression involves using the concentration of undissociated acetic acid, rather than the analytical concentration of acetic acid which is the sum of acetic acid and acetate. The analytical concentration is clearly the significant quantity. If we substitute

$$[CH_3COOH] = [CH_3COOH] \text{ analytical} \left/ \left(1 + \frac{K_a}{[H^+]}\right) \right. \tag{4-G-8}$$

we find that the equilibrium expression in terms of analytical concentrations is equal to

$$K\left(1 + \frac{K_a}{[H^+]}\right) = K' \tag{4-G-9}$$

and thus, is pH dependent. The standard free energy change, $\Delta G^{O'}$, distinguished by a prime, corresponds to K', and is customarily tabulated for pH 7.0. The activity of water is assumed to be constant at unit activity, and the temperature is often taken as 25°C. These conditions are more convenient than the physical chemist's standard states in the application of free energy changes to biological reactions.

It is clear that reaction (4-G-3) might not proceed very far, since there is only a modest decrease in the standard free energy. However, the story is not complete, for this biosynthesis illustrates still another type of coupling. The pyrophosphate formed in reaction (4-G-3) is hydrolyzed to inorganic phosphate, P_i.

$$PP_i + H_2O \rightleftharpoons 2P_i \qquad \Delta G^{O'} = -31.8 \text{ kJ} \tag{4-G-10}$$

Thus, the equilibrium of reaction (4-G-3) is pulled to the right. The complete reaction is now

$$\text{acetate} + \text{choline} + ATP \rightleftharpoons AMP + 2P_i + \text{acetylcholine} \quad \Delta G^{O'} = -36.4 \text{ kJ}$$

Biochemistry texts generally contain a table of standard free energies of hydrolysis, so that from relatively few hydrolytic reactions, a rather large number of coupled reactions can be put together and analyzed.

We have discussed a specific case of biosynthesis because the basic principles are quite general—a biosynthesis opposed to the spontaneous direction of the apparent reaction is accomplished by coupling the reaction to the hydrolysis of ATP. In many cases adenosine disphosphate (ADP) and P_i are the products, rather than AMP and PP_i.

Biosynthesis is one type of energy transformation in which the cell is engaged. As we have seen, the cell does chemical work, but there are other types of work that the cell also performs. In the case of muscle cells or motile cells, the performance of mechanical work is quite evident. Interestingly enough, the performance of mechanical work, too, is coupled to the hydrolysis of ATP to ADP and phosphate, but in this case, we still cannot describe the coupling adequately. Bioelectricity in the form of transport of ions is also related to the hydrolysis of ATP.

The free energy expended in transporting sodium from inside the cell to the outside fluid is

$$\Delta G^{o'} = RT \log \frac{[Na^+]_o}{[Na^+]_i} \approx 1.37 \log 20 \approx 7.5 \text{ kJ}$$

Can you suggest how many moles of Na^+ might reasonably be transported by the hydrolysis of one mole of ATP?

Cells, in fact, transport many substances. Indeed, few substances enter cells by a simple diffusion process, even in the direction of a concentration gradient. A rather dramatic example of active transport is the reabsorption of numerous substances by the kidney tubules. In the formation of urine, a "filtrate" of the blood is formed in the glomeruli, and as the filtrate moves through the renal tubules a multitude of substances, including inorganic ions, sugars, amino acids and even the protein serum albumin, are reabsorbed into the blood. We see, then, that ATP is a key substance in the performance of synthetic work, mechanical work, and transport work.

Energy Sources in Cells

The principal source of energy in aerobic cells is the oxidation of fuel substances such as glucose. Some aerobic cells can also derive energy from a non-oxidative process called glycolysis. This process is also the main source of energy for certain anaerobic organisms. How does a cell derive energy from the oxidation of glucose? First, let us consider glycolysis. Glycolysis involves the conversion of glucose to lactic acid in muscle cells, for example, or in other cases to other substances, notably ethanol in yeasts.

The reaction which results in lactic acid as a product is

$$C_6H_{12}O_6 \xrightleftharpoons{} 2CH_3\underset{OH}{\overset{H}{\underset{|}{\overset{|}{C}}}}-OOH \quad \Delta G^{o'} = -187.9 \text{ kJ} \qquad (4\text{-}G\text{-}11)$$

Since the hydrolysis of ATP is the source of energy for biological work, we can guess that the formation of lactic acid is coupled to the phosphorylation of ADP. Measurements show that for every molecule of glucose that is utilized in reaction (4-G-11), two molecules of ADP are phosphorylated.

$$2ADP + 2 P_i \xrightleftharpoons{} 2 ATP + 2 H_2O \quad \Delta G^{o'} = 2 \times 31.8 = 63.6 \text{ kJ} \qquad (4\text{-}G\text{-}12)$$

The complete reaction is

$$C_6H_{12}O_6 + 2 ADP + 2 Pi \xrightleftharpoons{} 2 ATP + 2 C_3H_6O_3 + 2 H_2O$$
$$\Delta G^{o'} = -124.2 \text{ kJ} \qquad (4\text{-}G\text{-}13)$$

Biochemists sometimes speak of the efficiency of a transformation. For example, the production of phosphate bond energy (ATP) in glycosis is said to be 63.6/187.9 = 38% efficient. However, this kind of evaluation is incorrect, because it corresponds to the reaction ocurring with each substance in its standard state at unit activity. In the cell, the substances are not at the same concentrations, and the general level for prominent substances is more on the order of 1×10^{-3} M than 1 M.

Nonetheless, by this procedure, one can get a general idea of how much free energy is conserved in the process, without making the detailed calculation which would require a knowledge of the actual concentrations. The concentrations will surely vary quite considerably, depending upon the kind of cell and the conditions of the cell.

Biochemists have determined how rections (4-G-11) and (4-G-12) are coupled or scrambled to produce reaction (4-G-13), and some fifteen steps are involved. These steps are needed to bring about the coupled phosphorylation process. Indeed, if a cell had an enzyme to catalyze reaction (4-G-11), no ATP would be produced. Under aerobic conditions, many cells oxidize glucose:

$$C_6H_{12}O_6 + 6O_2 \xrightleftharpoons{} 6\,CO_2 + 6\,H_2O$$
$$\Delta G^{o'} = -2{,}870 \text{ kJ} \tag{4-G-14}$$

and this reaction is coupled to the phosphorylation of 38 equivalents of ADP.

$$38 \text{ ADP} + 38 \text{ P}_i \xrightleftharpoons{} 38 \text{ ATP} + 38 \text{ H}_2O$$
$$\Delta G^{o'} = 38 \times 31.8 = 1{,}208 \text{ kJ} \tag{4-G-15}$$

The complete reaction is

$$C_6H_{12}O_6 + 6O_2 + 38 \text{ ADP} + 38 \text{ P}_i \xrightleftharpoons{} 6\,CO_2 + 44\,H_2O + 38 \text{ ATP}$$
$$\Delta G^{o'} = -1{,}662 \text{ kJ} \tag{4-G-16}$$

The mechanism by which reactions (4-G-14) and (4-G-15) are coupled to produce reaction (4-G-16) involves dozens of steps. A key substance involved in this coupling process is nicotinamide adenine dinucleotide, NAD^+, which acts as an oxidizing agent for a number of substances sequentially derived from glucose.

$$NADH + H^+ \xrightleftharpoons{} NAD^+ + 2H^+ + 2e^-$$

The reduced nucleotide, NADH, is in turn oxidized by the "electron transport chain," which consists of several sequential electron carriers, only the last of which, cytochrome oxidase, reacts with oxygen. Phosphorylation of ADP, three for every one of NADH oxidized, is coupled, in a yet unknown way, to the transport of electrons from carrier to carrier in the electron transport chain. Thus, almost all of the production of ATP attending the oxidation of glucose is derived from the oxidation of NADH. Oxygen reacts only with one carrier; it does not react directly with any of the "fuel" substances of the cell. Oxidation reactions are clearly important in biology, but they must be coupled to the formation of ATP. Indeed, if a cell had an enzyme that could catalyze reaction (4-G-14), no ATP would be produced and the cell would die. The importance of ATP in the biochemistry of cell processes cannot be overemphasized. It is, in fact, the link between the fuel substances received by the cell and the various functions performed by the cell. As illustrated in the previous examples, ATP is produced in the cell from ADP during the oxidation of substances such as glucose. The thermodynamically favored hydrolysis of ATP is then coupled with biosynthetic reactions which would otherwise not be favored because of the accompanying increases in free energy. The net result is that various biochemical reactions necessary for the cell's performance of chemical, mechanical, and osmotic work do occur. In short, ATP is the chemical agent, the hydrolysis of which acts as the driving force for many critical, but seemingly improbable, chemical reactions which occur in living cells.

Similarly the most versatile oxidizing agent in the cell is not oxygen but NAD^+. Indeed, oxygen reacts with only one substance in the cell. The couple NAD^+, NADH is the link between the fuel substances of the cell and the electron transport chain.

H. Summary

The following basic concepts of Solution Thermodynamics were presented in this chapter:

1. An ideal solution is one in which all components in the solution obey Raoult's Law, $P_i = X_i P_i'$. Ideal solutions usually result when the components have similar chemical properties; e.g. the same functional groups.

2. Real Solutions result from mixtures of dissimilar substances and may show positive deviations from Raoult's Law ($P_i > X_i P_i'$) or negative deviations ($P_i < X_i P_i'$). Generally Raoult's Law is closely behaved in real solutions for a component at high concentrations, say $X_i > 0.90$. We generally call the component with the highest concentration the solvent and we conclude that Raoult's Law is obeyed by the solvent even in real solutions.

3. In real solutions the vapor pressure of the solute obeys Henry's Law at low concentrations, say $X_i < 0.10$. Henry's Law can be expressed in terms of mole fraction or molality of the solute.

$$P_i = K_i^h X_i$$

$$P_i = (K_i^h)' m_i$$

4. The activity of the solvent is related to the molar free energy by

$$\overline{G}_1 = \overline{G}_1^o + RT \ln a_1$$

where $\overline{G}_1^o$ is the molar free energy of the pure solvent at standard conditions of $P = 1$ bar ~1 atm. The activity is related to the mole fraction of the solvent by

$$a_1 = \gamma_1 X_1$$

where γ_1 = activity coefficient of the solvent

$$\lim_{X_1 \to 1.0} \gamma_1 = 1.0$$

$$\lim_{X_1 \to 1.0} a_1 = X_1$$

5. The activity of the solute is related to the molar free energy by

$$\overline{G}_2 = \overline{G}_2^o + RT \ln a_2$$

where $\overline{G}_2^o$ is the molar free energy of an infinitely dilute solution at $m_2 = 1.0$. The activity is related to the molality by

$$a_2 = \gamma_2 \, m_2$$

where γ_2 = activity coefficient of the solute

$$\lim_{m_2 \to 0} \gamma_2 = 1$$

$$\lim_{m_2 \to 0} a_2 = m_2$$

6. An alternative name for molar free energy, $\overline{G}$, is chemical potential, μ. The change in chemical potential for a reaction is related to the activities of the reactants and products by

$$\Delta\mu = \Delta\mu^O + RT \ln \frac{a_C^c \, a_D^d}{a_A^a \, a_B^b}$$

The equilibrium constant is related to the change in standard chemical potential by

$$\Delta\mu^O = -RT \ln K$$

7. The activity of an ionic solute $A_x B_y$ is related to the chemical potential by

$$\mu_i = \mu_i^O + RT \ln a_i$$

where

$$\mu_i = x\mu_+ + y\mu_-$$

$$a_i = a_+^x \, a_-^y = a_\pm^{(x+y)}$$

where $a_\pm$ is the mean activity. The activity is related to concentration, m, by

$$a_i = \gamma_\pm^{(x+y)} (x^x y^y) \, m^{(x+y)}$$

Henry's Law for an ionic solute takes the form

$$P_2 = (K_2^h)' (x^x y^y) \, m^{(x+y)}$$

8. Biological process are frequently driven by coupling with other reactions which are exergonic such as

$$ATP + H_2O \rightleftharpoons AMP + PP_i \qquad \Delta G^O = -31.8 \text{ kJ}$$

$$PPi + H_2O \rightleftharpoons 2Pi \qquad \Delta G^O = -36.4 \text{ kJ}$$

ATP is produced by oxidation of fuel substances such as glucose, $C_6H_{12}O_6$:

$$C_6H_{12}O_6 + 6O_2 + 38\,ADP + 38P_i \rightleftharpoons 6\,CO_2 + 44\,H_2O + 38\,ATP$$
$$\Delta G^{O'} = -1{,}662 \text{ kJ}$$

Exercises

Data for Exercises 1, 2, 3, and 4. The following vapor pressure-composition data is available for mixtures of acetone (A) and ethyl ether (B) at 20°C. The P_A and P_B are in torr.

X_A	0	.052	.127	.249	.457	.612	.666	.842	.936	.979	.996	1.0
P_A	0	20	42	70	105	127	135	160	175	181	184	185
P_B	444	422	394	350	282	224	199	111	46	15	2	0

1. Prepare graphs of (a) partial vapor pressure of acetone versus the mole fraction of acetone, (b) partial vapor pressure of ethyl ether versus mole fraction of ethyl ether, and (c) the total vapor pressure of the solution versus composition. In the case of the graphs for the individual components, extrapolate each curve to zero mole fraction of the particular component and graphically evaluate the Henry's Law constants. Determine the modified Henry's Law constant for ethyl ether considering acetone the solvent. Answer: ethyl ether: $K_2^h = 724$ torr; acetone : 400 torr; $(K_2^h)' = 41$ torr/mol/kg.

2. What is the composition of the vapor above a solution of acetone and ethyl ether which contains 84.2 mole percent acetone? Answer: $Y_A = 0.590$, $Y_B = 0.410$.

3. Calculate the activities and activity coefficients of the solute, ethyl ether, and the solvent, acetone, for solutions having compositions in which the mole fraction of acetone equals 0.979, 0.936, and 0.842. (The modified Henry's Law constant for the solute was determined in Exercise 1 above.) Answer: $a_B = 0.366, 1.122, 2.709$, $\gamma_B = 0.988, 0.951, 0.836$; $a_A = 0.978, 0.946, 0.865$; $\gamma_A = 0.999, 1.011, 1.027$.

4. (a) Determine the change in the partial molar free energy of ethyl ether when a solution containing 93.6 mole percent acetone is diluted to 97.9 mole percent acetone. Activities for these solutions were calculated in Exercise 3.
   (b) Determine the change in the partial molar free energy of acetone when a solution containing 45.7 mole percent acetone is diluted with ethyl ether to 12.7 mol percent acetone. Values of the activities of acetone in these solutions are not necessary for solving this problem. Answer: (a) - 2.73 kJ, (b) - 2.23 kJ.

5. In Section C we derived an expression for the osmotic pressure, assuming the solvent obeyed Raoult's Law. A more precise expression can be derived using the activity of the solvent, a_1

$$\pi = - \frac{RT}{\overline{V}_1} \ln a_1$$

The osmotic pressure of human blood is 7.7 atm at 40°C. If the vapor pressure of pure water is 55.32 torr at 40°C, what is the vapor pressure of human blood at 40°C? Assume that only the water in blood contributes to its vapor pressure. Answer: 55.0 torr.

6. The modified Henry's Law constant for carbon dioxide in aqueous solutions is 29.5 atm/mol/kg water, and the standard free energy of formation of carbon dioxide, ΔG_f^o, at 25°C and one atmosphere is - 393.5 kJ/mol.
   (a) What pressure of carbon dioxide would be necessary in order to prepare an aqueous solution of carbon dioxide of unit activity?

(b) Assuming that carbon dioxide obeys the Ideal Gas Law, calculate the standard free energy of formation of carbon dioxide in aqueous solution at unit activity. Answer: (a) 29.5 atm, (b) - 385.1 kJ/mol.

7. Using ΔG^o values found in the Appendix, calculate the equilibrium constant for the dissociation of ammonium hydroxide.

$$NH_{3(aq)} + H_2O \rightleftharpoons NH_4^+{}_{(aq)} + OH^-{}_{(aq)}$$

Answer: 1.8×10^{-5}.

8. The oxidative decarboxylation of the malate ion,

$$malate^{2-}_{(aq)} + NADP^+_{(aq)} \rightleftharpoons pyruvate^-_{(aq)} + NADPH_{(aq)} + CO_{2(g)}$$

is catalyzed by malic enzyme which is found in the liver of mammals. If the equilibrium constant for this reaction has the value 1.83 at pH = 7 and 25°C, calculate the standard free energy change, $\Delta G^{o'}$, for the reaction. Answer: - 1.5 kJ.

9. Calculate the equilibrium constant (K_1) for the acid dissociation of protonated glycine at 25°C. The difference in the standard molar free energies of the products, glycine and hydrogen ion and protonated glycine is 13.41 kJ. See Exercise 10 for the chemical equation for the dissociation. Answer: 4.47×10^{-3}

10. Acid dissociation of protonated glycine occurs in two steps:

$$K_1$$
$$NH_3^+ CH_2COOH \rightleftharpoons NH_2CH_2COOH + H^+$$
$$K_2$$
$$NH_2CH_2COOH \rightleftharpoons NH_2CH_2COO^- + H^+$$

The pK = - log K values are 2.35 and 9.78. Calculate the pH of the solution which has an initial concentration of (a) 0.1 M glycine cation, (b) 0.1 M glycine anion. Hint: In part (b) consider the hydrolysis of glycine anion to give glycine and OH^-. Answer: (a) 1.72, (b) 11.39.

11. (a) Using $\Delta \overline{G}_f^o$ values found in the Appendix, and the value of $\Delta \overline{G}_f^o$ for glycine in aqueous solution (-373.6 kJ/mol), calculate ΔG^o for the oxidation of glycine in solution

$$NH_2CH_2COOH + \frac{3}{2} O_2 \longrightarrow 2\, CO_{2(aq)} + 2\, H_2O_{(l)}$$

(b) Is glycine stable with respect to oxidation in solution? What other factors must be taken into account in ascertaining the stability of glycine, or, for that matter, any other compound? Answer: (a) - 872.7 kJ, (b) No. The concentration of O_2 in solution.

12. The first acid ionization constant for ß-alanine is $10^{-3.55}$ at 25°C.

$$NH_3^+ CH_2CH_2COOH \rightleftharpoons NH_2CH_2CH_2COOH + H^+$$

The standard enthalpy of ionization for this process is equal to 5.1 kJ/mol. Assuming that the values of ΔH^o and ΔS^o do not change appreciably between 25°C and 75°C, determine the value of the first acid ionization constant of ß-alanine at 75°C. Answer: 6.0×10^{-4}.

13. The thermodynamic equilibrium constant for the conversion of glyceraldehyde 3-phosphate (G3-P) to dihydroxyacetone phosphate (DHAP)

$$\text{G3-P} \rightleftharpoons \text{DHAP}$$

in aqueous solution at 25°C and one atmosphere is 22. Calculate the free energy of the reaction, ΔG, when a solution 0.01 molal with respect to G3-P, and 0.44 molal with respect to DHAP is prepared. Assume that the activity coefficients of both solutes equal unity. In which direction will the reaction proceed? Answer: 1.72 kJ; Reaction will proceed from DHAP to G3-P.

14. According to Larry and Van Winkle, *J. Phys. Chem.*, 73: 570 (1969), chlorophylls a and b form one to one molecular complexes of a charge-transfer type with sym-trinitro-benzene. In the case of chlorophyll-b and trinitrobenzene in ethyl ether solutions,

$$\text{chlorophyll-b} + \text{trinitrobenzene} \rightleftharpoons \text{chlorophyll-b} \cdot \text{trinitrobenzene.}$$

The average values of the equilibrium constant determined at several temperatures are

T°C	5	10	15	20	25
K liter/mole	89±2	79±2	71±2	60±3	52±3

Plot log K versus the reciprocal absolute temperature. Determine ΔG^o, ΔH^o, and ΔS^o at 15°C. Answer: $\Delta G^o = -10.2$ kJ, $\Delta H^o = -19.2$ kJ, $\Delta S^o = -31.6$ J/K.

15. The mean ionic activity coefficients of Na_2SO_4 in aqueous solutions of 0.001, 0.010, and 0.100 molal at 0°C are 0.890, 0.719, and 0.449, respectively. Calculate the changes in the partial molar free energy of the solute for the successive dilutions of 0.100 molal to 0.010 molal to 0.001 molal. Compare the results for the two dilutions. Answer: 0.100 to 0.010 molal: -12.5 kJ; 0.01 to 0.001 molal: -14.2 kJ.

16. The standard molar free energy for $C_2H_5OH_{(l)}$ at 298K is $\overline{\Delta G}_f^o = -174.9$ kJ/mol. The vapor pressure of pure liquid ethanol is 56.5 torr. A 1.0 m solution of C_2H_5OH in H_2O has a partial vapor pressure due to ethanol of 3.65 torr. Calculate $\overline{\Delta G}_f^o$, $C_2H_5OH_{(aq)}$ using this data. Assume that the C_2H_5OH vapor obeys the Ideal Gas Equation of State, and the solute obeys Henry's Law in the 1.0 m solution. Answer: -181.7 kJ.

17. Describe how you could determine ΔG^o for

$$\underset{\text{calcite}}{CaCO_3} \longrightarrow \underset{\text{aragonite}}{CaCO_3}$$

using solubilities of these phases and activity measurements of $CaCO_3$ in aqueous solutions. Answer: $\Delta G^o = -RT \ln \dfrac{a_{CaCO_3(aq)} \text{ from aragonite}}{a_{CaCO_3(aq)} \text{ from calcite}}$

18. The solubility of tyrosine (HO - C_6H_4 - $CH_2\overset{\overset{\displaystyle NH_2}{|}}{CH}$ - COOH) in water has been measured as a function of temperature.

t^oC	0	25	50
solubility g/100ml	0.147	0.351	0.836

Calculate the heat of solution. Answer: 25.5 kJ/mol.

19. If the concentration of glucose is 0.050 M, lactic acid 0.029 M, ATP 0.018 M, ADP 0.0014 M, and P_i 0.010 M, calculate the free energy change per mole of glucose transformed in anaerobic glycolyses, when a small amount of reaction occurs. Assume T = 298K. What is meant by a small amount of reaction? Is this system in equilibrium? Answer: 98.8 kJ, an infinitesimal amount of reaction, No.

20. The standard free energy of hydrolysis of glucose-6-phosphate, $\Delta G^{o'}$, is - 13.8 kJ at 298K. Calculate the equilibrium constant for the phosphorylation of glucose by ATP. Answer: 1.4×10^3.

21. Write the equation for the oxidation of NADH by oxygen, corresponding to the transport of 2 equivalents of electrons, and calculate the value of $\Delta G^{o'}$ (in kJ). Hint: Recall that H^+ is assumed to be at 10^{-7} M in biological systems. How many moles of ADP might be phosphorylated in a coupled reaction?

$$\overline{G}^o_{NAD^+} - \overline{G}^o_{NADH} = - 22.4 \text{ kJ}$$

Answer: - 219.7 kJ, 7 moles could be phosphorylated.

22. (a) Calculate ΔG^o for the dissolution of $AgCl_{(s)}$.

$$AgCl_{(s)} \underset{\longrightarrow}{\longleftarrow} Ag^+_{(aq)} + Cl^-_{(aq)}$$

(b) Calculate the solubility product constant, K_{sp}, for $AgCl_{(s)}$.
Answer: (a) 55.66 kJ (b) 1.8×10^{-10}.

23. (a) Calculate the ΔG^o and K for the hydrolysis of Ac^-

$$Ac^-_{(aq)} + H_2O_{(l)} \underset{\longrightarrow}{\longleftarrow} HAc_{(aq)} + OH^-_{(aq)}$$

(b) Calculate the pH of a solution in which 0.100 moles of NaAc were added to sufficient water to make a 1L solution.
Answer: (a) 52.6 kJ, 6.1×10^{-10} (b) 8.9.

5
CHEMICAL KINETICS

*In order to understand the kinetics of a reaction it is necessary to elucidate the mechanism by which the reaction proceeds. The mechanism is a sequence of reaction steps which leads from reactants to products. In recent years lasers have been used to understand in more detail the nature of <u>each</u> reaction step. This information elevates the understanding of the kinetics to a higher level which describes not only the species involved, but also their orientation and energies required. Lasers that can be pulsed at speeds of 10–100 femtoseconds (10^{-15} s) have been developed that allow probing the reaction intermediates as they pass through the transition state. The time resolution is shorter than the vibrational and rotational motions in the molecule, and for all practical purposes the nuclei are frozen within this short time interval. Workers in this field have coined the term "Femtochemistry" to emphasize the short time frame of femtoseconds for studies in this area. There is a review article by Zewail, one of the active members working in this area [Ahmed H. Zewail, J. Phys. Chem., **1996**, 100, 1270]. The reaction of a $H^{\cdot}$ atom with CO_2 to give OH and CO has been described showing the unique formation of the intermediate complex and its change in geometry before going to the products of the reaction. The technique can also be used to study more complex reactions such as the decarbonylation of cyclopentanone, Diels-Alder reactions, intra- and intermolecular $H^{\cdot}$ atom transfers, and isomerizations.*

A. Objectives of Chemical Kinetics

The field of chemical kinetics is generally concerned with the rate at which reactions proceed. This is in contrast to thermodynamics, which deals only with the feasibility, or the spontaneity that the reaction may occur, with no guarantee that the reaction will proceed in any finite period of time. Chemical kinetics attempts to answer the question, "When will the reaction be complete?" This is essentially the same as determining, "How fast does the reaction proceed?" In other words, knowledge of the latter automatically answers the former question.

Although the general objective of chemical kinetics is the determination of the rate at which reactions proceed, there are other, more detailed, aspects of reactions which we shall investigate. Generally, these details must be known, if we ever hope to fully understand how reactions proceed, and eventually, to theoretically predict the rates of reactions. Thus, we will need specific information which will allow us to determine

   1. The mechanism by which the reaction proceeds

   2. The concentration dependence of the rate of the reaction

   3. The temperature dependence of the rate of the reaction

In contrast to thermodynamics, chemical kinetics is not based on the formulation of certain laws. Instead, various theories have been proposed to guide the chemist in understanding and predicting reaction kinetics.

B. Elementary Reaction Steps and Rate Expressions

Some reactions appear to proceed through a single reaction step. Reaction may occur between A and B as a result of a collision between them, forming an intermediate complex of short duration, followed by another reaction resulting in products. This could involve two molecules of the same or different kinds, or it could involve a single molecule that becomes excited through collisions to yield products. In a few cases, three molecules are thought to come together simultaneously in order to react; however, this seems to be a very unlikely event. It is difficult, and sometimes impossible, to distinguish a three-molecule collision from that of two molecules which collide and stay together for a very short period, then collide with a third molecule and subsequently react. The simultaneous collision of more than three bodies is extremely remote, and there is no evidence for a case where this occurs.

In this section, we shall examine the rate expressions which are applicable for various single reaction steps, frequently called elementary reaction steps. Complex reactions, which involve several elementary steps, will be considered in depth in the following section.

Bimolecular Reactions

For the present, let us consider the simple reaction of molecule A colliding with molecule B, to give products, P. The chemical reaction is

$$A + B \longrightarrow P \qquad (5\text{-}B\text{-}1)$$

P may be a single molecular species, or more than one species. However, all collisions between A and B do not necessarily lead to a reaction yielding products. There are several factors which must be satisfied in order for the reaction to proceed. As a result, only a certain fraction of the collisions will lead to reaction.

The rate at which A and B react will be the product of the rate at which the molecules collide, or come in contact, and a factor which expresses the probability that the molecules will react. The latter factor, expressing the probability that the molecules will react once they have come together, is clearly a function of the nature of the reacting molecules, and not the concentrations of the reacting species. The rate at which A and B collide, however, is a function of the concentrations of A and B, and can be determined through the collision frequency relationship.

The dependence of the collision frequency on concentration can be rationalized by first considering the collisions of a single molecule of A with molecules of B. If we follow the single molecule of A through the mixture, it will undergo a certain number of collisions in a given period of time. Dividing the total number of collisions by the time would give an average collision rate, collisions per second, for the single molecule of A. If the total number of molecules of B were doubled or tripled without changing the total volume of the system, the average collision rate for the single molecule of A would likewise double or triple. However, if the volume of the system were also doubled or tripled, as was the number of molecules of B, the average collision rate for the single molecule of A would remain the same. This reasoning suggests that the average collision rate of a single molecule of A with molecules of B is proportional to the concentration of molecules of B. Hence,

$$\text{Collision Rate (single molecule of A)} \; \alpha \; [B] \qquad (5\text{-}B\text{-}2)$$

Now, suppose that instead of a single molecule of A, two or three molecules of A are admitted to the system, and their collisions with molecules of B are considered collectively. If the average rate of collisions between molecules of A and molecules of B is again determined, including the collisions of all molecules of A, it will either double (in the case of two molecules of A), or triple (in the case of three molecules of A). Consequently, for N molecules of A present in the system, the average rate of collisions between molecules of A and molecules of B will be N times as great as the average rate of collisions for a single molecule of A. Thus, the total rate of collisions between molecules of A and molecules of B is also proportional to the total number of molecules of A in the system, N.

$$\text{Collision Rate (N molecules of A)} \propto N \qquad (5\text{-B-}3)$$

Combining equation (5-B-2) and (5-B-3) results in the expression

$$\text{Collision Rate (N molecules of A)} \propto N[B]$$

However, the desired rate is the average number of collisions per second per unit volume. Dividing both sides of the above equation by the total volume of the system, and utilizing the fact that molecules per unit volume is proportional to moles per liter, we obtain

$$\frac{\text{Collision Rate for N molecules of A}}{V} \propto \frac{N}{V}[B] \propto [A][B]$$

or

$$R \propto [A][B] \qquad (5\text{-B-}4)$$

where R is the rate per unit volume.

The rate of reaction can be expressed as the rate of *decrease in the concentration of the reactant*, A or B, which necessitates a *negative sign* to indicate this decrease in concentration. The rate of reaction can also be expressed as the rate of increase in the concentration of products, and correspondingly, this would be positive.

$$R = -\frac{d[A]}{dt} = -\frac{d[B]}{dt} = \frac{d[P]}{dt} \qquad (5\text{-B-}5)$$

The rate expression can then be written as

$$-\frac{d[A]}{dt} = -\frac{d[B]}{dt} = \frac{d[P]}{dt} = k[A][B] \qquad (5\text{-B-}6)$$

The proportionality constant, k, is called the **rate constant** for the reaction. In this elementary reaction step, a molecule of A and a molecule of B come together to give product, P. This is called a bimolecular reaction step, and the constant k is referred to as the bimolecular rate constant. In this specific case, where k is a bimolecular rate constant, the units are $(\text{concentration})^{-1}\,(\text{time})^{-1}$. Rate constants for other reactions will have different units.

Example 5-B-1. An example of an elementary bimolecular reaction is the gas phase reaction

$$Na + CH_3Cl \longrightarrow NaCl + CH_3^{\cdot}$$

The rate expression is

$$-\frac{d[Na]}{dt} = -\frac{d[CH_3Cl]}{dt} = \frac{d[NaCl]}{dt} = \frac{d[CH_3^{\cdot}]}{dt} = k[Na][CH_3Cl]$$

If the concentrations of Na and CH_3Cl are expressed in terms of moles/liter, and the rate is expressed in terms of moles/liter-sec, the units of the rate constant, k, are found by dimensional analysis.

$$\frac{moles}{liter\text{-}sec} = k \left(\frac{moles}{liter}\right)\left(\frac{moles}{liter}\right)$$

$$k = \frac{liters}{mole\text{-}sec} = \left(\frac{moles}{liter}\right)^{-1}(sec)^{-1}$$

Problem 5-B-1. The *cis-trans* isomerization of cis-ethylene-D_2 is catalyzed by nitric oxide. The bimolecular reaction step is

The rate constant for this bimolecular reaction step is 3.18 $cm^3mole^{-1}sec^{-1}$ at 561.2 K. If the concentration of cis-ethylene-D_2 is 10^{-3} moles/liter, and NO is at a concentration of 5×10^{-5} moles/liter, what is the rate of the cis-trans conversion which occurs at 561.2 K? Answer: 1.59×10^{-10} moles/L-sec.

If the species B were replaced by A, the other reactant, then the reaction would proceed only when two A molecules collide. Equation (5-B-1) would then become

$$2A \longrightarrow P \tag{5-B-7}$$

and equation (5-B-6) becomes

$$\frac{d[P]}{dt} = k[A][A] = k[A]^2 \tag{5-B-8}$$

However, since two A molecules must react to give one P molecule, A will decrease at twice the rate that P is produced. Hence, for this reaction,

$$-\frac{d[A]}{dt} = 2\frac{d[P]}{dt} \tag{5-B-9}$$

Substituting (5-B-9) into (5-B-8), the rate expression can be written as

$$-\frac{d[A]}{dt} = 2k[A]^2 = k'[A]^2 \tag{5-B-10}$$

The rate constant k' for the decrease of A is twice that of the rate constant for the increase of P.

Example 5-B-2. An example of a simple bimolecular reaction of this type is the dimerization of NO_2.

$$\underset{\substack{O \diagdown \\ \diagdown \\ O}}{\overset{\substack{\diagup O \\ \diagup}}{N}} + \underset{\substack{\diagup \\ O}}{\overset{\substack{O \diagdown \\ \diagdown}}{N}} \longrightarrow \underset{\substack{O \diagdown \\ \diagdown \\ O}}{\overset{\substack{\diagup O \\ \diagup}}{N}} - \underset{\substack{\diagdown \\ O}}{\overset{\substack{O \diagdown \\ \diagup}}{N}}$$

(5-B-11)

$$2\,NO_2 \longrightarrow N_2O_4 \qquad\qquad (5\text{-}B\text{-}12)$$

As one might expect from the structures of NO_2 and N_2O_4, this reaction could proceed by a simple bond formation process, since there appears to be little change in the O-N-O bond angle. Since the reaction occurs by a single reaction step, (5-B-12), the rate expression is:

$$\frac{d[N_2O_4]}{dt} = k[NO_2]^2$$

$$-\frac{d[NO_2]}{dt} = 2k[NO_2]^2 = k'[NO_2]^2 \qquad\qquad (5\text{-}B\text{-}13)$$

Unimolecular Reactions

If a reaction involves a single molecule which gives products, as

$$A \longrightarrow P \qquad\qquad (5\text{-}B\text{-}14)$$

the corresponding reaction rate would be given by

$$-\frac{d[A]}{dt} = k[A] \qquad\qquad (5\text{-}B\text{-}15)$$

The k in equation (5-B-15) is the unimolecular rate constant and has the units of reciprocal time.

Problem 5-B-2. Using dimensional analysis, show that the units for a unimolecular rate constant are independent of the concentration of the reacting species.

Problem 5-B-3.
(a) If a reaction involving three molecules occurred in a single step according to the equation

$$2A + B \longrightarrow P$$

write the rate expression for the formation of products.
(b) What is the relationship between $-d[A]/dt$ and $-d[B]/dt$?
Answer: (a) $\dfrac{d[P]}{dt} = k[A]^2[B]$ (b) $-\dfrac{d[A]}{dt} = -2\dfrac{d[B]}{dt}$

Problem 5-B-4. The reaction

$$2NO + O_2 \longrightarrow 2NO_2$$

may be a trimolecular (sometimes called termolecular) reaction.
(a) If it is assumed to be trimolecular, write the rate expression in terms of loss of reactants, and gain of products.

(b) If the [NO] were decreased by a factor of two (cut in half), how would the reaction rate be affected?

Answer: (a) $-\dfrac{d[NO]}{dt} = -2\dfrac{d[O_2]}{dt} = \dfrac{d[NO_2]}{dt} = k[NO]^2[O_2]$ (b) Decreased by a factor of 1/4.

C. Complex Reactions—Steady State and Equilibrium Approximations

Many reactions appear to be extremely simple according to their stoichiometric equation, but, in fact, are quite complex, frequently involving several elementary reaction steps. Therefore, the student must be very cautious in deducing a reaction rate expression from the stoichiometric equation. Although numerous examples could be given in this regard, the student should be convinced of this fact by the end of this section. As a matter of fact, probably more reactions occur through complex mechnisms, than by single elementary steps, especially reactions carried out in the gas phase.

An example of a reaction that appears to be quite simple according to the stoichiometric equation is the decomposition of ozone. The stoichiometric equation is given by

$$2O_3 \longrightarrow 3O_2 \qquad\qquad (5\text{-}C\text{-}1)$$

Ozone is a straight-chain molecule with an O-O-O angle of 116.8°.

It appears that two O_3 molecules could conceivably come together in the following manner:

An attraction could occur between the three O-atoms which, simultaneous with the bond ruptures, indicated by the wavy lines, could result in the formation of three O-O bonds. Another way in which two O_3 molecules could come together to produce three O_2 molecules in one reaction step is an end-on collision.

In this reaction step, one bond would be formed with the simultaneous breaking of two O-O bonds. However, these require rather unique arrangements of the two O_3 molecules, and according to existing data, the mechanism does not occur in this manner. If the reaction did take place like this, it would be a simple bimolecular rate expression, similar to that for NO_2 in equation (5-B-13). However, the rate of reaction is also dependent on the O_2 concentration, which would not be satisfied if the reaction proceeded as an elementary bimolecular reaction.

An alternate mechanism that appears to satisfy the experimental facts consists of the following series of elementary reaction steps:

$$O_3 \xrightarrow{\ k_1\ } O_2 + O \qquad\qquad (5\text{-}C\text{-}2)$$

$$O_2 + O \xrightarrow{\ k_{-1}\ } O_3 \qquad\qquad (5\text{-}C\text{-}3)$$

$$O + O_3 \xrightarrow{\ k_2\ } 2O_2 \qquad\qquad (5\text{-}C\text{-}4)$$

The first step involves the breaking of a single O-O bond, resulting in the very stable O_2 molecule and an O atom. The reverse reaction of the first step, giving back O_3, is considered in the second step. The rate constant for the second reaction step, k_{-1}, is designated with a subscript -1, to indicate that the reaction is the reverse of the first reaction. The third step can be thought of as an O abstraction from O_3 by an O atom. In each of these elementary reaction steps, a very simple bond breakage or bond formation is involved. On this basis, each is more likely to occur than the direct reaction of two O_3 molecules, which necessitates the breaking and formation of several bonds.

The reaction sequence (5-C-2) to (5-C-4), can be written more succinctly as

$$O_3 \underset{k_{-1}}{\overset{k_1}{\rightleftarrows}} O_2 + O \qquad\qquad (5\text{-}C\text{-}2)(5\text{-}C\text{-}3)$$

$$O + O_3 \xrightarrow{\ k_2\ } 2O_2 \qquad\qquad (5\text{-}C\text{-}4)$$

The reverse reaction (5-C-3) is indicated by an arrow in the reverse direction and the associated rate constant underneath the reverse arrow. The overall stoichiometric equation is satisfied by this reaction step, since product is obtained only when the back reaction, k_{-1}, does not occur. One O_3 molecule is lost and one O_2 molecule is produced in the first reaction step. In the third reaction step, the second O_3 molecule is reacted and the two additional O_2 molecules are formed to give a total of three O_2 molecules as product. Hence, the sum of reaction steps (5-C-2) and (5-C-4) yields the overall stoichiometric equation (5-C-1).

More than one rate expression can be written for a complex reaction, since there are more species present than just reactants and products, as in an elementary reaction step. In this case, there is an intermediate O atom, so there will be a rate expression for this species as well. For complex mechanisms, the rate of appearance or disappearance of a given species may involve several terms, since the species may be lost or produced in several elementary reaction steps. For example, the loss of O_3 in (5-C-2) to (5-C-4) is given by

$$-\frac{d[O_3]}{dt} = k_1\,[O_3] - k_{-1}\,[O_2][O] + k_2\,[O][O_3] \qquad\qquad (5\text{-}C\text{-}5)$$

Since O_3 is lost in (5-C-2) and (5-C-4), we have positive values for the k_1 and k_2 terms. On the other hand, O_3 is produced in (5-C-3), hence, a negative sign is assigned to the k_{-1} term. Similarly, for O_2,

$$\frac{d[O_2]}{dt} = k_1[O_3] - k_{-1}[O_2][O] + 2k_2[O][O_3] \tag{5-C-6}$$

2 is placed in the last term of (5-C-6), since two O_2 molecules are produced, and k_2 was specified in the third term of (5-C-5) in terms of the loss of a single O_3 molecule. In this case, the simple relationship between $-d[O_3]/dt$ and $d[O_2]/dt$ is not obtained directly, as in the previous section for equation (5-B-9), but will be shown later, after additional assumptions are made for simplification. The rate expression for O is given by

$$\frac{d[O]}{dt} = k_1[O_3] - k_{-1}[O_2][O] - k_2[O][O_3] \tag{5-C-7}$$

Equations (5-C-5), (5-C-6), and (5-C-7) rigorously define the rate of change of the concentrations O_3, O_2, and O with time, for the proposed reaction mechanism [equations (5-C-2)-(5-C-4)], however, their analytical solution is no simple task. For this reason, we seek a way to simplify the series of three equations to a single differential equation, a problem frequently encountered in solving the rate expressions for complex reactions. The techniques we shall use are commonly employed to reduce a system of differential equations to a single differential equation.

Differential equations can be simplified in two ways, each based on particular assumptions:

(a) reversible and fast reactions are assumed to be in equilibrium, which we refer to as the *equilibrium assumption*; and

(b) reactive intermediates are assumed to rapidly attain a steady state (constant concentration), which we refer to as the *steady state assumption*.

These assumptions are essentially equivalent when applied to the same problem. The former technique is simpler mathematically, but the latter shows the assumption of equilibrium explicitly, and obtains the alternate expression, not involving equilibrium, directly.

Equilibrium Assumption

The first two reaction steps, (5-C-2) and (5-C-3), can be assumed to be so fast that equilibrium is attained. Equilibrium in the first step necessitates that the forward and reverse reaction rates in (5-C-2) and (5-C-3) be equal:

forward rate = reverse rate

This is analogous to the assumption that $[O_3]$, $[O_2]$, and $[O]$ do not change significantly within the time of equilibrium in this first reaction step.

$$-\frac{d[O_3]}{dt} = \frac{d[O_2]}{dt} = \frac{d[O]}{dt} = 0 = k_1[O_3] - k_{-1}[O_2][O]$$

The two terms on the right side of this equation correspond to the forward and reverse rates, respectively.

$$k_1[O_3] = k_{-1}[O_2][O] \tag{5-C-8}$$

$$\frac{k_1}{k_{-1}} = \frac{[O_2][O]}{[O_3]} \tag{5-C-9}$$

The right side of equation (5-C-9) is the equilibrium expression for the first reaction step, hence, the ratio, k_1/k_{-1}, must be the equilibrium constant for the reaction.

$$K_1 = \frac{k_1}{k_{-1}} = \frac{[O_2][O]}{[O_3]} \qquad (5\text{-}C\text{-}10)$$

Since equilibrium is established in the first reaction step, the rate of production of products will be given by the rate expression for (5-C-4).

$$\frac{d[O_2]}{dt} = 2k_2[O][O_3] \qquad (5\text{-}C\text{-}11)$$

The concentration of the reactive intermediate, $[O]$, can be evaluated from (5-C-10), thus

$$[O] = \frac{K_1[O_3]}{[O_2]} \qquad (5\text{-}C\text{-}12)$$

Substitution of (5-C-12) into (5-C-11) yields

$$\frac{d[O_2]}{dt} = 2k_2 \frac{K_1[O_3]}{[O_2]} [O_3] = 2k_2 K_1 \frac{[O_3]^2}{[O_2]} \qquad (5\text{-}C\text{-}13)$$

Note in equation (5-C-13), that the reaction rate decreases with increasing $[O_2]$, in agreement with experimental observation. As stated earlier, this could not have been obtained if the reaction had been considered as a simple bimolecular reaction of two O_3 molecules.

Problem 5-C-1. The conversion of p-H_2 (para hydrogen) to o-H_2 (ortho hydrogen) is considered to go by the mechanism

$$\text{p-H}_2 \underset{\longleftarrow}{\overset{K}{\longrightarrow}} 2\text{H}^\bullet$$

$$\text{H}^\bullet + \text{p-H}_2 \xrightarrow{k_2} \text{o-H}_2 + \text{H}^\bullet$$

Derive the expression for the rate of production of o-H_2, assuming equilibrium in the first step. Answer: $\dfrac{d[\text{o-H}_2]}{dt} = k_2 K^{1/2}[\text{p-H}_2]^{3/2}$.

Probem 5-C-2. For the oxidation of NO to NO_2

$$2NO + O_2 = 2NO_2$$

the following mechanism has been proposed

$$\text{NO} + \text{O}_2 \underset{k_{-1}}{\overset{k_1}{\longrightarrow}} {}^\bullet\text{OONO}$$

$${}^\bullet\text{OONO} + \text{NO} \xrightarrow{k_2} 2\text{NO}_2$$

Assuming equilibrium in the first step, derive the rate equation for the production of NO_2.

Answer: $\dfrac{d[NO_2]}{dt} = \dfrac{k_1 k_2}{k_{-1}}$ $[NO]^2[O_2]$ where k_2 is defined as the rate constant for the formation of NO_2.

Problem 5-C-3. For the reaction

$$CO + Cl_2 = COCl_2$$

the following mechanism has been proposed

$$Cl_2 \xrightleftharpoons{K_1} 2Cl^{\cdot}$$

$$Cl^{\cdot} + CO \xrightleftharpoons{K_2} COCl^{\cdot}$$

$$COCl^{\cdot} + Cl_2 \xrightarrow{k_3} COCl_2 + Cl^{\cdot}$$

Assuming equilibrium in the first two steps, derive the rate expression for the formation of $COCl_2$. Answer: $\dfrac{d[COCl_2]}{dt} = K_1^{1/2} K_2 k_3 [Cl_2]^{3/2}[CO]$.

Steady State Approximation

The intermediate O atoms in the ozone decomposition reaction are diradicals and thus, are quite reactive. The steady state assumption can be employed frequently, since the concentration of highly reactive intermediates generally attains a low concentration quite rapidly. The steady state assumption assumes this low concentration is constant, or the rate of change of concentration of the intermediate is zero. For the mechanism of the decomposition of O_3, this simply equates equation (5-C-7) to zero:

$$\frac{d[O]}{dt} = 0 = k_1[O_3] - k_{-1}[O_2][O] - k_2[O][O_3] \qquad (5\text{-}C\text{-}14)$$

Solving for [O],

$$[O] = \frac{k_1[O_3]}{k_{-1}[O_2] + k_2[O_3]}$$

Substitution into equation (5-C-6),

$$\frac{d[O_2]}{dt} = k_1[O_3] - k_{-1}[O_2][O] + 2k_2[O][O_3],$$

gives

$$\frac{d[O_2]}{dt} = k_1[O_3] - \frac{k_{-1}[O_2]k_1[O_3]}{k_{-1}[O_2] + k_2[O_3]} + \frac{2k_2[O_3]k_1[O_3]}{k_{-1}[O_2] + k_2[O_3]} \qquad (5\text{-}C\text{-}15)$$

Simplifying, by finding a common denominator,

$$\frac{d[O_2]}{dt} = \frac{\cancel{k_1[O_3]k_{-1}[O_2]} + k_1k_2[O_3]^2 - \cancel{k_{-1}[O_2]k_1[O_3]} + 2k_2k_1[O_3]^2}{k_{-1}[O_2] + k_2[O_3]}$$

$$\frac{d[O_2]}{dt} = \frac{3k_1k_2[O_3]^2}{k_{-1}[O_2] + k_2[O_3]} \qquad (5\text{-C-}16)$$

Now, if we make the additional assumption that

$$k_{-1}[O_2] \gg k_2[O_3]$$

equation (5-C-16) becomes

$$\frac{d[O_2]}{dt} = \frac{3k_1k_2[O_3]^2}{k_{-1}[O_2]} = 3K_1k_2\frac{[O_3]^2}{[O_2]} \qquad (5\text{-C-}17)$$

where $K_1 = k_1/k_{-1}$, as shown earlier.

The dependence on $[O_3]$ and $[O_2]$ is identical in equations (5-C-17) and (5-C-13). The expressions are also identical with respect to the constants K_1 and k_2, and differ only in the constants 2 and 3. This difference can be acounted for, when we recall that the production of O_2 was neglected in the equilibrium step in the equilibrium assumption treatment. Only the O_2 produced as a result of the third reaction step was included in equation (5-C-11), in which the factor of 2 acounts for the production of two O_2 molecules. Since an O_2 molecule was produced in the previous equilibrium expression, (which is necessary before the third step can take place), the factor should be increased to 3, which will bring it into accord with the result given in (5-C-17). Frequently, a technique similar to that shown in the equilibrium assumption approach is carried out with little or no regard for the numerical coefficient. This is justifiable, since the primary interest in such an analysis is the resulting functional relationship with the concentration of the reactants, intermediates, and products. The expression given in equation (5-C-17) is correct under the steady state assumption, provided that $k_{-1}[O_2] \gg k_2[O_3]$.

Problem 5-C-4.
(a) Substitute (5-C-14) into (5-C-5) and derive an expression for the rate, $-d[O_3]/dt$.
(b) How is this related to $d[O_2]/dt$ given in (5-C-16)?

Answer: (a) $-\dfrac{d[O_3]}{dt} = \dfrac{2k_1k_2[O_3]^2}{k_{-1}[O_2] + k_2[O_3]}$ (b) $-\dfrac{1}{2}\dfrac{d[O_3]}{dt} = \dfrac{1}{3}\dfrac{d[O_2]}{dt}$

Problem 5-C-5. The following mechanism has been proposed for the oxidation of NO to NO_2:

$$NO + O_2 \xrightarrow{\;k_1\;} {}^{\bullet}OONO$$

$${}^{\bullet}OONO \xrightarrow{\;k_{-1}\;} NO + O_2$$

$${}^{\bullet}OONO + NO \xrightarrow{\;k_2\;} 2\,NO_2$$

Apply steady state to $\cdot$OONO and derive the rate expression for the formation of NO_2. Compare this result with that of Problem 5-C-2. Answer: $\dfrac{d[NO_2]}{dt} = \dfrac{2k_1k_2[NO]^2[O_2]}{k_{-1} + k_2[NO]}$

where k_2 is defined as the rate constant for the loss of $\cdot$OONO. This is the same form as the answer to Problem 5-C-2 except the denominator has an additional $k_2[NO]$.

It was previously shown that the rate expression derived using the steady state assumption could be made consistent with that derived using the equilibrium assumption. However, it should be emphasized that the actual steady state expression is given by equation (5-C-16), and only with the additional assumption, $k_{-1}[O_2] \gg k_2[O_3]$, is the result using the equilibrium approximation obtained. Thus, equation (5-C-16) is more general and could be valid even in the region where $k_{-1}[O_2]$ is on the order of $k_2[O_3]$.

D. Hydrogen + Halogen Reaction Mechanisms

However, in some cases, the rate expression cannot be used as evidence for or against one mechanism over another. A good example lies in the reaction

$$H_2 + I_2 \longrightarrow 2\,HI \tag{5-D-1}$$

This reaction has been investigated many times and evidence is given to suggest a complex mechanism. The elementary reaction step, however, produces the same rate expression.

As mentioned earlier, the results of (5-C-13) or (5-C-17) are significantly different from that predicted by a single elementary bimolecular reaction step. In this case, experimental evidence of the inhibiting nature of O_2 would support the complex mechanism over the simple bimolecular reaction step.

$H_2 + I_2 = 2HI$ Mechanism

In the past, investigators assumed reaction (5-D-1) went by a single elementary bimolecular reaction step. The intermediate complex was visualized as an H_2 molecule approaching an I_2 molecule as shown here. The attraction

$$
\begin{array}{ccc}
\text{H}\ldots\ldots\text{I} & & \text{H}-\text{I} \\
| \quad\quad | & \longrightarrow & + \\
\text{H}\cdots\cdots\text{I} & & \text{H}-\text{I}
\end{array}
$$

between H and I was thought to increase as the two molecules approached, until a bond was eventually formed with the simultaneous breaking of the H-H and the I-I bonds. The rate expression would be given by

$$-2\frac{d[H_2]}{dt} = -2\frac{d[I_2]}{dt} = \frac{d[HI]}{dt} = k[H_2][I_2] \tag{5-D-2}$$

The rate expression, given in equation (5-D-2), is in agreement with the experimentally determined concentration dependence.

Further investigations [J.H. Sullivan, *J. Chem.Phys.*, **1967**, *46*, 73], which involve the analogous photochemical reaction (reaction is initiated by radiation), suggest that the reaction is more complex. Two complex mechanisms have been proposed, and these, together with the corresponding rate expression derived for each, are given as follows:

Mechanism 1

$$I_2 \underset{\longleftarrow}{\overset{K_1}{\longrightarrow}} 2I^{\cdot}$$

$$2I^{\cdot} + H_2 \xrightarrow{k_2} 2HI$$

$$K_1 = \frac{[I^{\cdot}]^2}{[I_2]}$$

$$[I^{\cdot}]^2 = K_1[I_2]$$

$$\frac{d[HI]}{dt} = 2k_2[H_2][I^{\cdot}]^2$$

$$\frac{d[HI]}{dt} = 2k_2K_1[H_2][I_2] \qquad (4\text{-}D\text{-}3)$$

Mechanism 2

$$I_2 \underset{\longleftarrow}{\overset{K_1}{\longrightarrow}} 2I^{\cdot}$$

$$I^{\cdot} + H_2 \underset{\longleftarrow}{\overset{K_2}{\longrightarrow}} H_2I^{\cdot}$$

$$H_2I^{\cdot} + I^{\cdot} \xrightarrow{k_3} 2HI$$

$$K_1 = \frac{[I^{\cdot}]^2}{[I_2]}$$

$$[I^{\cdot}]^2 = K_1[I_2]$$

$$K_2 = \frac{[H_2I^{\cdot}]}{[I^{\cdot}][H_2]}$$

$$[H_2I^{\cdot}] = K_2[I^{\cdot}][H_2]$$

$$\frac{d[HI]}{dt} = 2k_3[H_2I^{\cdot}][I^{\cdot}]$$

$$= 2k_3K_2[I^{\cdot}][H_2][I^{\cdot}]$$

$$= 2k_3K_2[H_2][I^{\cdot}]^2$$

$$= 2k_3K_2[H_2]K_1[I_2]$$

$$\frac{d[HI]}{dt} = 2k_3K_1K_2[H_2][I_2] \qquad (4\text{-}D\text{-}4)$$

As one can see from equations (5-D-3) and (5-D-4), the mechanism cannot be specified on the basis of the rate expression. In all three rate expressions, (5-D-2), (5-D-3), and (5-D-4), the rate depends on the first power of $[H_2]$ and $[I_2]$. In a case such as this, additional information will be required to distinguish between the different proposed mechanisms. It is a general rule that a mechanism cannot be proved from a rate expression. Though it can be disproved if the rate expression derived from the proposed mechanism does not agree with the experimental facts, the reverse case of unequivocally proving a mechanism is not possible.

Problem 5-D-1. Let us assume the mechanism for the oxidation of NO to be

$$O_2 \underset{\longleftarrow}{\overset{K_1}{\longrightarrow}} 2O^{\cdot}$$

$$O^{\cdot} + NO \xrightarrow{k_2} NO_2$$

Assume equilibrium in the first step and derive the rate expression for the production of NO_2. Is it different from the results of Problems 5-B-4 and 5-C-2?

Answer: $\dfrac{d[NO_2]}{dt} = k_2 K_1^{1/2} [NO][O_2]^{1/2}$; yes.

$H_2 + Br_2 = 2HBr$ Mechanism

A final example of a classic reaction mechanism, which appears to be extremely simple from the stoichiometric equation, is

$$H_2 + Br_2 \longrightarrow 2HBr \qquad\qquad (5\text{-}D\text{-}5)$$

The reaction mechanism, which has been rather well established, involves several elementary reaction steps:

$$
\begin{array}{lll}
Br_2 \xrightarrow{\ k_1\ } 2Br\cdot & & \text{initiation} \\[4pt]
Br\cdot + H_2 \xrightarrow{\ k_2\ } HBr + H\cdot & & \text{propagation} \\[4pt]
\uparrow\ \underleftarrow{\hspace{5cm}} & & \text{chain reaction} \qquad (5\text{-}D\text{-}6) \\[4pt]
& \uparrow & \\[4pt]
H\cdot + Br_2 \xrightarrow{\ k_3\ } HBr + Br\cdot & & \text{propagation} \\[4pt]
H\cdot + HBr \xrightarrow{\ k_{-2}\ } H_2 + Br\cdot & & \text{inhibition} \\[4pt]
Br\cdot + Br\cdot \xrightarrow{\ k_{-1}\ } Br_2 & & \text{termination}
\end{array}
$$

This reaction is unique compared to previous examples, in that it involves what is called a *chain mechanism*. The reaction is initiated in the first step by breaking the Br-Br bond to give Br atoms. The Br atoms can react with H_2 in the second step to give product HBr plus H atoms. The H atoms produced can react with Br_2, as shown in the third step, to give more of the product HBr plus Br atoms. However, Br atoms are responsible for reaction step 2, so the reaction at this point can be started again. In other words, a chain reaction is started in steps 2 and 3, producing product HBr from H_2 and Br_2, the atoms being only intermediates. The long left pointing and two up pointing arrows in the reaction scheme shows the return of the Br atoms from step 3 to step 2.

This chain reaction would proceed indefinitely except that other reactions may compete for the radicals formed. This is shown by the inhibiting step 4, where the product HBr can scavenge H atoms to give H_2 and Br. Since Br is regenerated, it can return to step 2 to continue the chain mechanism. Because it does not stop the chain reaction, it is referred to as an inhibiting step.

However, two Br atoms can also combine to give Br_2, as shown in step 5. Since Br_2 is a reactant, it must go back to step 1 and will produce Br atoms again. In this case, the chain mechanism, which is propagated by Br and H atom formation, is terminated, and reaction step 5 is appropriately called a termination step. There are many important gas phase reactions which undergo chain mechanisms by radical intermediates, and the $H_2 + Br_2$ reaction serves as an example of one of the many possibilities.

The rate expression for this mechanism can be derived using the steady state assumption on the reactive H and Br atom intermediates. The details of the calculation will not be carried out in this text. However, if the steady state assumption is applied to the atom intermediates H and Br, the final rate expression is given by

$$\frac{d[HBr]}{dt} = \frac{2k_2\left(\frac{k_1}{k_{-1}}\right)^{1/2}[H_2][Br_2]^{1/2}}{1+\frac{k_{-2}}{k_3}\frac{[HBr]}{[Br_2]}}$$

(5-D-7)

This is a rather unique rate expression and is in agreement with existing experimental data. The difference between this expression and that for the $H_2 + I_2$ reaction mechanisms in (5-D-2), (5-D-3), and (5-D-4), should be noted. In particular, when [HBr] is low, as it would be initially with only H_2 and Br_2 as reactants, equation (5-D-7) can be simplified to

$$\frac{d[HBr]}{dt} = 2k_2\left(\frac{k_1}{k_{-1}}\right)^{1/2}[H_2][Br_2]^{1/2}$$

(5-D-8)

In this case, the rate depends on $[Br_2]$ to the 1/2 power. Also, the inhibition or slow-down of the rate is obvious from equation (5-D-7), since [HBr] is in the denominator.

One principal difference between the H_2-I_2 and H_2-Br_2 reaction mechanisms should be noted. Off hand, one might expect that the reaction mechanisms would be similar if they were complex, however, this is not the case. The complex mechanisms proposed for I_2 and Br_2 are initiated by halogen atom formation, but this happens more readily with I_2 than Br_2, since the bond dissociation energy for I_2 is lower than the corresponding value for Br_2. When we consider the second step, the two mechanisms are different, apparently because I atoms cannot abstract a H atom from H_2 very readily. The H_2 bond is quite strong (435.1 kJ), and the HI bond, which is formed while the H-H bond is being broken, is not exceptionally large (295.4 kJ). We can visualize the reaction intermediate as follows:

$$X^{\cdot} \cdots H\text{---}H \longrightarrow X\text{----}H \cdots H^{\cdot}$$

where $X^{\cdot}$ represents the halogen atom. The Br-H bond, on the other hand, is considerably stronger (362.8 kJ), and apparently the H atom abstraction reaction can take place.

The reaction mechanisms for the reaction between H_2 and Cl_2 or F_2 are similar to that for the H_2+Br_2 reaction. Again the halogens, Cl_2 and F_2, dissociate into Cl and F, respectively, and then proceed to react with H_2.

$$Cl^{\cdot} + H_2 \longrightarrow HCl + H^{\cdot}$$

$$F^{\cdot} + H_2 \longrightarrow HF + H^{\cdot}$$

The bond enthalpies for HCl and HF are quite large, 431.3 and 564.8 kJ, respectively, both of which are even greater than the H-Br bond enthalpy of 362.8 kJ. Consequently, these reaction steps can proceed even more favorably than the Br + H_2 reaction step.

After discussing these examples of reaction mechanisms, we can make the following conclusions:

1. The stoichiometric equation cannot be used to predict the reaction mechanism.

2. The rate expression can be derived for a proposed reaction mechanism consisting of elementary steps, and for convenience, the steady state approximation can be used for reactive intermediates.

3. The rate expression, and its agreement with experimental data, is useful in supporting a proposed mechanism, but it cannot be used as unequivocal evidence for a reaction mechanism.

Problem 5-D-2. In the following reaction scheme, identify the initiation step or steps, the chain mechanism, and the termination step.

$$C_2H_6 \xrightarrow{k_1} 2\,CH_3^{\cdot}$$

$$CH_3^{\cdot} + C_2H_6 \xrightarrow{k_2} CH_4 + C_2H_5^{\cdot}$$

$$C_2H_5^{\cdot} \xrightarrow{k_3} C_2H_4 + H^{\cdot}$$

$$H^{\cdot} + C_2H_6 \xrightarrow{k_4} H_2 + C_2H_5^{\cdot}$$

$$H^{\cdot} + C_2H_5^{\cdot} \xrightarrow{k_5} C_2H_6$$

What are the principal products? Answer: Initiation is step 1; propagation steps 2, 3, 4; Termination is step 5. Chain mechanism is steps 3 and 4. Principal products are C_2H_4 and H_2.

Problem 5-D-3. For the reaction

$$CO + Cl_2 = COCl_2$$

the following mechanism has been proposed:

$$Cl_2 \underset{k_{-1}}{\overset{k_1}{\rightleftharpoons}} 2\,Cl^{\cdot}$$

$$Cl^{\cdot} + CO \underset{k_{-2}}{\overset{k_2}{\rightleftharpoons}} COCl^{\cdot}$$

$$COCl^{\cdot} + Cl_2 \xrightarrow{k_3} COCl_2 + Cl^{\cdot}$$

(a) Identify the nature of each reaction step and a chain mechanism.
(b) Assuming steady state for Cl and COCl, derive the rate expression for the formation of $COCl_2$. Hint: The steady state equation must be solved simultaneously to obtain equations for [Cl] and [COCl]. This is accomplished most easily by adding the steady state equations which eliminates [COCl]. Solving for [Cl] and substitution into the steady state equation for [COCl] gives the equation for [COCl].
(c) Show that the steady state solution in (b) is equivalent to the equilibrium solution in Problem 5-C-3 if one assumes

$$k_{-2} \gg k_3\,[Cl_2]$$

Answer: (a) initiation step 1; termination step -1; propagation steps 2, -2, 3; chain mechanism steps 2 and 3; (b) $\dfrac{d[COCl_2]}{dt} = \left(\dfrac{k_1}{k_{-1}}\right)^{1/2} \dfrac{k_2 k_3 [Cl_2]^{3/2}[CO]}{k_{-2} + k_3[Cl_2]}$

E. General Rate Expression and Order of Reaction

The **reaction order** is formally defined for a reaction with the stoichiometric equation:

$$aA + bB \longrightarrow cC + dD \qquad (5\text{-}E\text{-}1)$$

This reaction is not an elementary reaction step. We can assume there is a general rate expression of the form

$$\frac{d[P]}{dt} = k[A]^{\alpha}[B]^{\beta}[C]^{\gamma}[D]^{\delta} \qquad (5\text{-}E\text{-}2)$$

where α = reaction order with respect to component A, β = reaction order with respect to component B, etc., and P represents either of the products C or D.

$$\frac{d[P]}{dt} = \frac{1}{c}\frac{d[C]}{dt} = \frac{1}{d}\frac{d[D]}{dt} = -\frac{1}{a}\frac{d[A]}{dt} = -\frac{1}{b}\frac{d[B]}{dt}$$

In some cases, for example the H_2-Br_2 reaction in equation (5-D-7), the reaction rate expression in its most general form is not in the form of equation (5-E-2), and hence, the order of the reaction is not a meaningful term. However, generally rate expressions of this type can be put in the form of equation (5-E-2), as shown in equation (5-D-8) for the H_2-Br_2 reaction.

The overall order of the reaction (n) is defined as

$$n = \alpha + \beta + \gamma + \delta.$$

The k in equation (5-E-2) is the overall rate constant for the reaction. It can be related to the rate constants of the elementary reaction steps if the reaction is complex.

Referring to the examples which we considered in the previous two sections, we can identify the reaction order from the resultant rate expression.

Equation (5-B-6):

$$\frac{d[P]}{dt} = k[A][B] \qquad (5\text{-}B\text{-}6)$$

1st order with respect to A;
1st order with respect to B;
2nd order for the overall reaction;
the overall rate constant is equal to the bimolecular rate constant for the elementary step.

Equation (5-B-8):

$$\frac{d[P]}{dt} = k[A]^2 \qquad (5\text{-}B\text{-}8)$$

2nd order with respect to A;
2nd order for the overall reaction;
the bimolecular rate constant is equal to the overall rate constant.

Equation (5-C-17):

$$\frac{d[O_2]}{dt} = 3K_1k_2 \frac{[O_3]^2}{[O_2]} \qquad (5\text{-}C\text{-}17)$$

2nd order with respect to O_3;
inverse first order with respect to O_2;
1st order for the overall reaction $k = 3K_1k_2$.

Equations (5-D-2), (5-D-3), (5-D-4):

$$\frac{d[HI]}{dt} = k[H_2][I_2] \qquad (5\text{-}D\text{-}2)$$

$$\frac{d[HI]}{dt} = 2k_2\,K_1[H_2][I_2] \qquad (5\text{-}D\text{-}3)$$

$$\frac{d[HI]}{dt} = 2k_3\,K_1K_2[H_2][I_2] \qquad (5\text{-}D\text{-}4)$$

1st order with respect to H_2;
1st order with respect to I_2;
2nd order for overall reaction
for (5-D-2) the bimolecular rate constant is equal to the overall or general rate constant;
for (5-D-3) $k = 2k_2K_1$;
for (5-D-4) $k = 2k_3K_1K_2$.

Equation (5-D-7):

$$\frac{d[HBr]}{dt} = \frac{2k_2\left(\dfrac{k_1}{k_{-1}}\right)^{1/2}[H_2][Br_2]^{1/2}}{1 + \dfrac{k_{-2}}{k_3}\dfrac{[HBr]}{[Br_2]}} \qquad (5\text{-}D\text{-}7)$$

undefined.

Equation (5-D-8):

$$\frac{d[HBr]}{dt} = 2k_2\left(\frac{k_1}{k_{-1}}\right)^{1/2}[H_2][Br_2]^{1/2} \qquad (5\text{-}D\text{-}8)$$

1st order with respect to H_2;
1/2 order with respect to Br_2;
3/2 order for the overall reaction;
$k = 2k_2(k_1/k_{-1})^{1/2}$.

Problem 5-E-1. In equation (5-D-7), we could consider the situation where

$$\frac{k_{-2}[HBr]}{k_3[Br_2]} \gg 1.$$

The rate expression then becomes

$$\frac{d[HBr]}{dt} = \frac{2k_2 \left(\dfrac{k_1}{k_{-1}}\right)^{1/2} [H_2][Br_2]^{1/2}}{\dfrac{k_{-2}\,[HBr]}{k_3\,[Br_2]}}$$

$$\frac{d[HBr]}{dt} = 2k_2 \left(\frac{k_1}{k_{-1}}\right)^{1/2} \frac{k_3}{k_{-2}} \frac{[H_2][Br_2]^{3/2}}{[HBr]} \tag{5-E-3}$$

What is the order of reaction with respect to H_2, Br_2, HBr? What is the overall order for the reaction, and what is the relationship between the general rate constant and the elementary rate constants?
Answer: H_2 first order; Br_2 3/2 order; HBr inverse first order; $n = 3/2$ overall order;

$k = 2k_2 \left(\dfrac{k_1}{k_{-1}}\right)^{1/2} \dfrac{k_3}{k_{-2}}$.

From these examples and the discussion in sections B and C, we can conclude: The molecularity of an elementary reaction is the order of that elementary reaction. The reverse is not necessarily true; that is, the order of the reaction does not specify the molecularity of an elementary reaction, since the reaction could be complex.

Problem 5-E-2. The thermal decomposition of ethane according to the stoichiometric equation

$$C_2H_6 \longrightarrow C_2H_4 + H_2$$

is 1st order with respect to C_2H_6. What can be said of the order of the reaction? Answer: nothing.

F. Initial Rate Method for the Determination of Reaction Order and Overall Rate Constant

There are several techniques for determining the order of a reaction, and the interested student should consult a text in chemical kinetics for a more complete discussion of the subject. We will reconsider this topic later in the chapter, but at this time, one of these methods will be presented. This technique, which is quite simple in principle and straightforward in mathematical evaluation, is called the initial rate method.

The technique involves the determination of the initial rate for the reaction from data of concentration versus time, for different initial concentrations of reactants. The specific concentrations are selected so that the orders of the reaction with respect to each reactant can be evaluated readily. For example, let us refer to the general reaction given in (5-E-1) and the general rate expression given in equation (5-E-2). If we are to determine the order of the reaction with respect to component A, then we will keep the concentration of the other species constant. Then equation (5-E-2) can be written as

$$\frac{d[P]}{dt} = \underset{\text{constant}}{(k[B]^\beta \cdots)} [A]^\alpha$$

or,

$$\frac{d[P]}{dt} = k'[A]^{\alpha} \tag{5-F-1}$$

where the $[B]^{\beta}$ and similar terms for other species are included in k', that is, $k' = (k[B]^{\beta}\cdots)$.

In order to determine α, we simply measure the initial rate, dP/dt, at various initial concentrations of A. Equation (5-F-1) can be put into linear form by taking the logarithm:

$$\ln\frac{d[P]}{dt} = \ln k' + \alpha \ln [A] \tag{5-F-2}$$

A graph of ln(dP/dt) versus ln[A] should then be linear, with the slope = α and the intercept equal to ln k'. (See Figure 5-F-1.) The logarithms in (5-F-2) can be either natural logarithms to the base e or base 10 logarithms. Since all three terms in (5-F-2) are logarithms, the factor 2.303 relating these two logarithms will cancel. This method of evaluating the rate expression is less precise than other methods, since initial rate measurements are difficult to determine. Also the data must span a large range of concentration in order to obtain a precise slope.

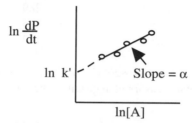

Figure 5-F-1. The Initial Rate Method for Determining the Order of Reaction

If the rate has been determined at only two different concentrations of A, a graph is not necessary to evaluate the order. We can represent the data at the different concentrations by subscripts 1 and 2 and substitute these two sets of values into equation (5-F-2).

$$\ln \left(\frac{d[P]}{dt} \right)_1 = \ln k' + \alpha \ln[A]_1$$

$$\ln \left(\frac{d[P]}{dt} \right)_2 = \ln k' + \alpha \ln [A]_2$$

We can solve these two equations simultaneously by subtracting to eliminate ln k'.

$$\ln \left(\frac{d[P]}{dt} \right)_1 - \ln \left(\frac{d[P]}{dt} \right)_2 = \alpha(\ln[A]_1 - \ln[A]_2)$$

Solving for α,

$$\alpha = \frac{\ln\left(\dfrac{d[P]}{dt} \right)_1 - \ln\left(\dfrac{d[P]}{dt} \right)_2}{\ln[A]_1 - \ln[A]_2} \tag{5-F-3}$$

After determining the order of the reaction with respect to each of the substituents, the rate constant can be determined from a measured value of the rate at known concentrations of the reactants. These values are then substituted into equation (5-E-2) and the rate constants calculated.

Example 5-F-1. Determine the reaction order of NO in the reaction

$$2\ NO + H_2 \longrightarrow N_2O + H_2O$$

using the technique of initial rates with the following data. The superscript zero refers to the initial value.

$P^o_{H_2}$ = 400 torr = constant		P^o_{NO} = 400 torr = constant	
	Rate of reaction		Rate of reaction
P^o_{NO} (torr)	(torr/100 sec)	$P^o_{H_2}$ (torr)	(torr/100 sec)
359	150	289	160
300	103	205	110
152	25	147	79

The temperature was held constant at 826°C for all reactions. The rate of reaction was followed by measuring the total pressure change with time. From the stoichiometry,

$$-\frac{d(\Delta P)}{dt} = \frac{dP_{N_2O}}{dt}$$

We proceed by taking logarithms of the initial concentrations, or in this case, the pressures, since volume is constant, and also the logarithms of the rates of reaction. These are shown in the following tables:

$P^o_{H_2}$ = 400 torr		P^o_{NO} = 400 torr	
$\ln P^o_{NO}$	$\ln \dfrac{d(-\Delta P)}{dt}$ (torr/sec)	$\ln P^o_{H_2}$	$\ln \dfrac{d(-\Delta P)}{dt}$ (torr/sec)
5.883	0.405	5.666	0.470
5.704	0.0296	5.323	0.0953
5.024	-1.386	4.990	-0.235

A graph of $\ln d(-\Delta P)/dt$ versus $\ln P^o_{NO}$, at constant $P^o_{H_2}$ = 400 torr, gives the order of the reaction with respect to [NO]. (See Figure 5-F-2.) The slope is close to 2.0, therefore, the reaction is assumed to be exactly second order with respect to NO; that is,

$$\frac{d(-\Delta P)}{dt} = k'[NO]^2$$

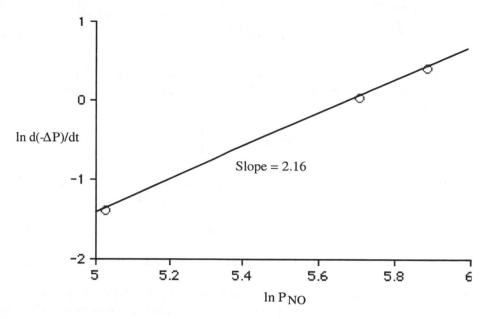

Figure 5-F-2. Order of Reaction for NO

Problem 5-F-1.

(a) From the data given previously on the reaction between NO and H_2, determine the order of the reaction with respect to H_2.

(b) Write the general rate expression in the form of equation (5-E- 2).

Answer: (a) one (b) $\dfrac{dP_{N_2O}}{dt} = \dfrac{d(-\Delta P)}{dt} = k(P_{NO})^2\, P_{H_2}$.

After determining the form of the general rate expression and the order of the reaction with respect to each of the substituents, the rate constant can be evaluated from the measured rates at known initial concentrations. In the previous problem involving the reaction between NO and H_2, we obtained

$$\frac{d(-\Delta P)}{dt} = k[NO]^2[H_2].$$

When $P_{NO}^{o} = 359$ torr and $P_{H_2}^{o} = 400$ torr, the rate is 150 torr/100 sec or 1.50 torr/sec. Substituting this information into our general rate expression and solving for k,

$$k = (1.50\ \text{torr/sec})/(359\ \text{torr})^2(400\ \text{torr}) = 2.91 \times 10^{-8}\ \text{torr}^{-2}\text{sec}^{-1}$$

Problem 5-F-2. Calculate the rate constant for the previous reaction from the data P_{NO}^{o} = 400 torr and the various values for $P_{H_2}^{o}$ and rates of reaction. Find the average of the three rate constants. Answer: 3.45×10^{-8} torr^{-2}/sec, 3.35×10^{-8} torr^{-2}/sec, 3.36×10^{-8} torr^{-2}/sec, average = 3.386×10^{-8} torr^{-2}/sec.

G. Temperature Dependence of Reaction Rates

With very few exceptions, the rate of a reaction increases with temperature. We express this increase in rate by the increase in the overall rate constant for the reaction. Generally, the reaction is run at some constant temperature and the rate constant determined. Then the temperature is changed, the reaction is rerun, and the rate constant determined once again. This procedure may be repeated at several temperatures to obtain several sets of T and k.

An empirical relationship between k and T which is generally satisfied by most reactions over some limited temperature region is the **Arrhenius Equation**. It is given by

$$k = Ae^{-E^*/RT} \qquad (5\text{-}G\text{-}1)$$

where A = a constant which is referred to as the pre-exponential constant, E^* = activation energy, and R = gas constant. Equation (5-G-1) can be put into a linear relationship by taking the logarithm:

$$\ln k = \ln A - \frac{E^*}{R}\left(\frac{1}{T}\right) \qquad (5\text{-}G\text{-}2)$$

This equation is linear, if ln k is plotted versus 1/T. The negative slope is equal to E^*/R, and using this relationship, the activation energy can be evaluated.

$$E^* = R\,(-\text{ slope}) \qquad (5\text{-}G\text{-}3)$$

The intercept of such a plot is ln A, so that A can be readily evaluated.

Example 5-G-1. To illustrate an example of a calculation of the Arrhenius parameters for a reaction, E^* and A, let us consider the homogeneous thermal decomposition of N_2O into its elements.

$$2N_2O \longrightarrow 2N_2 + O_2$$

This reaction was investigated by C. N. Hinshelwood and R. E. Burk, [*Proc. Roy. Soc.*, A106: 284 (1924)] and was found to be second order. The rate constant was determined as a function of temperature, with the following data obtained:

T(K)	k(liters/mole-sec)	$1/T \times 10^4$	$\ln(k \times 10^3)$
1125	11.59	8.89	9.36
1085	3.76	9.22	8.23
1053	1.67	9.50	7.42
1030	0.871	9.71	6.77
1001	0.380	9.99	5.94
967	0.136	10.34	4.91
838	0.0011	11.93	0.095

A graph of ln k versus 1/T is required to evaluate E^* and A, and these logarithms and reciprocals are calculated to give the values shown above. Figure 5-G-1 illustrates the graph.

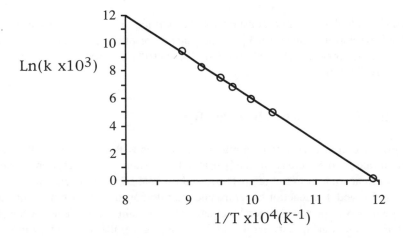

Figure 5-G-1. Calculation of the Arrhenius Parameters for the Reaction: E^* and A

The slope is evaluated from the graph and found to be -2.99×10^4. Since

$$\ln k = \ln A - \frac{E^*}{RT}$$

$$-\text{slope} = \frac{E^*}{R}$$

$$E^* = -(-2.99 \times 10^4 \text{ K})(8.3142 \text{ J/K-mol})$$

$$E^* = 2.49 \times 10^5 \text{ J/mol} = 249 \text{ kJ/mol}$$

It is convenient to drop the unit of $(\text{mol})^{-1}$ since it is assumed that the activation energy corresponds to the reaction of molar quantities of the reactants. For example, for the reaction

$$2A + B \longrightarrow P$$

the activation energy corresponds to 2 moles of A reacting with 1 mole of B.

The pre-exponential constant, A, is also evaluated from the equation:

$$\ln k = \ln A - \frac{E^*}{RT}.$$

The $\ln k$ and $1/T$ are read off the graph from a single point on the straight line, for example, $1/T \times 10^4 = 12.0 \text{ K}^{-1}$, or $1/T = 12.0 \times 10^{-4} \text{ K}^{-1}$; and $\ln(k \times 10^{+3}) = 0.00$, or $\ln k = -6.908$.

$$-6.908 = \ln A - \frac{2.49 \times 10^5 \text{ J/mol} (12.0 \times 10^{-4} \text{K}^{-1})}{(8.3142 \text{ J/K-mol})}$$

$$-6.908 = \ln A - 35.9$$

$$\ln A = 29.0$$

$$A = 4.0 \times 10^{12} \text{ liters/mol-sec}$$

It should be noted, in equation (5-G-1), that the units of k determine the units of A, and likewise, in equation (5-G-2), where they determine the units of $\ln A$ or the intercept. However, the units of k have no effect on the slope of a $\ln k$ versus $1/T$ graph.

Problem 5-G-1. The decomposition of OF_2 in the presence of He, presumably according to

$$OF_2 + He \longrightarrow OF + F + He$$

has been studied in the temperature range 773 K to 973 K in 50 degree intervals [L. Dauerman et al., *J. Phys. Chem.*, **1967**, *71*, 3999]. The following bimolecular rate constants were measured at various temperatures.

TK	k(liters/mol-sec)	TK	k(liter/mol-sec)	TK	k(liters/mol-sec)
773	120	873	1309	923	3451
773	126	873	1228	973	9291
823	465	873	1495	973	9394
823	472	923	4500	973	9213
823	574	923	4387		

Calculate the activation energy (E^*) and the pre-exponential constant (A) for this reaction. Answer: $E^* = 136.0$ kJ/mol; $A = 2 \times 10^{11}$ L/mol-sec.

If the overall rate constant for a reaction has been determined at only two temperatures, the graphical solution is unnecessary. We can let the subscripts 1 and 2 represent the values at the two different temperatures, then substitute them into equation 5-G-2) to give

$$\ln k_1 = \ln A - \frac{E^*}{RT_1}$$

$$\ln k_2 = \ln A - \frac{E^*}{RT_2}$$

Subtracting these to eliminate A,

$$\ln k_1 - \ln k_2 = -\frac{E^*}{RT_1} + \frac{E^*}{RT_2}$$

Solving for E^*,

$$(\ln k_1 - \ln k_2) = \left(\frac{1}{T_2} - \frac{1}{T_1}\right)\frac{E^*}{R}$$

$$E^* = \frac{R(\ln k_1 - \ln k_2)}{\left(\frac{1}{T_2} - \frac{1}{T_1}\right)} = \frac{R \ln \frac{k_1}{k_2}}{\left(\frac{1}{T_2} - \frac{1}{T_1}\right)} \tag{5-G-4}$$

Knowing the value of E^*, the value of A can be evaluated using equation (5-G-5) and the value of k and T at either of the two temperatures.

$$A = k_1 e^{[+E^*/RT_1]} \quad \text{or} \quad A = k_2 e^{[+E^*/RT_2]} \tag{5-G-5}$$

Problem 5-G-2.
(a) Referring to the data in Example 5-G-1, for T = 1125K and T = 838K, calculate E^* and A, using equation (5-G-4) and (5-G-5).
(b) Now, referring to the same data for points T = 1125K and T = 1085K, calculate E^* and A.
(c) Which of the results, (a) or (b), is more precise, and why?
Answer: (a) $E^* = 253.6$ kJ/mol, $A = 6 \times 10^{12}$ L/mol-sec (b) $E^* = 285.8$ kJ/mol, $A = 2 \times 10^{14}$ L/mol-sec (c) Answer to (a) should be more precise since the data is over a larger temperature range.

Problem 5-G-3. For reactions involving biological and some organic compounds, a rule of thumb estimate is that the reaction rate constant doubles for a temperature increase of 10°C. What activation energy does this correspond to? (Hint: If $k_1 = 2k_2$, then $T_1 = T_2 + 10$. Assume $T_2 = 300$ K and calculate E^* according to equation (5-G-4)). Answer: $E^* = 53.6$ kJ/mol.

H. Activation Energy for Complex Reactions

In Section E of this chapter, we showed the relationship between the overall rate constant and the rate constant for the elementary steps of a specific mechanism. Using this relationship, we can also obtain the relationship between the overall activation energy and the activation energies for the elementary reaction steps.

The Arrhenius Equation was defined in equation (5-G-1), and its logarithmic form in equation (5-G-2). We can obtain the differential form of equation (5-G-2), by taking the derivative with respect to temperature, assuming $E^* = $ constant.

$$\frac{d\ln k}{dT} = \frac{E^*}{RT^2} \tag{5-H-1}$$

Frequently, the reaction rate constant can be conveniently related to the equilibrium constant for some intermediate reaction step. For example, in equation (5-C-17),

$$\frac{d[O_2]}{dt} = \frac{3K_1k_2[O_3]^2}{[O_2]}$$

the overall rate constant is given by

$$k = 3\frac{k_1k_2}{k_{-1}} = 3K_1k_2 \tag{5-H-2}$$

where the equilibrium constant, K_1, is equal to k_1/k_{-1}. Since k_1 and k_{-1} are expressed in concentration units, so is K_1.

From information in a previous chapter, we can readily obtain a general relationship between K_1 and T. Equation (3-F-4) gives the temperature dependence of K_p:

$$\frac{d\ln K_p}{dT} = \frac{\Delta H^o}{RT^2}$$

It can be shown that the temperature dependence of K is

$$\frac{d\ln K}{dT} = \frac{\Delta U^o}{RT^2} \tag{5-H-3}$$

where K is expressed in terms of molar concentrations, the standard state now refers to unit molarity, and ΔU^o is the change in standard internal energy.

For a gas phase reaction, assuming ideal gases, K_p is related to K, the equilibrium constant expressed in units of moles/liter, by

$$K_P = \frac{(P_C)^c (P_D)^d}{(P_A)^a (P_B)^b}$$

$$K_P = \frac{\left(\dfrac{n_C RT}{V}\right)^c \left(\dfrac{n_D RT}{V}\right)^d}{\left(\dfrac{n_A RT}{V}\right)^a \left(\dfrac{n_B RT}{V}\right)^b}$$

$$K_P = \frac{\left(\dfrac{n_C}{V}\right)^c \left(\dfrac{n_D}{V}\right)^d}{\left(\dfrac{n_A}{V}\right)^a \left(\dfrac{n_B}{V}\right)^b} (RT)^{(c+d-a-b)} \qquad (5\text{-}H\text{-}4)$$

Since $[C] = n_C/V$, $[D] = n_D/V$, etc.

$$K_p = \frac{[C]^c [D]^d}{[A]^a [B]^b} (RT)^{\Delta n} = K(RT)^{\Delta n} \qquad (5\text{-}H\text{-}5)$$

where $\Delta n = c + d - a - b$ = change in number of moles for the reaction. Recalling Section D of Chapter 3, the partial pressure term in the K_p expression, e.g., $(P_C)^c$, is actually a ratio of the equilibrium partial pressure to the pressure of the standard state ($P_C = 1$ atm). For this reason, K_p is a unitless quantity.

The equilibrium constant (K), expressed in concentration units, is also unitless, since each molar concentration is referred to a standard state of unit molarity. The relationship $\Delta G^o = -RT \ln K$ can be derived in a manner analogous to that in Section 3-D, however, in this case, the standard state for ΔG^o is unit molarity and the specified temperature. The RT term in (5-H-5) is unitless, since it should be multiplied by moles/liter-atm.

If we take the logarithm of equation (5-H-5), then

$$\ln K_p = \ln K + \Delta n \ln RT \qquad (5\text{-}H\text{-}6)$$

Differentiating (5-H-6),

$$\frac{d \ln K_p}{dT} = \frac{d\ln K}{dT} + \frac{\Delta n R}{RT} = \frac{d\ln K}{dT} + \frac{\Delta n RT}{RT^2}$$

Rearranging,

$$\frac{d \ln K}{dT} = \frac{d\ln K_p}{dT} - \frac{\Delta n RT}{RT^2} \qquad (5\text{-}H\text{-}7)$$

Substituting in (3-F-4),

$$\frac{d \ln K}{dT} = +\frac{\Delta H^o}{RT^2} - \frac{\Delta n RT}{RT^2} = \frac{\Delta H^o - \Delta n RT}{RT^2} = \frac{\Delta U^o}{RT^2} \qquad (5\text{-}H\text{-}3)$$

where ΔU^o = change in internal energy for the reaction step. The standard state for ΔU^o is now unit molarity.

We can now find the relationship between the overall activation energy and that of the reaction steps, using (5-H-3) and (5-H-1). For example, let us consider the ozone decomposition reaction, where the rate expression is given by (5-C-17), and k by (5-H-2). Taking the logarithm, we get

$$\ln k = \ln 3 + \ln K_1 + \ln k_2$$

Differentiating with respect to temperature,

$$\frac{d\ln k}{dT} = 0 + \frac{d\ln K_1}{dT} + \frac{d\ln k_2}{dT}$$

Substituting the appropriate terms from (5-H-1) and (5-H-3) into this equation gives

$$E^* = 0 + \_U\ EQ\ \backslash S(o,1)\ + E\ EQ\ \backslash S(*,2)$$

$$E^* = \_U\ EQ\ \backslash S(o,1)\ + E\ EQ\ \backslash S(*,2)$$

In an analogous manner, we can obtain the expression for the activation energy for equation (5-D-8):

$$EQ\ \backslash F(d[HBr],dt)\ = 2k2\ EQ\ \backslash B\backslash BC\backslash((\backslash F(k1,k\text{-}1))\ 1/2\ [H2][Br2]1/2$$

$$k = 2k2(k1/k\text{-}1)1/2 = 2k2K\ EMBED\ Equation.DSMT36$$

$$\ln k = \ln 2 + \ln k_2 + \frac{1}{2}\ \ln K_1$$

$$\frac{d\ln k}{dT} = 0 + \frac{d\ln k_2}{dT}\ + \frac{1}{2}\frac{d\ln K_1}{dT}$$

$$E^* = E_2^*\ + \frac{1}{2}\ \Delta U_1^o$$

According to this mechanism E^* depends only on $1/2\Delta U_1^o\ = 1/2D_{Br-Br}$ and E_2^*. The other E_i^* do not affect the activation energy for the reaction.

Problem 5-H-1. Obtain the expression for the activation energy for the rate constant in (5-D-4):

$$\frac{d[HI]}{dt}\ = 2k_3K_1K_2[H_2][I_2]$$

Answer: $E^* = E_3^*\ + \Delta U_1^o\ + \Delta U_2^o$.

Problem 5-H-2. Obtain the expression for the activation energy for the rate constant in (5-E-3)

$$\frac{d[HBr]}{dt}\ = 2k_2\left(\frac{k_1}{k_{-1}}\right)^{\frac{1}{2}}\frac{k_3}{k_{-2}}\ \frac{[H_2][Br_2]^{3/2}}{[HBr]}$$

Answer: $E^* = E_3^*\ + \Delta U_2^o\ + \frac{1}{2}\ \Delta U_1^o$

I. Relationship Between Chemical Kinetics and Thermodynamics

In Section C of this chapter, equations (5-C-8) through (5-C-10), we illustrated how the equilibrium constant can be related to the forward and reverse rate constants for an elementary reaction step. In an analogous manner, we can treat the overall forward and reverse expressions and obtain a relationship with the overall equilibrium constant. For a complex reaction, this derivation is not obvious.

General rate expressions for the forward rate and the reverse rate can be set up similar to equation (5-E-2),

$$\frac{d[P]}{dt} = k[A]^{\alpha}[B]^{\beta}[C]^{\gamma} \cdots \tag{5-E-2}$$

If we let f represent the forward direction and r the reverse direction, the general rate expressions can then be expressed as:

$$\frac{dP}{dt} = k_f \text{ (concentration terms)}$$

$$\frac{dR}{dt} = k_r \text{ (concentration terms)}$$

where P represents products and R represents reactants. At equilibrium, these two rate expressions must be equal, since there is no resulting net change in concentration for reactants and products at equilibrium.

$$\frac{dP}{dt} = \frac{dR}{dt}$$

$$k_f \text{ (concentration terms)} = k_r \text{ (concentration terms)}$$

$$K = \frac{k_f}{k_r} \tag{5-I-1}$$

The K in equation (5-I-1) is numerically equal to the equilibrium constant (K) in equation (5-H-5), providing the same concentration units are used.

Taking the logarithm of (5-I-1),

$$\ln K = \ln k_f - \ln k_r$$

Differentiating with respect to temperature,

$$\frac{d \ln K}{dT} = \frac{d \ln k_f}{dT} - \frac{d \ln k_r}{dT}$$

From equations (5-H-1) and (5-H-3),

$$\frac{\Delta U^o}{RT^2} = \frac{E_f^*}{RT^2} - \frac{E_r^*}{RT^2} \tag{5-I-2}$$

$$\Delta U^o = E_f^* - E_r^*$$

Equation (5-I-1) and (5-I-2) represent the relationship between chemical kinetics and thermodynamics. This is shown more descriptively in Figure 5-I-1, which represents the change in internal energy as a function of some generalized reaction coordinate. The reaction coordinate is a coordinate representing the extent to which the reacting molecule or molecules have reacted in the direction of the product molecule(s). In Figure 5-I-1, the reaction is exothermic, and the internal energy of the products is lower than the energy of the reactants. In this case, the forward activation energy is less than the reverse activation energy $(E^*_f < E^*_r)$. For an exothermic reaction, $E^*_r \geq \Delta U^o$ and $E^*_f \geq 0$.

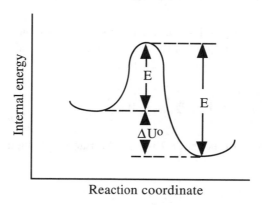

Figure 5-I-1. Change in Internal Energy During Reaction

Problem 5-I-1. Draw an energy diagram similar to Figure 4-I-1 for an endothermic reaction. What is the relationship between E^*_f and ΔU^o and the magnitude of E^*_r? Answer: $E^*_f \geq \Delta U^o$; $E^*_r \geq 0$.

It should be emphasized that equations (5-I-1) and (5-I-2) are *independent of the reaction mechanism*.

To illustrate the applications of (5-I-1) and (5-I-2), we can consider the decomposition of O_3 into O_2. The steady state assumption given in (5-C-17) was derived, but now an expression for the reverse rate must be obtained in order to apply (5-I-1) and (5-I-2). The reverse rate can be represented by a series of reaction steps similar to those in (5-C-2) to (5-C-4). The nomenclature of the rate constants in the latter equations will be retained. In addition, the back reaction of (5-C-4) must now be considered.

$$2\,O_2 \underset{k_2}{\overset{k_{-2}}{\rightleftharpoons}} O + O_3$$

$$O + O_2 \xrightarrow{k_{-1}} O_3$$

The rate of production of O_3 is then given by

$$\frac{d[O_3]}{dt} = k_{-2}\,[O_2]^2 - k_2[O][O_3] + k_{-1}[O][O_2] \qquad (5\text{-}I\text{-}3)$$

Applying the steady state treatment to O as before,

$$\frac{d[O]}{dt} = 0 = k_{-2}[O_2]^2 - k_2[O][O_3] - k_{-1}[O][O_2]$$

Solving for [O],

$$[O] = \frac{k_{-2}[O_2]^2}{k_2[O_3] + k_{-1}[O_2]}$$

Substituting into (5-I-3) and simplifying gives

$$\frac{d[O_3]}{dt} = \frac{2k_{-1}k_{-2}[O_2]^3}{k_2[O_3] + k_{-1}[O_2]} \tag{5-I-4}$$

In this case, $k_{-1}[O_2] \gg k_2[O_3]$ since k_{-1} pertains to the simple bond formation which produces O_3. This should require little activation energy. In addition, the initial $[O_2]$ concentration would be relatively large. Equation (5-I-4) then reduces to

$$\frac{d[O_3]}{dt} = 2k_{-2}[O_2]^2 = k_r[O_2]^2 \tag{5-I-5}$$

From the stoichiometric equation (5-C-1), the rate of production of O_3 and O_2 are equated at equilibrium by

$$-3\frac{d[O_3]}{dt} = 2\frac{d[O_2]}{dt} \tag{5-I-6}$$

Substituting in (5-I-5) and (5-C-17) gives

$$3(2)k_{-2}[O_2]^2 = 2(3)K_1 k_2 \frac{[O_3]^2}{[O_2]} \tag{5-I-7}$$

$$\frac{[O_2]^3}{[O_3]^2} = K_1 \frac{k_2}{k_{-2}} = K_1 K_2 = K$$

where $K_2 = k_2/k_{-2}$, and K is the overall equilibrium constant for the stoichiometric equation (5-C-1). From (5-I-5) and (5-C-17),

$$k_r = 2k_{-2},$$

$$k_f = 3K_1 k_2.$$

Using the coefficients in (5-I-6) derived from the stoichiometric equation and (5-I-1),

$$K = \frac{2k_f}{3k_r} = \frac{6K_1 k_2}{6k_{-2}} = \frac{K_1 k_2}{k_{-2}} = K_1 K_2 \tag{5-I-8}$$

Obviously, (5-I-8) is equivalent to (5-I-7).

As shown in the previous section, the activation energy for the forward rate equation was related to the energies of the individual reaction steps. Likewise,

$$E_f^* = \Delta U_1^o + E_2^* \tag{5-I-9}$$

From (5-I-5),

$$E_r^* = E_{-2}^* \tag{5-I-10}$$

Substituting (5-I-9) and (5-I-10) into (5-I-2),

$$\Delta U^o = \Delta U_1^o + E_2^* - E_{-2}^*$$

Since

$$\Delta U_2^o = E_2^* - E_{-2}^*$$

$$\Delta U^o = \Delta U_1^o + \Delta U_2^o \tag{5-I-11}$$

Obviously (5-I-11) is in agreement with the principles of thermodynamics. E is a state function, and ΔE^o for the overall reaction must equal the sum of the ΔU^os for the two elementary steps.

Problem 5-I-2. A reaction was investigated in both the forward and reverse directions. The rate constant for the forward reaction was found to be 5.1×10^{-8} liters/mol-sec, with an activation energy of 161.1 kJ/mol. The rate constant for the reverse reaction was found to be 4.0×10^{-5} sec^{-1}, with an activation energy of 66.1 kJ/mol. Calculate the equilibrium constant for the reaction and the change in internal energy using equations (5-I-1) and (5-I-2). Answer: 1.275×10^{-3} L/mol; $\Delta E = 95.0$ kJ/mol.

J. Integrated Rate Expressions

In Section D of this chapter, we discussed the general rate expression and order of a reaction, with respect to certain components involved in the reaction. A technique involving initial rate data was used to determine both the order of the reaction and the overall rate constant.

In this section, we will consider the integration of the rate expression for some assumed order for the reaction. The resulting integrated equation will give a relationship between the concentration of a species involved in the reaction, either reactant or product, as a function of time. As you recall, this was our general objective in kinetics—to answer at what rate the concentration of a reactant will decrease, or that of a product will increase.

In addition to the explicit relationship between concentration and time, the integrated equation permits the investigator to establish the order of the reaction and the rate constant with greater certainty. With the initial rate method, only concentration values over a limited time interval can be used to obtain the initial slope or rate. The error in the slope can be considerable, if the concentration values are not known with great precision. This, of course, projects greater uncertainty in the determination of the orders of the reaction, and greater error in the rate constant. Using integrated rate expressions, concentration data over large time intervals can generally be employed.

First-Order Reaction

A reaction which is first order with respect to one reactant, A, has the rate expression

$$-\frac{d[A]}{dt} = k[A] \tag{5-J-1}$$

In order to integrate (5-J-1), it must first be written in differential form.

$$-d[A] = k[A]\,dt \tag{5-J-2}$$

After separating the variables [A] and t, as discussed in Appendix C, the integration is accomplished. Dividing by [A] and integrating,

$$\int_{[A]^O}^{[A]} -\frac{d[A]}{[A]} = \int_0^t k\,dt \qquad (5\text{-}J\text{-}3)$$

The lower limits of integration are the initial values t = 0, [A] = [A]^O. The upper limits are the concentration of A, [A], at time, t. Using the integration formulas in Appendix C.

$$-\ln[A] + \ln[A]^O = kt - k{\cdot}0 \qquad (5\text{-}J\text{-}4)$$

Rearranging (5-J-4) gives

$$\mathbf{\ln[A] = - kt + \ln[A]^O} \qquad (5\text{-}J\text{-}5)$$

Equation (5-J-5) is written in the form of a linear equation, where the variables are ln[A] and t, the slope is -k, and the intercept is ln[A]^O.

Example 5-J-1. As an example of the application of equation (5-J-5) in testing a first-order rate and in determining the first-order rate constant, we can consider the following rate data for the decomposition of trinitrobenzoic acid in anisole as a solvent, at t = 80.1^{O}C. The reaction is

$$C_6H_2(NO_2)_3COOH \longrightarrow C_6H_3(NO_2)_3 + CO_2$$

The reaction was followed by titration of the trinitrobenzoic acid in samples of the reaction mixture taken at periodic intervals. The data is shown in the following table.

time x10^{-3}(min)	V (mL)	log [A] + 3
0	18.70	1.273
1.590	14.03	1.147
3.055	10.73	1.031
4.220	9.05	.957
5.665	7.05	.848
7.115	5.50	.740
8.565	4.76	.678
∞	1.80	.255

The data in column two give mL of 0.005M Ba(OH)$_2$ reacted with reaction samples of 10mL. The concentration of trinitrobenzoic acid, A, is then calculated by

$$[A] = \frac{(0.005 \text{ moles Ba(OH)}_2/L)(2\text{mol A}/1\text{mol Ba(OH)}_2)(V \text{ mL})}{10 \text{ mL}} = 10^{-3} \text{ V mol A/L}$$

In column three, (log[A] + 3) is represented by log V. Since only relative values of [A] are of significance in a first-order reaction, we can plot (log[A] +3) versus t, as shown in Figure 4-J-1. Since ln[A] = 2.303 log[A],

$$k = - 2.303 \text{ (slope)} = 2.303 \,(.73 \times 10^{-4} \text{ min}^{-1}) = 1.66 \times 10^{-4} \text{ min}^{-1}$$

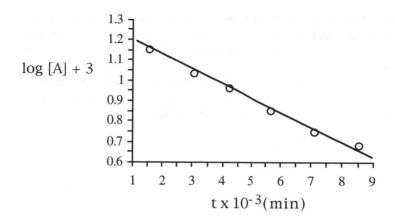

Figure 5-J-1. Sample Test for a First-order Rate of Reaction

Problem 5-J-1. An isotope of phosphorus is radioactive, giving off a positron (positive electron).

$$\begin{smallmatrix}30\\15\end{smallmatrix} P \longrightarrow \begin{smallmatrix}30\\14\end{smallmatrix} Si + e^+$$

The following measurements of activity as a function of time were recorded.

time (sec)	0	10	30	50	80	120	160
activity x 10^{-4} counts/sec	4.5	3.8	3.5	3.1	2.7	2.3	1.9

Since activity is proportional to concentration, plot the logarithm of the activity versus time, and determine the first order rate constant. Answer: $k = 4.7 \times 10^{-3}$ sec^{-1}.

Equation (5-J-5) can be arranged to give a linear graph with a positive slope and an intercept of zero. Combining the ln[A] and ln[A]^0 terms in (5-J-5)

$$\ln \frac{[A]^0}{[A]} = kt \qquad (5\text{-}J\text{-}6)$$

The graph of ln[A]^0/[A] versus t, yields a positive slope of k and a zero intercept.

Problem 5-J-2. Plot the data in Problem 5-J-1, in accordance with equation (5-J-6), as ln[A]^0/[A] versus t. Evaluate the rate constant. Note the apparent nonlinearity of the plot in this problem compared to Problem 5-J-1, if a zero intercept is assumed. Answer: $k = 7.4 \times 10^{-3}$ sec^{-1}.

The graph in Problem 5-J-2 points out a disadvantage in using equation (5-J-6) to test a first order rate, which occrs when the initial data point (that is, t = 0, [A] = [A]^0) is poor. This can frequently arise, not because [A]^0 is not sufficiently known but, rather, because the initial time is difficult to establish. The reaction should be started instantaneously, but experimentally this is impossible. Mixing of solutions, dissolution of solids, and so forth take a finite period of time, and the reaction will actually begin during this mixing and dissolution period. Unfortunately, a graph of ln[A]^0/[A] involves this initial [A]^0 value in all data points. Therefore, all data points are displaced on the y-axis as a result of this error. Problem 5-J-2 should convince the student of this effect.

Frequently, the rate of a first-order reaction is expressed in terms of half-life ($\tau_{1/2}$), or mean lifetime (τ). The advantage of using these terms is that an order of magnitude, or "feeling" for the rate at which the reaction proceeds, is more readily interpreted from a half-life or mean lifetime value. The units for $\tau_{1/2}$, or τ are time—something we all have a direct concept of. On the other hand, the first order rate constant has units of reciprocal time, and these are not immediately meaningful with respect to the actual rate at which the concentration decreases with time.

The relationships between $\tau_{1/2}$ or τ and k can be derived by specifying the limits of integration in (5-J-3) in terms of the definition of $\tau_{1/2}$ and τ. $\tau_{1/2}$ is the time when the concentration of A has diminished to 1/2 its original value, that is, when

$$\begin{cases} t = 0 \\ [A] = [A]^o \end{cases}$$

and when

$$\begin{cases} t = \tau_{1/2} \\ [A] = \dfrac{[A]^o}{2} \end{cases}$$

Using these limits in (5-J-3),

$$\int_{[A]^o}^{[A]^o/2} -\frac{d[A]}{[A]} = \int_0^{\tau_{1/2}} kdt \qquad (5\text{-}J\text{-}7)$$

If we integrate (5-J-7), and rearrange the equation as in (5-J-6),

$$\ln\frac{[A]^o}{[A]^o/2} = k\tau_{1/2}$$

Simplifying,

$$\ln 2 = k\tau_{1/2}$$

or

$$\tau_{1/2} = \frac{\ln 2}{k} = \frac{0.693}{k} \qquad (5\text{-}J\text{-}8)$$

Problem 5-J-3. From the rate constant obtained in Problem 5-J-1, calculate $\tau_{1/2}$, the half-life for the reaction. Answer: $\tau_{1/2} = 1.5 \times 10^2$ sec.

The relationship between τ and k can be derived in a similar manner. The definition of τ is the time for the concentration of the reactant to diminish to 1/e of its value (e is the base of our natural logarithms). Equation (5-J-3) becomes

$$\int_{[A]^o}^{[A]^o/e} -\frac{d[A]}{[A]} = \int_0^{\tau} kdt$$

$$\ln\frac{[A]^o}{[A]^o/e} = \ln e = 1 = k\tau$$

$$\tau = \frac{1}{k} \qquad (5\text{-}J\text{-}9)$$

Problem 5-J-4. Calculate the mean lifetime for the first-order reaction in Problem 5-J-1. Answer: $\tau = 2.1 \times 10^2$ sec.

If the first order reaction is followed experimentally by measuring the concentration of product with time, we must convert equation (5-J-5), ln [A] = -kt + ln[A]^O, into an equation involving concentration of product. If the stoichiometric equation is

$$A = P$$

a molecule of P is produced for every molecule of A which reacts. Therefore, the concentration of A plus P should equal the initial concentration of A; that is,

$$[A]^O = [A] + [P] \qquad (5\text{-}J\text{-}10)$$

Solving (5-J-10) for [A] and substituting into (5-J-5),

$$\ln([A]^O - [P]) = - kt + \ln[A]^O \qquad (5\text{-}J\text{-}11)$$

A graph of ln([A]^O - [P]) versus t should be linear, with slope equal to k.

If the stoichiometric equation was of a different form, a new equation must replace (5-J-10). For example,

$$A \longrightarrow 2 P_1 + P_2$$

where P$_1$ and P$_2$ are products. Then (5-J-10) would take the form

$$[A]^O = [A] + \frac{1}{2} [P_1] \qquad (5\text{-}J\text{-}12)$$

if [P$_1$] were followed as a function of time. Solving (5-J-12) for [A] and substituting into (5-J-5)

$$\ln\left([A]^O - \frac{1}{2} [P_1]\right) = - kt + \ln[A]^O$$

We can now plot $\ln\left([A]^O - \frac{1}{2} [P_1]\right)$ versus t.

Problem 5-J-5. The stoichiometric equation for the thermal decomposition of N$_2$O$_5$ in CCl$_4$ is

$$2N_2O_5 \longrightarrow 2N_2O_4 + O_2$$
$$\downarrow \uparrow$$
$$4NO_2$$

The reaction at 40°C can be followed by the evolution of O$_2$. The following table gives the values as a function of time.

t (sec)	V_{O_2} (mL)
0	0
300	3.42
600	6.30
900	8.95
1200	11.40
1500	13.55
1800	15.52
∞	34.75

Calculate the first-order rate constant for the reaction. (Hint: $[N_2O_5]^o = [N_2O_5] + 2[O_2]$. At infinite time, $[N_2O_5] = 0$ and $_{\infty}[N_2O_5]^o = 2[O_2]^{\infty}$; hence, $[N_2O_5] = 2[O_2]^{\infty} - 2[O_2]$. Since $V_{O_2} \propto [O_2]$, $[N_2O_5] = c(V_{O_2} - V_{O_2})$, and $\ln[N_2O_5] = \ln c + \ln(V_{O_2} - V_{O_2})$. Substitute into (4-J-5)). Answer: $k = 3.3 \times 10^{-4} \ sec^{-1}$.

Second-Order Reaction

A second-order reaction can be first order with respect to two reactants, for instance, A and B, or second order, with respect to a single species, such as A. The latter is simpler from a mathematical viewpoint, and will be considered first. The rate equation is

$$-\frac{d[A]}{dt} = k[A]^2 \tag{5-J-13}$$

Proceeding as before,

$$\int_{[A]^o}^{[A]} \frac{d[A]}{[A]^2} = -\int_0^t kdt$$

$$-\frac{1}{[A]} + \frac{1}{[A]^o} = -kt$$

$$\frac{1}{[A]} = kt + \frac{1}{[A]^o} \tag{5-J-14}$$

According to (5-J-14), a graph of $1/[A]$ versus t will give a linear plot with

$$slope = k,$$

$$intercept = \frac{1}{[A]^o}$$

If a product concentration is followed as a function of time, the relationship between the reactant concentration, [A], and product concentration must be found as before, when we considered first-order reactions. Then, solving for [A], and substituting into (5-J-14), yields an expression for product concentration as a function of time.

The second order reaction involving two reactants, A and B, can also arise from a reaction with the general stoichiometric equation

$$aA + bB \longrightarrow cC + dD.$$

The rate expression for the second order reaction is given by

$$\frac{1}{c}\frac{d[C]}{dt} = -\frac{1}{a}\frac{d[A]}{dt} = k[A][B] \tag{5-J-15}$$

In order to carry out this integration it is imperative that

$$\frac{[A]^o}{a} \neq \frac{[B]^o}{b} \tag{5-J-16}$$

It will be shown later, if $[A]^o/a = [B]^o/b$, then the rate equation given by equation (5-J-15) can be reduced to that given by equation (5-J-13). The solution to the differential equation given by equation (5-J-15) becomes more difficult since we must first change [A] and [B] into a single concentration variable. This is accomplished most readily by expressing the rate expression in terms of [C]. The mass balance expressions for A and B are given by

$$[A]^o = [A] + \frac{a}{c} [C]$$

$$[B]^o = [B] + \frac{b}{c} [C]$$

where $\frac{a}{c}$ [C] and $\frac{b}{c}$ [C] represent the amount of [A] [B] reacted. Solving these expressions for [A] and [B] and substituting into equation (5-J-15) gives

$$\frac{1}{c} \frac{d[C]}{dt} = k([A]^o - \frac{a}{c} [C]) ([B]^o - \frac{b}{c} [C])$$

Using the differential expression for d[C] we can obtain the differential equation and separate the variables, as before

$$\frac{\frac{1}{c} d[C]}{([A]^o - \frac{a}{c}[C]) ([B]^o - \frac{b}{c}[C])} = kdt \qquad (5\text{-J-18})$$

In order to integrate the left hand side of equation (5-J-18) we must break up the fraction into partial fractions of the form

$$\frac{1}{([A]^o - \frac{a}{c}[C]) ([B]^o - \frac{b}{c}[C])} = \frac{C_1}{[A]^o - \frac{a}{c}[C]} + \frac{C_2}{[B]^o - \frac{b}{c}[C]} \qquad (5\text{-J-19})$$

We solve for C_1 and C_2 by multiplying both sides by the denominator on the left hand side of equation (5-J-19), collecting the constant term and those involving the variable [C]. Since there is only the constant 1 on the left side, the constant term must equal 1 and the variable term must be equal to zero; i.e.,

$$C_1[B]^o + C_2[A]^o = 1$$

$$\frac{b}{c} C_1 + \frac{a}{c} C_2 = 0$$

Solving these equations simultaneously, we obtain

$$C_1 = \frac{a}{a[B]^o - b[A]^o}$$

$$C_2 = \frac{-b}{a[B]^o - b[A]^o}$$

C_1 and C_2 are substituted into equation (5-J-19) and our differential equation becomes

$$\frac{-1}{a[B]^o - b[A]^o} \int_0^{[C]} \frac{-\frac{a}{c} d[C]}{[A]^o - \frac{a}{c}[C]} + \frac{1}{a[B]^o - b[A]^o} \int_0^{[C]} \frac{-\frac{b}{c} d[C]}{[B]^o - \frac{b}{c}[C]} = \int_0^t kdt$$

The two integrals on the left are log functions since

$$d([A]^o - \frac{a}{c}[C]) = -\frac{a}{c} d[C]$$

$$d([B]^o - \frac{b}{c}[C]) = -\frac{b}{c} d[C]$$

Integrating and substituting in the upper and lower limits of the integration we obtain

$$\frac{-1}{a[B]^o - b[A]^o}\left(\ln([A]^o - \frac{a}{c}[C]) - \ln[A]^o\right) + \frac{1}{a[B]^o - b[A]^o}\left(\ln([B]^o - \frac{b}{c}[C]) - \ln[B]^o\right) = kt.$$

Combining the log functions we have

$$\ln \frac{[B]^o - \frac{b}{c}[C]}{[A]^o - \frac{a}{c}[C]} = (a[B]^o - b[A]^o)\ kt + \ln \frac{[B]^o}{[A]^o} \qquad (5\text{-}J\text{-}21)$$

Since

$$[B] = [B]^o - \frac{b}{c}[C] \qquad (5\text{-}J\text{-}22)$$

$$[A] = [A]^o - \frac{a}{c}[C] \qquad (5\text{-}J\text{-}23)$$

Equation (5-J-21) can be written in terms of [A] and [B]

$$\ln \frac{[B]}{[A]} = (a[B]^o - b[A]^o)\ kt + \ln \frac{[B]^o}{[A]^o} \qquad (5\text{-}J\text{-}24)$$

Equation (5-J-24) is in the form of a linear equation as indicated above. A graph of $y = \ln[B]/[A]$ versus t will give a slope, m,

$$m = (a[B]^o - b[A]^o)\ k$$

from which the rate constant, k, can be evaluated. In order to use equation (5-J-24) the concentrations of both [A] and [B] need to be known.

We can also express the integrated equation in terms of a single reactant variable, say [A], by eliminating [C] in equations (5-J-22) and (5-J-23)

$$\ln \frac{[B]^o + \frac{b}{a}[A] - \frac{b}{a}[A]^o}{[A]} = (a[B]^o - b[A]^o)\ kt + \ln \frac{[B]^o}{[A]^o}$$

As mentioned earlier, if

$$\frac{[A]^o}{a} = \frac{[B]^o}{b}$$

then equation (5-J-15) is reduced to the rate equation (5-J-13). Previously we derived the relationship between [A] and [B]

$$[B] = [B]^o + \frac{b}{a}[A] - \frac{b}{a}[A]^o$$

However, if

$$\frac{[A]^o}{a} = \frac{[B]^o}{b}$$

then

$$[B]^o = \frac{b}{a}\ [A]^o$$

and

$$[B] = \frac{b}{a}\ [A]^o + \frac{b}{a}\ [A] - \frac{b}{a}\ [A]^o = \frac{b}{a}\ [A] \qquad (5\text{-J-}25)$$

Substitution of (5-J-25) into equation (5-J-15) gives

$$-\frac{1}{a}\frac{d[A]}{dt} = k[A]\frac{b}{a}\ [A]$$

$$-\frac{d[A]}{dt} = kb[A]^2$$

which is equivalent to equation (5-J-13) since kb is a constant.

Problem 5-J-6. Consider the second order reaction

$$H_{2(g)} + I_{2(g)} \longrightarrow 2HI_{(g)}.$$

(a) If $[H_2]^o \neq [I_2]^o,$ write the integrated rate equation if [HI] is followed as a f(t).
(b) If $[H_2]^o \neq [I_2]^o,$ write the integrated rate equation if $[I_2]$ is followed as a f(t).
(c) If $[H_2]^o = [I_2]^o,$ write the integrated rate equation if $[H_2]$ is followed as a f(t).
(d) If $[H_2]^o = [I_2]^o,$ write the integrated rate equation if [HI] is followed as a f(t).

Answers: (a) $\ln \dfrac{[I_2]^o - \frac{1}{2}[HI]}{[H_2]^o - \frac{1}{2}[HI]} = \left([I_2]^o - [H_2]^o\right)\ kt + \ln \dfrac{[I_2]^o}{[H_2]^o}$

(b) $\ln \dfrac{[H_2]^o + [I_2] - [I_2]^o}{[I_2]} = \left([H_2]^o - [I_2]^o\right)\ kt + \ln \dfrac{[H_2]^o}{[I_2]^o}$

(c) $\dfrac{1}{[H_2]} = kt + \dfrac{1}{[H_2]^o}$ (d) $\dfrac{1}{\left([H_2]^o - \frac{1}{2}[HI]\right)} = kt + \dfrac{1}{[H_2]^o}$

nth-order Reaction

The first order and second order reactions, with respect to a single species, were readily integrated to give equations (5-J-5) and (5-J-14). In some cases, fractional orders such as 1/2, 3/2, and so forth can occur, as we observed in Sections C and D of this chapter. The integrated expression could be obtained for every specific order, however, it is more beneficial to obtain a general nth order integrated expression. Then, when a specific order is encountered (n ≠ 1), one needs simply to substitute for n in the nth-order integrated expression.

The rate expression for the nth-order reaction is

$$-\frac{d[A]}{dt} = k[A]^n \qquad (n \neq 1) \qquad (5\text{-J-}26)$$

Separating variables, as we have previously done,

$$\int_{[A]^o}^{[A]} \frac{d[A]}{[A]^n} = - \int_0^t kdt.$$

Referring to equation C-4 in the Appendix, and integrating, yields

$$\frac{[A]^{-n+1}}{-n+1} - \frac{([A]^o)^{-n+1}}{-n+1} = - kt \qquad (n \neq 1) \qquad (5\text{-}J\text{-}27)$$

$$[A]^{-n+1} = (n - 1) kt + ([A]^o)^{-n+1}$$

If a product is followed as a function of time, the relationship between the product concentration and reactant concentration can be obtained from the stoichiometric equation. Substituting in terms of product concentration for the reactant concentration in (5-J-27), yields the integrated expression in terms of product concentration. The procedure described earlier in this section should be followed.

Problem 5-J-7. If a reaction is suspected of being $\frac{3}{2}$ order with respect to A, what type function would you plot to test this hypothesis? Answer: $[A]^{-1/2}$ versus t should be linear.

Problem 5-J-8. If a stoichiometric equation was

$$A \longrightarrow B + C$$

and the reaction was $\frac{3}{2}$ order with respect to A, what type plot would you make if the product [C] were followed with time? Answer: $([A]^o - [C])^{-1/2}$ versus t should be linear.

Isolation Method

With equations (5-J-5) and (5-J-27), any order with respect to a single reactant can be tested over the entire concentration range covered in the experimental data. A general reaction involving more than one reactant can also be examined with respect to one of the components using these equations, if the isolation method is used. The general rate expression given by equation (5-E-2) can be reduced to

$$- \frac{d[A]}{dt} = k'[A]^\alpha \qquad (5\text{-}J\text{-}28)$$

if the concentrations of B, C, and D are held constant, where $k' = k[B]^\beta [C]^\gamma [D]^\delta$. Since we wish to integrate (5-J-28), this means [B] is also constant with time. Previously, we derived (4-F-1) by holding the initial concentration of B constant; now, we must hold [B] constant at all times. This is accomplished by means of the isolation method, in which we make the concentrations of B, C, and D far in excess of [A]. Then, during the course of the reaction, [A] will decrease, but [B], [C], and [D] will remain approximately constant. Thus, the order of the reaction is isolated as to the order with respect to A—hence, the name *isolation method*.

Equation (5-J-28) is identical to (5-J-1) and (5-J-26), with $\alpha = 1$ and $\alpha = n$, respectively. (The second-order reaction considered in (5-J-13) is a special case of (5-J-26) for the general nth order ($n \neq 1$) reaction.) Hence, equations (5-J-5) and (5-J-26) can be used for the integrated expression of (5-J-28). Thus, the order in a general reaction with respect to component A can be examined over the entire concentration region investigated. An analogous procedure for the determination of the order with respect to reactant B can be carried out. In this case, the concentration of A is made in excess. Extension to reactions involving three reactants is obvious, where the concentration of two species must be kept in excess at the same time, in order to establish the order with respect to the third component.

K. Enzyme Kinetics

Biochemical reactions are involved in energy transformations. A cell carries out several thousand chemical reactions along sequential pathways that have numerous points of inter-connection. These reactions must be ordered in space and time, and they must be so inte-grated as to produce a self regulating, viable system, capable of sustaining and reproducing itself. It is immediately evident that the basic reactions associated with living cells cannot occur at a significant rate under physiological conditions of an aqueous medium, neutral pH, and low temperature unless catalyzed, because reactions that occur "of their own" can-not be controlled. Thus, with respect to oxidation, glucose is kinetically stable, although it is thermodynamically labile. To carry out the reactions within cells, cells synthesize pro-teins, which are called enzymes, that catalyze specific reactions. Every significant reaction in the living cell is mediated by an enzyme, and most enzymes catalyze only one reaction. Others may mediate a number of reactions of the same type. The activity of key enzymes is controlled by activators and inhibitors. For example, the rate of anaerobic glycolysis, which produces ATP, is controlled in part by the ratio of ATP to AMP. AMP activates an enzyme, phosphofructokinase, which catalyzes one of the steps in glycolysis, but this en-zyme is inhibited by ATP. Thus, the existing level of phosphorylation controls the rate at which glycolysis proceeds, and therefore the rate at which AMP is converted to ATP.

In a slower acting control mechanism, the cell regulates the rate of synthesis of all the enzymes involved in a particular sequence of reactions. The cell can also regulate the ac-cess of a substrate to its enzyme, when they exit in different subcellular structures. Let us now turn our attention to enzymes. To be specific, we shall consider acetyl-cholinesterase, an enzyme that catalyzes the hydrolysis of acetylcholine

$$(CH_3)_3N^+C_2H_4)O\overset{\overset{\displaystyle O}{\displaystyle \|}}{C}CH_3 + H_2O \underset{\longleftarrow}{\longrightarrow} (CH_3)_3N^+C_2H_4OH + CH_3COOH \qquad (5\text{-}K\text{-}1)$$

This enzyme will also catalyze the synthesis of acetylcholine by reaction (5-K-1), but the equilibrium lies very far to the right. Enzymes are very efective catalysts which are generally present in very much lower concentrations than the reactants. To illustrate this, one mole of acetylcholinesterase can hydrolyze 7×10^5 moles of acetylcholine per minute. However, this enzyme is especially fast, and 10^4 per minute is a more typical value. This is the enzyme turnover number as will be defined later from the derived rate expression. The rate of an enzyme catalyzed reaction is proportional to the concentration of enzyme, but it is only proportional to the concentration of the substrate at low concentration. At higher concentrations the rate of reaction is <u>not</u> proportional to substrate concentration. When the substrate concentration is increased, the rate increases, but a limiting maximum velocity, V_{max}, is approached.

Enzyme Catalysis

This manner of the dependence of the rate upon substrate concentration can be explained by the scheme

$$E + S \underset{k_{-1}}{\overset{k_1}{\rightleftarrows}} E \cdot S \xrightarrow{k_2} E + P_1 + P_2 \qquad (5\text{-}K\text{-}2)$$

where in the case of acetylcholinesterase, S, the substrate, is acetylcholine, and the products are choline and acetic acid. Water is a second substrate, but since its con-centration is fixed, we need not explicitly consider this substance in the scheme at this time. The rate of the reaction is given by

$$V = -\frac{d[S]}{dt} = \frac{d[P_1]}{dt} = \frac{d[P_2]}{dt} = k_2[E \cdot S] \qquad (5\text{-}K\text{-}3)$$

The total enzyme concentration, $[E]^o$, which can be controlled by varying the amount we added to our reaction vessel, is given by

$$[E]^o = [E] + [E \cdot S] \qquad (5\text{-}K\text{-}4)$$

Since the substrate concentration is very large compared to the enzyme concentration, there is no significant decrease in [S] when E· S forms. It is customary to determine experimentally the initial velocities of the reaction as we discussed in Section F. For example, in reaction (5-K-1), measurements of the increasing concentration of acetic acid as a function of time, are made during a period of time in which [S] changes very little. The reason for this restriction is that sometimes one of the products is an inhibitor, and the scheme we have written will become invalid if the concentration of this product should reach a kinetically significant level during the course of the experiment. Thus, for the usual type of experiment, [S] changes by only a few percent. When enzyme and substrate are mixed, [E· S] very quickly rises from zero to a steady state level (see Section C). After a few milliseconds, we have

$$\frac{d[E \cdot S]}{dt} = k_1[E][S] - (k_{-1} + k_2)[E \cdot S] = 0 \qquad (5\text{-}K\text{-}5)$$

from which

$$\frac{[E][S]}{[E \cdot S]} = \frac{k_{-1} + k_2}{k_1} = K_S \qquad (5\text{-}K\text{-}6)$$

Solving for [E] and substituting into (5-K-4),

$$[E]^o = [E \cdot S]\left(1 + \frac{K_S}{[S]}\right) \qquad (5\text{-}K\text{-}7)$$

Solving for [E· S] and substituting into (5-K-3),

$$V = \frac{k_2[E]^o}{1 + \dfrac{K_S}{[S]}} \qquad (5\text{-}K\text{-}8)$$

This last equation is called the **Michaelis-Menten Equation**. The velocity of most enzyme catalyzed reactions has the Michaelis-Menten form, although the mechanism of the reaction may be more complicated than the scheme given here. It is therefore customary to write

$$V = \frac{k_{cat}[E]^o}{1 + \dfrac{K_M}{[S]}} \qquad (5\text{-}K\text{-}9)$$

The constant k_{cat} is called the *turnover number* and K_M the *Michaelis constant*. For the mechanism we have considered,

$$k_{cat} = k_2 \qquad (5\text{-}K\text{-}10)$$

$$K_M = K_S \qquad (5\text{-}K\text{-}11)$$

but, in general, k_{cat} and K_M will be more complicated expressions, involving the kinetic constants of a more extensive scheme. Examination of equation (5-K-9) shows that as [S] increases, V approaches $k_{cat}[E]^o$ as a limit. This limit is often symbolized by V_{max}. When [S] is small compared to K_M, there is very little E· S, and therefore, $[E] = [E]^o$. Under this condition, the reaction is second order; and, the second order rate constant is k_{cat}/K_M.

From Figure (5-K-1), or equation (5-K-9) we see that K_M is the value of [S], when the velocity is $\frac{1}{2} V_{max}$. These constants cannot be evaluated from Figure (5-K-1) in the graphical manner indicated, because we usually cannot have concentrations of S sufficiently large to reach the maximum velocity. However, this is not necessary because these constants can be evaluated easily from a double reciprocal plot (or some variant) of V^{-1} versus $[S]^{-1}$.

$$\frac{1}{V} = \frac{1}{k_{cat}[E]^o} + \frac{K_M}{k_{cat}[E]^o} \frac{1}{[S]} \tag{5-K-12}$$

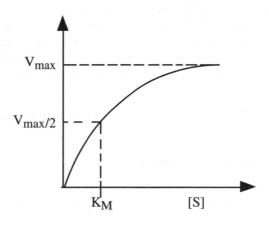

Figure 5-K-1. Dependence of Rate of Enzyme Catalyzed Reaction on Substrate Concentration

In using this equation, V is approximated by

$$V = \frac{\Delta[P_1]}{t} = \frac{\Delta[P_2]}{t} = -\frac{\Delta[S]}{t}$$

and [S] should be replaced by the arithmetic average value of [S] during the time, t. If there is no product inhibition, this equation can be used not only for initial velocities, but with the above substitutions, it can be used for measurements in which as much as fifty percent of the substrate is utilized. If this equation is obeyed, a plot of V^{-1} versus $[S]^{-1}$ is linear.

Problem 5-K-1. Show how the slope and intercept of the linear equation (5-K-12) are related to the constants k_{cat} and K_M, and describe how k_{cat} and K_M can be evaluated.

Answers: m = slope = $K_M/ k_{cat} [E^o]$, b = intercept = $1/k_{cat} [E]^o$, $K_M = \frac{m}{b}$, $k_{cat} = \frac{1/b}{[E]^o}$.

What is the nature of the interactions between enzyme and substrate in the enzyme substrate complex? There may be coulombic interaction, as between the positively charged ammonium function of acetylcholine and a suitably located negatively charged group in the enzyme. Hydrophobic interactions are often very important. The methyl groups of acetylcholine contribute to binding, because there is a negative free energy change attending the removal of nonpolar groups out of the aqueous medium, and into contact with nonpolar amino acid residues in the region of the active site of the protein. Other types of interaction include hydrogen bonds and reversible weak covalent bonds, as for example, between a carbonyl group and a nucleophile.

The spatial orientation of these interactions determines the specificity of the enzyme. Thus, in comparison with acetylcholine, ethyl acetate, which does not contain the quaternary ammonium head with its coulombic and hydrophobic bonding capabilities, is an extremely poor substrate for acetylcholinesterase. It is immediately apparent from this discussion that a substance such as tetramethyl ammonium ion, which has many of the binding features of acetylcholine, must combine with acetylcholinesterase at its active site, and thereby preclude the binding of acetylcholine. This substance, then, is an **inhibitor**. In some cases, substances may bind at sites other than the active site of an enzyme, and may, or may not, exclude the simultaneous binding of substrate. Such substances may be **activators** or inhibitors.

Enzyme Inhibition

We will consider two ways in which a substance may act as a reversible inhibitor. We have already noted that a substance which resembles the substrate will bind to the enzyme at the active site and prevent the simultaneous binding of the substrate. Since the substrate S and inhibitor I are competing in their binding to the enzyme, the process is called *competitive inhibition*. In addition to the enzyme catalysis reaction given in equation (5-K-2),

$$E + S \underset{k_{-1}}{\overset{k_1}{\rightleftarrows}} E \cdot S \xrightarrow{k_2} E + P_1 + P_2 \quad \text{(enzyme catalysis)} \qquad (5\text{-}K\text{-}4)$$

we have the reaction of enzyme with inhibitor

$$E + I \rightleftarrows E \cdot I \qquad \text{(competitive inhibition)} \qquad (5\text{-}K\text{-}13)$$

It is also apparent that an inhibitor might bind to a site other than the active site so that it is possible for both substrate and inhibitor to bind simultaneously to the enzyme. This process is called *noncompetitive inhibition*. To allow for this possibility, we must add another equation for the formation of the inactive ternary complex

$$E \cdot S + I \rightleftarrows E \cdot S \cdot I \qquad \text{(noncompetitive inhibition)} \qquad (5\text{-}K\text{-}14)$$
$$\text{(inactive)}$$

It is apparent that the complex $E \cdot S \cdot I$ could arise in a different way, namely

$$E \cdot I + S \rightleftarrows E \cdot S \cdot I \qquad (5\text{-}K\text{-}15)$$

We must use the same symbol, $E \cdot S \cdot I$, for the ternary complex, no matter how it is formed. The rate of the reaction is still given by equation (5-K-3), but it will be lower than before because the concentration of $E \cdot S$ is decreased by the formation of the complexes $E \cdot I$ and $E \cdot S \cdot I$.

In addition to the steady state equation, we now have three equilibrium equations so that we may write four equilibrium expressions and the stoichiometric equation for the total concentration of enzyme $[E]^0$.

$$\frac{[E][S]}{[E \cdot S]} = K_S \qquad\qquad \frac{[E \cdot I][S]}{[E \cdot S \cdot I]} = K'_S \qquad (5\text{-}K\text{-}16)$$

$$\frac{[E][I]}{[E \cdot I]} = K_I \qquad\qquad \frac{[E \cdot S][I]}{[E \cdot S \cdot I]} = K'_I$$

$$[E]^0 = [E] + [E \cdot S] + [E \cdot I] + [E \cdot S \cdot I] \qquad (5\text{-}K\text{-}17)$$

There are five equations and four unknowns. Evidently there is some redundancy; the equations are not independent. Thus by inspection, we note that

$$\frac{K_S}{K'_S} = \frac{K_I}{K'_I} \qquad (5\text{-}K\text{-}18)$$

This is an important result to remember. It tells us that the effect of the prior binding of S on the binding of I is precisely the same as the effect of the prior binding of I on the binding of S. Since there is nothing so far that requires S to be a substrate, this conclusion applies to the simultaneous binding of any two substances to a third. This result is fundamental to the understanding of biological phenomena.

Because of the relationship (5-K-18), we can omit one equation. Let us omit the equation corresponding to K'_S of equation (5-K-15). The rate expression given in (5-K-3) is

$$V = k_2[E \cdot S]$$

To obtain $[E \cdot S]$, we solve equations (5-K-16) for $[E]$, $[E \cdot I]$, and $[E \cdot S \cdot I]$ in terms of $[E \cdot S]$, and substitute the result in equation (5-K-17) to obtain

$$[E]^0 = [E \cdot S]\left(1 + \frac{[I]}{K'_I} + \left(1 + \frac{[I]}{K_I}\right)\frac{K_S}{[S]}\right) \qquad (5\text{-}K\text{-}19)$$

Therefore,

$$V = \frac{k_2[E]^0}{1 + \dfrac{[I]}{K'_I} + \left(1 + \dfrac{[I]}{K_I}\right)\dfrac{K_S}{[S]}} \qquad (5\text{-}K\text{-}20)$$

and

$$\frac{1}{V} = \left(1 + \frac{[I]}{K'_I}\right)\frac{I}{k_2[E]^0} + \left(1 + \frac{[I]}{K_I}\right)\left(\frac{K_S}{k_2[E]^0}\right)\left(\frac{1}{[S]}\right) \qquad (5\text{-}K\text{-}21)$$

Thus if an inhibitor is present, the slope and intercept of the double reciprocal plot may be affected. If the inhibitor is a competitive inhibitor, its prior binding prevents the binding of S, as we have already discussed. This means that K'_S is infinite. This also means that K'_I is infinite (see equation (5-K-18)). Thus the intercept is not changed by the presence of a competitive inhibitor, as shown in Figure 5-K-2. Note that if the substrate concentration is increased in the presence of a fixed concentration of a competitive inhibitor, the degree of inhibition will decrease until, in the limit of very high substrate concentrations, there is no inhibition. Competitive inhibition is a quite common phenomenon.

On the other extreme, it is possible for the prior binding of I to have no effect at all upon the binding of S. In this case $K'_S = K_S$ and, therefore, $K'_I = K_I$. This is noncompetitive inhibition and both the slope and intercept are affected by precisely the same factor, $(1 + [I]/K_I)$. In this case, the degree of inhibition is not affected by changes in substrate concentration. Sometimes an inhibitor will affect the intercept and slope, but by different

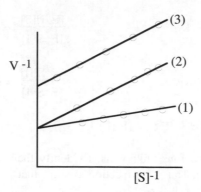

Figure 5-K-2. Effect of Inhibition on Enzyme Catalysis Reaction Rate: (1) Without Inhibitor (I) = 0); (2) Competitive Inhibition ($K_I' = \infty$); (3) Noncompetitive Inhibition ($K_I' = K_I$)

factors, as indicated by the general result above. In this case, the inhibition is said to have competitive and noncompetitive components. If only the intercept is affected, the inhibition is said to be noncompetitive.

Although the basic scheme we have used to introduce the subject is oversimplified, with respect to being applicable to all but a few enzymes, the essential ideas are easily presented by this illustration. We have presented a formal treatment, but, except for the case of competitive inhibition, we have not indicated why a substance acts as an inhibitor or activator. This problem remains for the scientists who are investigating the mechanism of a particular enzyme, and, in many cases, the "why" of this phenomenon is not understood.

In order to better appreciate enzyme kinetics, we will briefly consider a slightly more complicated mechanism which is applicable to acetylcholinesterase and other hydrolytic enzymes. In the enzymatic hydrolysis of acetylcholine, choline is "split" out, and the enzyme is acetylated; that is, the enzyme acts as a nucleophile. The acetyl enzyme then rapidly reacts with water to form acetic acid and the free enzyme. The scheme is

$$E + S \underset{k_{-1}}{\overset{k_1}{\rightleftarrows}} E \cdot S \xrightarrow{k_2} E' + P_1 \tag{5-K-22}$$

$$k_3 \downarrow + H_2O$$

$$E + P_2$$

where E' is the acetyl enzyme intermediate.

Assuming the steady state relationship for [E· S] and [E']

$$[E]^0 = [E] + [E \cdot S] + [E'] \tag{5-K-23}$$

$$\frac{d[E \cdot S]}{dt} = k_1[E][S] - (k_{-1} + k_2)[E \cdot S] = 0 \tag{5-K-24}$$

$$\frac{d[E']}{dt} = k_2[E \cdot S] - k_3[E'] = 0 \tag{5-K-25}$$

Substituting [E] in terms of [E· S] from equation (5-K-24) and [E'] from equation (5-K-25) into equation (5-K-23), we obtain

$$[E]^0 = [E \cdot S]\left(\frac{K_S}{[S]} + 1 + \frac{k_2}{k_3}\right) \tag{5-K-26}$$

The rate is

$$V = k_2[E \cdot S] = k_3[E']$$ (5-K-27)

Substituting $[E \cdot S]$ from equation (5-K-26) into equation (5-K-27), we obtain the Michaelis form shown in equation (5-K-9),

$$V = \frac{k_{cat}[E]^o}{1 + \dfrac{K_M}{[S]}}$$

where

$$k_{cat} = \frac{k_2}{1 + \dfrac{k_2}{k_3}}$$ (5-K-28)

$$K_M = \frac{k_{-1} + k_2}{k_1\left(1 + \dfrac{k_2}{k_3}\right)}$$ (5-K-29)

The form of K_M is rather interesting. Since $k_2 < k_{-1}$, K_M equals the dissociation constant of the Michaelis complex, divided by the kinetic factor $(1 + k_2/k_3)$. If $k_2 > k_3$, K_M may be very small.

When $[S]$ is small compared to K_M, the rate controlling process is the acetylation of the enzyme. The quantity, k_{cat}/K_M, is the second-order rate constant for this process, and is very large. Its value for acetylcholinesterase, 7×10^9 min^{-1}, is about an order of magnitude slower than a diffusion-controlled reaction. The second order rate constant is an important quantity, because it relates the transition state to the reactants.

L. Summary

The following concepts were used to understand the rate at which chemical reactions proceed.

1. A *mechanism* consists of a series of *elementary reaction steps* that describe the pathway for the reaction. An elementary reaction step is one which specifes which species actually come together to react upon a single encounter.

2. The *equilibrium approximation* assumes that one or more reversible reactions occur sufficiently fast that equilibrium is established for these reversible steps. This approximation simplifies the rate expressions for the mechanism so that the concentration of intermediates can be evaluated from the equilibrium constant expressions for these reversible reactions.

3. The *steady state approximation* assumes that an intermediate is sufficiently reactive that its concentration is small and remains essentially constant; i.e., the rate of reaction of an intermediate is essentially zero. This approximation converts the rate expressions for the intermediates into simple algebraic equations that can be solved for the concentration of intermediates.

4. The general rate expression for a reaction is

$$\frac{d[P]}{dt} = k\,[A]^{\alpha}\,[B]^{\beta}\,[C]^{\gamma}\cdots$$

where A, B, C consist of all reactants and products. The α, β, γ, ... must be determined experimentally by measuring the rate of change of a reactant or product at different concentrations of the reactants and products. α, β, γ, ... are the *orders* of the reaction with respect to the species A, B, C,$\cdots$.

5. Agreement between the experimentally established rate expression and the rate expression derived from the proposed mechanism specifies that the mechanism *may be correct*. Disagreement specifies that the mechanism (or approximations) *is incorrect*.

6. The temperature dependence for a reaction is dependent upon the temperature dependence of the rate constant, which is assumed to obey the Arrhenius equation

$$k = A\,e^{-E^{*}/RT}$$

where E^{*}= activation energy and A = pre-exponential constant. E^{*} and A can be evaluated from measurements of k as a function of temperature, T. The logarithmic form of the Arrhenius Equation is helpful in the evaluation of E^{*} and A from the resulting linear relationship.

$$\ln k = -\frac{E^{*}}{R}\left(\frac{1}{T}\right) + \ln A$$

7. Integrated rate expressions for the first and n-th order rates are useful in evaluating precise values for the rate constants. These are

$$\ln [A] = - kt + \ln [A]^{o}$$
$$[A]^{-n+1} = (n-1)\,kt + ([A]^{o})^{-n+1}$$

Exercises

1. In Exercise 19, at the end of Chapter 3, the ΔH^{o} and ΔG^{o} were calculated for the
   reaction $N_2O_4 \longrightarrow 2\,NO_2$.
   If the activation energy for the reaction $2\,NO_2 \longrightarrow N_2O_4$ is 92 kJ/mol, what is
   the activation energy for the reaction,

$$N_2O_4 \longrightarrow 2\,NO_2\,?$$

   Assume T = 298K. Answer: 147 kJ/mol.

2. The decomposition of A is catalyzed by

$$A + C \underset{k_{-1}}{\overset{k_1}{\rightleftarrows}} AC$$

$$AC \xrightarrow{k_2} P + C$$

(a) Derive the rate expression and show the dependence of the rate on the catalyst concentration. Assume steady state for AC.

(b) What order of reaction, with respect to A and C, would this mechanism predict?

Answer: (a) $\dfrac{d[P]}{dt} = -\dfrac{d[A]}{dt} = \dfrac{k_1 k_2 [A][C]}{k_{-1} + k_2}$ (b) First order with respect to [A] and [C].

3. The thermolysis of ethyl acetate has been studied by a flow technique [H. Kwart, S. F. Sarner, and J. H. Olson, *J. Phys Chem.*, **1969**, *73*, 4056] in the temperature range 472 to 561°C. The following Arrhenius equation was determined from the study:

$$k = 1.87 \times 10^{12}\ e^{(-195{,}400\ \text{J/mole})/RT}\ \text{sec}^{-1}.$$

Calculate the reaction rate constant at 500°C. Answer: $k = 0.12\ \text{sec}^{-1}$.

4. The value of A in the Arrhenius equation is often approximated as $10^{13}\ \text{sec}^{-1}$ for unimolecular reactions. Calculate the half-life ($\tau_{1/2}$) for the reaction (assumed unimolecular),

$$CH_3I \longrightarrow CH_3^{\cdot} + I^{\cdot}$$

if the C-I bond dissociation energy is 235.6 kJ/mole. Assume the temperature is 200 °C. Since this reaction is a single reaction step, the bond dissociation energy, ΔU, is equal to the activation energy. This is generally assumed for any single, bond-breaking reaction step. Answer: $\tau_{1/2} = 0.7 \times 10^{13}$ sec.

5. Previously a mechanism was proposed for the reaction

$$CO + Cl_2 = COCl_2$$

which involved an intermediate $COCl^{\cdot}$ (Problems 5-C-3 and 5-D-4) . Another mechanism that has been proposed for this reaction is

$$Cl_2 \underset{k_{-1}}{\overset{k_1}{\rightleftarrows}} 2Cl^{\cdot}$$

$$Cl^{\cdot} + Cl_2 \underset{k_{-2}}{\overset{k_2}{\rightleftarrows}} Cl_3^{\cdot}$$

$$Cl_3^{\cdot} + CO \xrightarrow{k_3} COCl_2 + Cl^{\cdot}$$

(a) Identify the nature of each reaction step and a chain mechanism if one exists.

(b) Assume equilibrium in the first two steps and derive the rate expression for the formation of $COCl_2$. (Hint: Since there are two intermediates, $Cl^{\cdot}$ and $Cl_3^{\cdot}$, you should use the two equilibrium expressions to solve for $Cl_3^{\cdot}$.)

(c) Assuming steady state for $Cl^{\cdot}$ and $Cl_3^{\cdot}$, derive the rate expression for the formation of $COCl_2$. (Hint: The steady state equations must be solved simultaneously to obtain equations for $[Cl^{\cdot}]$ and $[Cl_3^{\cdot}]$. This is accomplished most easily by adding the steady state equations similar to the procedure used in Problem 5-D-4.)

(d) Show that the steady state solution in (c) can be made equivalent to the rate expression assuming equilibrium in part (b).

(e) Compare the rate expressions derived from this mechanism with those in problems 5-C-3 and 5-D-3.

Answers: (a) k_1: initiation; k_{-1}: termination; k_2, k_{-2}, k_3: propagation

(b) $\dfrac{d[COCl_2]}{dt} = k_3\, K_2\, K_1^{1/2}\, [Cl_2]^{3/2}\, [CO]$

(c) $\dfrac{d[COCl_2]}{dt} = k_3\, k_2 \left(\dfrac{k_1}{k_{-1}}\right)^{1/2} \dfrac{[Cl_2]^{3/2}[CO]}{k_{-2} + k_3[CO]}$ (e) The rate expressions obtained in (b)

and (d) are identical with those in 5-C-3 and 5-D-3.

6. The following rate data were obtained for the reaction

$$NH_4^+ + NO_2^- = N_2 + 2\, H_2O \text{ (acid solution)}$$

[J. H. Dusenbury and R. E. Powell, *J Am. Chem. Soc.*, **1951**, *73*, 3266]

HNO_2 (mol/liter)	0.0092	0.0092	0.0488	0.0249
NH_4^+ (mol / liter)	0.098	0.049	0.196	0.196
rate (mol/liter/sec) x 10^{-8}	34.9	16.6	335.	156.

(a) Express the rate law.
(b) Calculate the rate constant.
(c) On this basis only, can you definitely state that this reaction is bimolecular? Explain briefly.
(d) If you wanted to determine a more accurate rate constant, how would you go about doing so?

Answers: (a) $\dfrac{d[N_2]}{dt} = k[NH_4^+]\,[NO_2^-]$ (b) $k = 3.6 \times 10^{12}$ L/mol-sec

(c) No; the order of the reaction does not prove the mechanism; hence, one cannot be sure the reaction is second order. (d) Measure $[NH_4^+]$ or $[NO_2^-]$ as a function of time and use equation (5-J-21), or measure $[N_2]$ as a function of time and use equation (5-J-24).

7. The oxidation of Br^- by BrO_3^- in fused alkali nitrates has been studied by J. M. Schlegel [*J. Phys. Chem.*, **1969**, *73*, 4152]. When BrO_3^-, Br^-, and CrO_4^- are in excess, compared to $Cr_2O_7^{2-}$, the overall reaction is

$$Cr_2O_7^{2-} + BrO_3^- + Br^- = 2CrO_4^{2-} + Br_2 + O_2.$$

This reaction was found to be first order for BrO_3^- and second order for Br^-. Show that these reaction orders are consistent with the mechanism:

$$Cr_2O_7^- + BrO_3^- \underset{k_{-1}}{\overset{k_1}{\rightleftarrows}} BrO_2^+ + 2CrO_4^{2-}$$

$$BrO_2^+ + 2Br^- \underset{k_{-2}}{\overset{k_2}{\rightleftarrows}} BrO_2^- + Br_2$$

$$BrO_2^- \overset{k_3}{\longrightarrow} Br^- + O_2$$

Assume the first two steps are rapid, and that the last step is the slow, rate determining step. (Hint: Express the rate of the reaction in terms of $d[O_2]/dt$.)

8. The overall reaction (stoichiometric equation) for $H_2 + NO$ is

$$2H_{2(g)} + 2NO \longrightarrow N_{2(g)} + 2H_2O$$

The following mechanism has been proposed for this reaction:

$$2\,NO \underset{k_{-1}}{\overset{k_1}{\rightleftarrows}} N_2O_2$$

$$H_2 + N_2O_2 \xrightarrow{k_2} N_2O + H_2O$$

$$H_2 + N_2O \xrightarrow{k_3} N_2 + H_2O$$

(a) Write the rate expression for the loss of NO.
(b) Write the rate expression for N_2O_2, and solve for $[N_2O_2]$, using the steady state approximation. The result should be

$$[N_2O_2] = \frac{k_1[NO]^2}{k_{-1} + 2k_2[H_2]}$$

(c) Apply the steady state treatment to N_2O.
(d) Show that the mechanism is, or can be, in agreement with the experimentally determined rate law

$$\frac{-d[NO]}{dt} = k[NO]^2[H_2]$$

This was the result found in Example 5-F-1.
(e) State the molecularity of the reactions represented by the four rate constants.
(f) The k in (d) can be shown to be

$$k = \frac{2k_1 k_2}{k_{-1}}$$

What is the relationship between E^* and the activation energies for the various steps?

Answers: (a) $-\dfrac{d[NO]}{dt} = k_1\,[NO]^2 - k_{-1}\,[N_2O_2]$ (b) $[N_2O_2] = \dfrac{k_1[NO]^2}{2k_2[H_2] + k_{-1}}$

(c) $[N_2O] = \dfrac{k_2[N_2O_2]}{k_3}$ (e) The reverse of the first step, represented by the rate constant k_{-1} is unimolecular, whereas, the other three reaction steps k_1, k_2, k_3 are bimolecular. (f) $E^* = E_1^* + E_2^* - E_{-1}^* = \Delta U_1^o + E_2^*$.

9. At high temperatures, dimethyl ether decomposes as follows:

$$(CH_3)_2O \longrightarrow CH_4 + H_2 + CO$$

When the reaction was run at $504^o C$ in a closed container, the pressure increased with time as follows:

Time (sec)	0	390	665	1195	2240	3155	∞
Pressure (mm)	312	408	468	562	714	779	938

Determine the order of the reaction and the rate constant. Propose a mechanism for the reaction. What other information might one desire to determine the mechanism?

Answer: First order reaction; graph of $\ln \dfrac{3P^o - P}{2}$ versus t is linear; $k = 4.5 \times 10^{-4}$ sec^{-1}.

10. The kinetic mechanism for thermal electron attachment can be represented by the following series of fundamental reaction steps. The process is carried out at approximately atmospheric pressure in helium with a low concentration (<1%) of a dopant such as CH_4. The thermal electrons are produced by some high energy source such as photons or β particles from a radioactive source. Photons with an average intensity or flux, represented by I_{hv}, are used in the proposed mechanism and k_P is the rate constant for the production of electrons.

$$hv + D \xrightarrow{\ k_P\ } D^+ + e^-$$

$$e^- + AB \underset{k_{-1}}{\overset{k_1}{\rightleftharpoons}} AB^-$$

$$AB^- \xrightarrow{\ k_2\ } A + B^-$$

$$e^- + D^+ \xrightarrow{\ k_D\ } D$$

$$AB^- + D^+ \xrightarrow{\ k_N\ } neutrals$$

It is assumed in this mechanism that e^- attachment forms a molecular negative ion, AB^-, which then may undergo subsequent dissociation into $A + B^-$ or e^- detachment. The energy of detachment is approximately the electron affinity of the molecule, AB.

(a) Write the rate expressions for the formation of e^- and AB^-.
(b) Assume steady state for e^- and AB^-. Solve for $[AB^-]$ from the steady state expression for AB^- and substitute into the steady state expression for e^- and solve for $[e^-]$.
(c) Experimentally a potential is applied and the e^- are collected and measured as a current. The current measured should be proportional to the $[e^-]$. Measurements are made when there is no AB present and we designate the e^- concentration by $[e^-]^o$. Then with AB present the current measurement when the e^- concentration is reduced to $[e^-]$. From the previous expression for $[e^-]$ show that

$$[e^-]^o = \frac{k_P\, I_{hv}\, [D]}{k_D[D^+]}$$

Solve for the ratio $[e^-]^o/[e^-]$.
(d) Note that

$$\frac{[e^-]^o}{[e^-]} - 1 = K[AB]$$

where K is a combination of rate constants. Show how K can be found graphically.

Answers: (a) $\dfrac{d[e^-]}{dt} = k_P I_{hv}[D] - k_1[e^-][AB] + k_{-1}[AB^-] + k_2[AB^-] - k_D[e^-][D^+]$

$\dfrac{d[AB^-]}{dt} = k_1[e^-][AB] - k_{-1}[AB^-] - k_2[AB^-] - k_N[AB^-][D^+]$

(b) $[e^-] = \dfrac{k_P I_{hv}[D]}{k_1[AB] - \dfrac{(k_{-1})k_1[AB]}{k_{-1} + k_2 + k_N[D^+]} + k_D[D^+]}$

(c) $\dfrac{[e^-]^o}{[e^-]} = \dfrac{1}{k_D[D^+]}\left[k_1[AB] - \dfrac{(k_{-1})k_1[AB]}{k_{-1} + k_2 + k_N[D^+]} \right] + 1$

(d) Plot $([e^-]^o/[e^-] - 1)$ versus $[AB]$; slope $= K = \dfrac{1}{k_D[D^+]}\left[\dfrac{k_1(k_2 + k_N[D^+])}{k_{-1} + k_2 + k_N[D^+]} \right]$.

11. Consider the reaction of ICl with H_2 which gives the following stoichiometric equation

$$2\,ICl + H_2 = 2\,HCl + I_2$$

The following mechanism has been proposed for this reaction

$$ICl \xrightarrow{\ k_1\ } I^\cdot + Cl^\cdot$$

$$Cl^\cdot + H_2 \xrightarrow{\ k_2\ } HCl + H^\cdot$$

$$H^\cdot + ICl \xrightarrow{\ k_3\ } HCl + I^\cdot$$

$$I^\cdot + ICl \xrightarrow{\ k_4\ } I_2 + Cl^\cdot$$

$$I^\cdot + I^\cdot \xrightarrow{\ k_5\ } I_2$$

(a) Identify the nature of each reaction step in terms of: initiation, propagation, branching, inhibition, and termination.

(b) Identify a chain mechanism if one exists.

(c) Apply steady state to the intermediates $Cl^\cdot$, $H^\cdot$, $I^\cdot$ and obtain expressions for $[Cl^\cdot]$, $[H^\cdot]$, and $[I]^\cdot$. (Hint: solve the steady state equation for $Cl^\cdot$ to get $[Cl^\cdot]$ and substitute into the steady state equation for $H^\cdot$. Then substitute for $[H^\cdot]$ into the steady state equation for $I^\cdot$.

(d) Using the results of part c, show that the rate equation is

$$\frac{1}{2}\frac{d[HCl]}{dt} = \frac{d[I_2]}{dt} = k_1[ICl] + k_4\left(\frac{2k_1}{k_5}\right)^{1/2}[ICl]^{3/2}$$

(e) Discuss the relative magnitudes of the two terms in the rate equation in regards to the nature of the mechanism. Answers: (a) k_1: initiation; k_2, k_3, k_4: propagation; k_5: termination (b) k_2, k_3, k_4: chain mechanism (c) $I^\cdot = 2^{1/2}\left(\dfrac{k_1}{k_5}\right)^{1/2}[ICl]^{1/2}$ (e)

The second term should be much greater since it only involves $k_1^{1/2}$ versus k_1. k_1 should be small since it is the rate constant for the initiation step.

12. The kinetics of the alkylhydroperoxide-tetranitromethane reaction was studied by Sager and Hoffsommer [W. F. Sager and J. C. Hoffsommer, *J. Phys. Chem.*, **1969**, 73, 4155]. The reaction is

$$C(NO_2)_4 + ROO^- + H_2O \longrightarrow C(NO_2)_3^- + NO_2^- + H^+ + ROH + O_2$$

The rates of reaction were followed spectrophotometrically for the $C(NO_2)_3^-$ ion concentration with time. In aqueous solution, ROO^- is in equilibrium with $ROOH$.

$$ROOH \underset{\longleftarrow}{\overset{K_A}{\longrightarrow}} ROO^- + H^+$$

$$[ROO^-] = \frac{K_A[ROOH]}{[H^+]}$$

The rate equation is then of the general form

$$-\frac{d[C(NO)_4]}{dt} = \frac{d[C(NO_2)_3^-]}{dt} = k[C(NO_2)_4]^a\,[ROO^-]^b = kK_A^b\,[C(NO_2)_4]^a\frac{[ROOH]^b}{[H^+]^b}$$

If [ROOH] is in excess, it will remain constant, and $[H^+]$ can be kept constant by proper buffers. The rate expression then becomes

$$-\frac{d[C(NO_2)_4]}{dt} = k'\,[C(NO_2)_4]^a$$

(a) Show that integration of the rate equation, assuming $a = 1$, yields the expression

$$\ln[C(NO_2)_4] = -k't + \ln[C(NO_2)_4]^o$$

where $[C(NO_2)_4]^o$ is the initial concentration.

(b) Experimentally, the $[C(NO_2)_3^-]$ was measured as a function of time. In order to test the integrated equation, we must find the relationship between $[C(NO_2)_3^-$ and $[C(NO_2)_4]$. From the stoichiometry of the reaction, the $[C(NO_2)_3^-]$ at infinity, is the sum

$$[C(NO_2)_3^-]^\infty = [C(NO_2)_3^-] + [C(NO_2)_4]$$

or

$$[C(NO_2)_4] = [C(NO_2)_3^-]^\infty - [C(NO_2)_3^-]$$

The $[C(NO_2)_3^-]$ was found from absorbance, A, measurements,

$$A = \varepsilon \ell\,[C(NO_2)_3^-]$$

where ℓ = path length = 1 cm., ε = molar extinction coefficient. Show that

$$[C(NO_2)_4] = \frac{A^\infty - A}{\varepsilon}$$

The following A measurements, as a function of time, were observed for the reaction with R = ethyl.

$$[C_2H_5OOH] = 2.43 \times 10^{-2}\,M,\ pH = 7.00,\ t = 25.14^oC$$

t (min)	A
1.49	0.015
4.11	0.034
11.45	0.089
21.19	0.150
39.05	0.239
81.57	0.384
104.86	0.431
137.54	0.487
182.35	0.519
∞	0.559

Show that this data satisfied the liner equation derived in part (a), thus proving the rate is first order with respect to $C(NO_2)_4$. Evaluate k'.

(c) Using data similar to that described in part (b), k' was determined at various [ROOH] and pHs.

R	[ROOH] x 10^3 M	pK_A	$[C(NO_2)_4]^0$ x 10^5M	pH	k'(sec^{-1}) x 10^3
H	9.64	11.54	3.06	7.2	0.482
	18.6		4.42	7.2	0.904
	21.6		3.95	7.2	1.12
C_2H_5	25.5	11.35	2.18	7.4	0.625
	12.1		4.47	7.4	0.307
t-butyl	10.8	12.54	4.58	7.8	0.137
	15.7		4.07	7.8	0.202

Determine the order of the reaction with respect to [ROOH]; that is, determine the b in the general rate expression in part (a). Evaluate k for the three R groups given in the above table.

Answers: (b) k' = 1.4 x 10^{-2} min^{-1},

(c)	R	b	k(L/mol-sec)
	H	1.00	1.10 x 10^3
	C_2H_5	0.96	0.22 x 10^3
	t - butyl	1.04	0.70 x 10^3

13. The reaction

$$2\ Ce(IV) + As(III) \longrightarrow 2\ Ce(III) + As(V)\ \text{(catalyzed by } I_2)$$

was followed as a function of time by titration to determine the remaining Ce(IV) concentration. The Ce(IV) was titrated with standard iron(II) ammonium sulfate. The logarithm of the volume of titrant as a function of time is shown in the following graph:

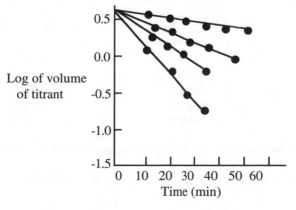

Kinetic study for (A) 0.0, (B) 5.0, (C) 10.0, and (D) 15.0 g/100ml concentrations of iodine. These solutions were diluted 1:5 in arsenite in the final analysis.

(a) If the concentration of As(III) is in excess, compared to Ce(IV), what is the order of the reaction with respect to [Ce(IV)]?

(b) From the data in this figure, it is obvious that the reaction is catalyzed by I_2. It is proposed that the catalysis occurs by the following mechanism:

$$2Ce(IV) + 2I^- \xrightarrow{k_1} 2Ce(III) + I_2$$

$$As(III) + I_2 \underset{k_{-2}}{\overset{k_2}{\rightleftarrows}} As(v) + 2I^-$$

$$2Ce(IV) + As(III) \longrightarrow 2Ce(III) + As(V)$$

From the results in this graph, what is the order of the reaction with respect to $[I_2]$? Could this result be consistent with the proposed mechanism? Assume steady state for I_2. For more details, see the original paper by E. Gabinski and B. Zak, *J. Chem. Ed.*, **1962**, *39*, 634. Answers: (a) The reaction should be first order with respect to Ce(IV). [Ce(IV)] α volume titrant and [As(III)] is in excess. (b) First order with respect to I_2. Plot $(k - k_0)$ versus $[I_2]$, where k_0 is the rate constant for the uncatalyzed reaction. Yes, the first order dependence of I_2 is consistent with the mechanism.

14. The quantity, $1/(1 + K_M/[S])$, is the fraction of total enzyme that exists as enzyme-substrate complex.

(a) If there are two substrates, S_1 and S_2, in a reaction, and if the binding of these are independent, write the expression for the velocity of the enzyme-catalyzed reaction.

(b) If one substrate concentration is held constant, will the velocity expression have the Michaelis-Menten form?

Answer: (a) $v = \dfrac{k_{cat}[E]^o}{\left(1 + \dfrac{K_{M_1}}{[S_1]}\right)\left(1 + \dfrac{K_{M_2}}{[S_2]}\right)}$ (b) yes.

15. To 100 mL of 0.1 M NaCl, thermostatted at 25°C, and held at constant pH = 7.00 with an automatic titrator, 8.09 μg of acetylcholinesterase (mol. wt = 55,000) were added, along with varying concentrations of substrate. The following rates of hydrolysis were observed, with and without 3×10^{-7} M inhibitor, 3-hydroxyphenyl-trimethylammonium bromide, averaged over a period of 10 minutes.

$[S] \times 10^5$M	v (μmol/min)	v, with I (μmol/min)
5.00	0.0340	
6.66	0.0405	0.0278
10.0	0.0500	0.0370
14.3	0.0600	0.0448
25.0	0.0710	0.0601
50.0	0.0870	0.0740

Using reciprocal plots, evaluate K_M, K_I, K'_I, and k_{cat}. What type of inhibition does this problem illustrate?
Answers: $k_{cat} = 6.8 \times 10^2$ min^{-1}, $K_M = 9.9 \times 10^{-5}$M, $K_I = 4.0 \times 10^{-7}$M; competitive inhibition.

16. The following mechanism has been proposed for the thermal decomposition of acetaldehyde:

$$CH_3CHO \xrightarrow{k_1} \cdot CH_3 + \cdot CHO$$

$$CH_3 + CH_3CHO \xrightarrow{k_2} CH_4 + CH_3CO \cdot$$

$$CH_3CO \cdot \xrightarrow{k_3} \cdot CH_3 + CO$$

$$2 \cdot CH_3 \xrightarrow{k_4} C_2H_6$$

(a) Identify the nature of the reaction steps; e.g. initiation, propagation, etc.

(b) Identify a chain mechanism. Is this consistent with the stoichometric equation

$$CH_3CHO = CH_4 + CO$$

Briefly explain.

(c) Write the rate expression for the rate of formation of $\cdot CH_3$.

(d) Write the rate expression for the formation of C_2H_6. Make this rate consistent with the rate expression for $\cdot CH_3$ in question 3.

(e) Apply the steady state approximation to the intermediates $\cdot CH_3$ and $CH_3CO \cdot$ and derive the rate expression

$$\frac{d[CH_4]}{dt} = \frac{k_2(k_1)^{1/2}}{(k_4)^{1/2}} [CH_3CHO]^{3/2}$$

(f) Experimentally it is found that the rate expression for the decomposition of acetaldehyde is

$$-\frac{d[CH_3CHO]}{dt} = k[CH_3CHO]^{3/2}$$

Based on this information and the rate expression derived in question (e), what can you conclude about the proposed mechanism?

(g) For the rate expression given in question (f), what is the integrated equation expressing $[CH_3CHO]$ as a function of time. Explain the graph that you would make in order to determine an accurate rate constant.

(h) If total pressure were measured as a function of time, what graph of pressure versus time would you make in order to test the validity of the rate expression for CH_3CHO in question (f) Let P^o represent the initial pressure of pure CH_3CHO and P the total pressure at any time, t. Answers: (a) k_1: initiation; k_2 and k_3: propagation; k_4: termination (b) steps 2 and 3 make up a chain mechanism; yes, adding steps 2 and 3 gives the stoichiometric equation. (c) $\dfrac{d[CH_3^{\cdot}]}{dt} =$

$k_1[CH_3CHO] - k_2[CH_3^{\cdot}] [CH_3CHO] + k_3 [CH_3CO^{\cdot}] - k_4[CH_3^{\cdot}]^2$ (d) $\dfrac{d[C_2H_6]}{dt} =$

$\frac{1}{2} k_4[CH_3^{\cdot}]^2$. (f) The mechanism may be correct since there is agreement between the experimental rate expression and the derived rate expression based upon the proposed mechanism. (g) $[CH_3CHO]^{-1/2} = \frac{1}{2} kt + ([CH_3CHO]^o)^{-1/2}$;

Graph $[CH_3CHO]^{-1/2}$ versus t, slope $= \frac{1}{2} k$ or $k = 2(slope)$

(h) $(2P^o_{CH_3CHO} - P)^{-1/2}$ versus t and slope $= \frac{1}{2}(RT)^{-1/2}k$.

17. The half-life for the first-order radioactive decay of ^{14}C is 5730 yr. An archeological sample contained wood that had only 72% of the ^{14}C found in living trees. What is its age? Answer: 2715 yr.

18. The reaction

$$n\text{-}C_7H_{16(g)} \longrightarrow 3\,C_2H_{4(g)} + CH_{4(g)}$$

is first order with respect to C_7H_{16}. If $[C_2H_4]$ is followed as a function of time, give the integrated rate expression which you would use to obtain an accurate rate constant for this reaction. Answer: $\ln([n\text{-}C_7H_{16}]^0 - \frac{1}{3}[C_2H_4]) = -kt + \ln[n\text{-}C_7H_{16}]^0$.

19. Consider the following reaction

$$SO_2 + \frac{1}{2}\,O_2 = SO_3$$

which is catalyzed by NO. The following mechanism has been proposed for this reaction

$$NO + O_2 \underset{k_{-1}}{\overset{k_1}{\rightleftharpoons}} O_2\cdot NO$$

$$O_2\cdot NO + NO \overset{k_2}{\longrightarrow} 2NO_2$$

$$NO_2 + SO_2 \overset{k_3}{\longrightarrow} NO + SO_3$$

(a) Write the rate expressions for $O_2\cdot NO$, NO_2, and SO_3. Be certain all three rate expressions are consistent.
(b) Apply steady state to $O_2\cdot NO$ and NO_2, eliminate $[O_2\cdot NO]$ and solve for $[NO_2]$.
(c) Derive the rate expression for the formation of SO_3.

Answers: (a) $\dfrac{d[O_2\cdot NO]}{dt} = k_1[NO][O_2] - k_{-1}[O_2\cdot NO] - k_2[O_2\cdot NO][NO]$

$\dfrac{d[NO_2]}{dt} = 2k_2[O_2\cdot NO][NO] - k_3[NO_2][SO_2]$, $\quad \dfrac{d[SO_3]}{dt} = k_3[NO_2][SO_2]$

(b) $[NO_2] = \dfrac{2k_2k_1[NO]^2[O_2]}{k_3(k_{-1} + k_2[NO])[SO_2]}$ (c) $\dfrac{d[SO_3]}{dt} = \dfrac{2k_1k_2[NO]^2[O_2]}{k_{-1} + k_2[NO]}$.

20. The following mechanism has been proposed for the reaction of hydrogen with organic chlorine compounds, represented by the general formula RCl.

$$He^* + H_2 \overset{k_1}{\longrightarrow} He + 2H^\cdot$$

$$H^\cdot + RCl \overset{k_2}{\longrightarrow} HCl + R^\cdot$$

$$R^\cdot + H_2 \overset{k_3}{\longrightarrow} RH + H^\cdot$$

$$H^\cdot + H^\cdot \overset{k_4}{\longrightarrow} H_2$$

where He^* represents an excited state of He.

(a) Identify the nature of each reaction step such as propagation, initiation, etc.
(b) Identify a chain mechanism if it is present and give the stoichiometric equation for the reaction.
(c) Write the rate expressions for the production of $H^{\bullet}$ and $R^{\bullet}$.
(d) Apply steady state conditions for $H^{\bullet}$ and $R^{\bullet}$ and show the rate expression for the loss of RCl is given by

$$-\frac{d[RCl]}{dt} = \left(\frac{k_1}{k_4}\right)^{1/2} k_2 [He^*]^{1/2} [H_2]^{1/2} [RCl]$$

(e) Derive the relationship between the activation energies for the individual reaction steps and the overall activation energy for the reaction.
(f) If the concentration of H_2 were much greater than the concentration of RCl and $[He^*]$ were constant, derive the equation which represents the concentration of RCl as a function of time.
(g) If the concentration of RCl were much greater than the concentration of H_2 and $[He^*]$ were constant, derive the equation which represents the concentration of H_2 as a function of time.
(h) Assuming the conditions of part (g), derive the equation which represents the concentration of HCl as a function of time.
(i) If we were to establish experimentally that the rate expression is

$$-\frac{d[RCl]}{dt} = k [He^*]^{1/2}[H_2]^{1/2} [RCl]$$

what may we conclude about the mechanism?
(j) Describe how you would determine the rate expression for this reaction of H_2 plus RCl in the presence of He^* using the initial rate method.

Answers: (a) k_1: initiation; k_2 and k_3: propagation; k_4 termination (b) k_2 and k_3 make up a chain mechanism; $RCl + H_2 = HCl + RH$ (c) $\frac{d[H^{\bullet}]}{dt} = k_1[He^*][H_2] - k_2[H^{\bullet}][RCl] + k_3[R^{\bullet}][H_2] - k_4[H^{\bullet}]^2$; $\frac{d[R^{\bullet}]}{dt} = k_2[H^{\bullet}][RCl] - k_3[R^{\bullet}][H_2]$
(e) $E^* = E_2^* + \frac{1}{2}E_1^* - \frac{1}{2}E_4^*$ (f) $[RCl] = [RCl]^0 e^{-kt}$ (g) $[H_2]^{1/2} = -\frac{1}{2}kt + ([H_2]^0)^{1/2}$
(h) $([H_2]^0 - [HCl])^{1/2} = -\frac{1}{2}kt + ([H_2]^0)^{1/2}$ (i) Since the experimental rate expression is of the same form as the derived rate expression based on the mechanism, the mechanism may be correct. (j) Measure the initial rates of formation of product (HCl or RH) or loss of reactant (RCl or H_2), providing it is not in excess, as a function of $[H_2]$, keeping the concentrations of all other reactants and products constant. Graph $\ln R$ versus $\ln[H_2]$ to find slope $= \alpha$. Repeat this procedure by varying each reactant and product, keeping the other concentrations constant.

21. The composition of a liquid phase reaction

$$2A \longrightarrow B$$

was followed by a spectrophotometric method with the following results:

t(min)	0	10	20	30	40	∞
[B](mol/L)	0	0.089	0.153	0.200	0.230	0.312

Determine the order of the reaction and its rate constant.

Answer: First order with $k = 0.034$ min^{-1}.

22. The reaction between ICl and H_2 follows the stoichiometric equation

$$2ICl_{(g)} + H_{2(g)} \longrightarrow 2HCl_{(g)} + I_{2(g)}$$

The rate expression for this reaction has been experimentally found to be second order

$$-\frac{d[ICl]}{dt} = \frac{[HCl]}{dt} = k[ICl][H_2]$$

Write the integrated rate expression that you would use if you were to follow the reaction by measuring the:
(a) [HCl] with time.
(b) [ICl] with time.

Answer: (a) $\ln \dfrac{[H_2]^0 - \frac{1}{2}[HCl]}{[ICl]^0 - [HCl]} = \left(2[H_2]^0 - [ICl]^0\right) \; kt + \ln \dfrac{[H_2]^0}{[ICl]^0}$

(b) $\ln \dfrac{[H_2]^0 + \frac{1}{2}[ICl] - \frac{1}{2}[ICl]^0}{[ICl]} = \left(2[H_2]^0 - [ICl]^0\right) \; kt + \ln \dfrac{[H_2]^0}{[ICl]^0}.$

6

MOLECULAR MOTIONS AND ENERGIES

Biological processes generally involve large molecules and the study of small molecules would not seem to be pertinent to this field. However, over the past 10–15 years it has been found that nitric oxide, NO, plays an important role as a messenger molecule in several biological processes. In 1980 Robert F. Furchgott, a recent recipient of a Nobel Award in Chemistry, rationalized that acetylcholine acting on receptors on endotheliel cells provokes the release of a small molecule that diffuses to the adjacent muscle layer and relaxes it. The small molecule could not be identified at that time, although it was later identified as NO. One reason for the difficulty of identification is that NO is a reactive radical with a lifetime of only 6–10 seconds. Ferid Murad, also a recipient of the Nobel Prize for work in this area, explained the effectiveness of nitroglycerin as a treatment of heart attacks by the metabolic conversion to nitric oxide which subsequently migrates to the muscle, causing it to relax. In 1986 Furchgott and Louis J. Ignarro independently predicted that the small molecule released by the acetycholine reaction was also NO, and this was confirmed experimentally the following year. Louis J. Ignarro also shared the Nobel Prize with Furchgott and Murad which was awarded for this work involving NO. It is presently thought that the role of NO as a messenger may be much more prevalent even in its function in the brain and the immune system. A review of the role of NO in biological processes can found in an article by S.H. Snyder and D.S. Bredt in the May 1992 issue of Scientific American. *There is also evidence that CO could play a similar role in other processes.*

In Chapters 2 and 3, we introduced certain thermodynamic variables (U, H, S, and G) to accomplish our basic objectives; namely, criterion for spontaneity and calculation of the maximum yield. Each of these thermodynamic variables describes a macroscopic property of the substance; that is, the magnitude of the variable is a measure of the entire substance as it exitsts in the laboratory. For example, the molar internal energy of $NaCl_{(s)}$ at 25°C is a measure of the energy of one mole of $NaCl_{(s)}$ as it exists in the laboratory at 25°C. At that time it was not necessary to know why the substance had a certain internal energy or free energy. We now ask this question: "What properties of the substance determine the magnitude of the thermodynamic variables U, H, S, and G?" In order to obtain an answer, we must look at the properties of the molecules which make up the substance, rather than the macroscopic properties of the substance.

In this chapter, we will consider individual molecules and the energies associated with the motions of these molecules. These individual molecular energy levels can be related to all the thermodynamic parameters, though at the present time, we will be concerned exclusively with their relationship with molecular structure. The knowledge of molecular energy levels, along with additional basic principles of Statistical Thermodynamics, allows a derivation of the equilibrium constant for ideal gas mixtures in terms of fundamental molecular parameters.

A. Objectives

The principal topics which are the objectives for disussion in this chapter are:

Molecular Motions. The nuclei of a molecule are continually undergoing motion. We wish to describe this motion and understand its relationship to the geometric structure of the molecule.

Quantum Mechanics. The principles of quantum mechanics will be introduced, and solutions to some simple problems will be shown, which illustrate quantized states and quantized energies.

Molecular Energies. The motions of the nuclei of a molecule are associated with the energy of the molecule. Various formulas (arising from quantum mechanical derivations) will be given, which describe the various quantized energies which a molecule can have, each of which is related to a specific nuclear motion.

B. Molecular Motions—Classical Equipartition Principle

In view of the classical molecular model, in this section we will attempt to discover what kinds of motion are possible for a molecule, and how these motions can best be described. We can begin by stating that the various types of motion which a molecule can have are best described in terms of the motion of its component parts, that is, the individual atoms which compose it. Each atom can move in any direction, the motion being resolvable in 3 coordinates. Each atom is said to have *3 degrees of freedom*. Since it requires three coordinates to describe the position of each atom, it will require $3n$ coordinates to describe completely the spatial orientation of a molecule containing n atoms. The molecule is said to have $3n$ degrees of freedom, which means that any motion whatsoever that the molecule undergoes will be completely described by the changes in $3n$ coordinates. The problem, then, is how the molecule as a whole moves as a result of the collective motion of the individual atoms. We can resolve it by considering the result of certain combinations of the motions of the individual atoms.

For example, let us suppose that all the atoms of the molecule move simultaneously in the x direction. The "center of mass" of the molecular system moves in the x direction. Therefore the molecule is said to have *translational* motion in the x direction. This motion can be completely described by expressing the change in the x coordinate of the center of

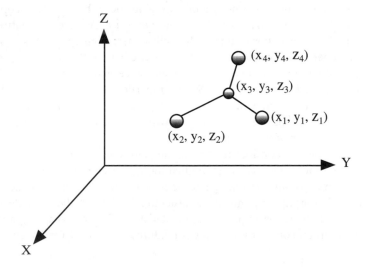

Figure 6-B-1. Designation of Atomic Location by Three Cartesian Coordinates (x_i, y_i, z_i)

mass of the molecule, which is really some combination of the various x coordinates of the n individual atoms. Similarly, if all the individual atoms were to move simultaneously in the y or the z direction, then the center of mass of the molecule would likewise move in the y or the z direction. *Translational motion is defined as the motion of the center of mass of the system.* Any translational motion may be resolved into 3 components in the x, y and z directions. Thus, a molecule is said to have 3 degrees of translational freedom. By describing the 3 coordinates of the center of mass of the molecular system, the translational motion can be completely described.

Whether we choose to consider the motions of each individual atom, or the motions of the molecule as a whole, there will be 3n unique modes of motion, or degrees of freedom, as they are called. Subtracting the 3 degrees of translational freedom, leaves 3n - 3 modes of motion to be considered.

Since any motion which involves a change in the position of the center of mass of the system can be resolved into the various translational components, the 3n - 3 remaining kinds of motion do not involve any change in the position of the center of mass of the system in space. It is possible that all n atoms move in such a manner that the center of mass remains fixed, while the system as a whole revolves about an imaginary line passing through the center of mass, in either the x, y, or z directions. This type of motion is known as *rotational motion* and, in general, accounts for 3 of the remaining degrees of freedom. These motions are usually referred to as "rotations about x, y, or z axes." Linear molecules, however, constitute a special case.

Any rotational motion of a linear molecule can be resolved into 2 components of rotation about two mutually perpendicular axes, which are also perpendicular to the internuclear axis. Thus, linear molecules have only 2 rotational degrees of freedom.

There remain, then, (3n - 6) modes of motion or degrees of freedom, other than translational and rotational, for nonlinear molecules, and (3n - 5) such modes for linear molecules. The nature of these remaining modes of motion will now be described. It is possible for the atoms to move in such a manner that the center of mass remains fixed in space, while the molecule as a whole also remains fixed with regard to rotation about any axis. The motion in such a case is that of the atoms, which move to and fro about their equilibrium positions. This type of motion is known as *vibrational motion*. There are generally 3n - 6 modes of vibrational motion, referred to merrily as "vibrations," for a nonlinear molecule. This means that (3n - 6) unique patterns of vibrational motion may occur, and that any others which may appear to be unique can be resolved into combinations of the (3n - 6).

For a diatomic molecule (n = 2), the 3n = 6 total degrees of freedom can be represented by the following motions:

    3 translational degrees of freedom,
    2 rotational degrees of freedom,
    1 vibrational degree of freedom.

Since a diatomic molecule is necessarily linear, there are only 2 rotational degrees of freedom. Consequently, 3n - 5 = 1 vibrational degree of freedom.

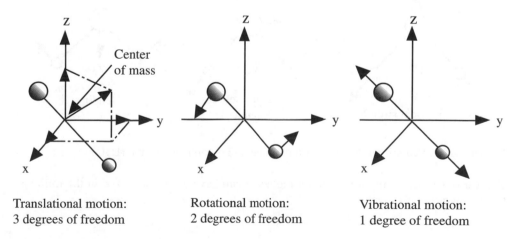

Translational motion: Rotational motion: Vibrational motion:
3 degrees of freedom   2 degrees of freedom   1 degree of freedom

Figure 6-B-2. Nuclear Motions for a Diatomic Molecule

The vibrational mode can only be the stretching mode in which there is a periodic change in internuclear separation. The various motions or degrees of freedom for a diatomic molecule are shown graphically, in Figure 6-B-2. There are 2 rotational degrees of freedom, but for simplicity only one is shown in Figure 6-B-2.

For a polyatomic molecule containing n atoms, a similar mathematical decomposition into translational and internal coordinates can be accomplished, and further decomposition into rotational and vibrational motion can be made, assuming small internuclear displacements. Since translational motion involves only the motion of the center of mass, again, only 3 degrees of freedom are ascribed to this motion for a polyatomic molecule. If the polyatomic molecule is linear, the rotational motion can be defined in 2 rotational degrees of freedom. Thus, there remain 3n - 5 vibrational degrees of freedom for the linear polyatomic molecule. These polyatomic vibrational motions are called *normal modes*, and they are uniquely defined mathematically. The mathematical treatment insists that the motions for all the atoms in the molecule oscillate about a given mean position, with the *same frequency for a specific vibrational mode*. Therefore, there is a *characteristic frequency of vibration* associated with each normal mode. As an example, the motions for the linear triatomic CO_2 are shown in Figure 6-B-3.

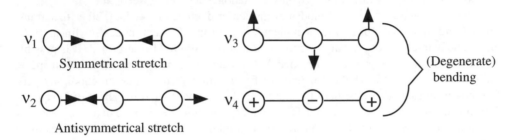

Figure 6-B-3. Vibrational Modes for a Linear Triatomic Molecule (CO_2)
(+), (-) Refer to In and Out of Plane Displacement

A nonlinear polyatomic molecule again has 3 translational degrees of freedom, but has 3 rotational degrees of freedom. Since there are 3 possible independent moments of inertia for a nonlinear molecule, 3 components are necessary to completely describe the rotational motion, (hence, the 3 rotational degrees of freedom). This leaves a nonlinear polyatomic molecule with 3n - 6 vibrational modes or degrees of freedom. An example of the degrees of freedom associated with a nonlinear, triatomic molecule is shown for the triatomic H_2O, in Figure 6-B-4.

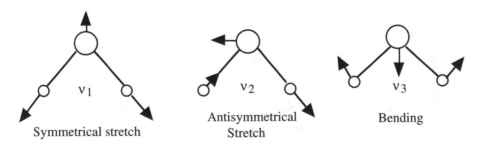

Figure 6-B-4. Vibrational Modes for a Nonlinear, Triatomic Molecule (H_2O)

The various types of motion which a molecule can have are summarized in the following block diagram.

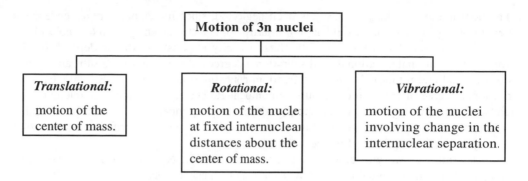

Each of these 3n modes of motion involves a certain amount of energy, which can increase if the molecule absorbs energy, or decrease if the molecule loses energy.

Problem 6-B-1. Calculate the number of translational, rotational, and vibrational degrees of freedom of (a) Xe, (b) HCl, (c) O_3 (bent, nonlinear structure), and (d) acetylene (linear). Answer: (a) 3 trans, 0 rot, 0 vib (b) 3 trans, 2 rot, 1 vib (c) 3 trans, 3 rot, 3 vib (d) 3 trans, 2 rot, 7 vib.

The classical **equipartition principle** states that, on the average, the internal energy absorbed by a molecule is distributed equally among the translational, rotational, and vibrational degrees of freedom. Each translational and rotational degree of freedom contains (1/2)RT energy per mole of molecules, while each vibrational mode or degree of freedom is considered to have RT energy per mole of molecules. The vibrational mode is given RT energy/mole, since its total energy contains both the classical kinetic energy term $(1/2mv^2)$, and the potential energy term $[1/2k(r - r_e)^2]$ for a harmonic oscillator, in which the term $(r - r_e)$ is the displacement of the internuclear separation from the equilibrium separation, r_e. Based on the equipartition principle, the molar internal energy can be written in terms of the average energies of motion of the molecules. The bar over the symbol means the *molar quantity*. Since we will only be concerned with pure substances in the remainder of this text, the solid dot nomenclature will be dropped.

$$\overline{U} = \overline{U}_{\text{translational}} + \overline{U}_{\text{rotational}} + \overline{U}_{\text{vibrational}} + \overline{U}_{\text{other}} \qquad (6\text{-}B\text{-}1)$$

For a linear molecule,

$$\overline{U} = \underset{\text{(translational)}}{\frac{3}{2}RT} + \underset{\text{(rotational)}}{RT} + \underset{\text{(vibrational)}}{(3n\text{-}5)RT} + U_{\text{other}} \qquad (6\text{-}B\text{-}2)$$

For a nonlinear molecule,

$$\overline{U} = \underset{\text{(translational)}}{\frac{3}{2}RT} + \underset{\text{(rotational)}}{\frac{3}{2}RT} + \underset{\text{(vibrational)}}{(3n\text{-}6)RT} + U_{\text{other}} \qquad (6\text{-}B\text{-}3)$$

Since the heat capacity at constant volume (C_V) is equal to $(\partial U/\partial T)_V$, by differentiating equations (6-B-2) and (6-B-3), we obtain

$$\text{linear molecule: } C_v = \underset{\text{(translational)}}{\frac{3}{2}R} + \underset{\text{(rotational)}}{R} + \underset{\text{(vibrational)}}{(3n\text{-}5)R} \qquad (6\text{-}B\text{-}4)$$

$$\text{nonlinear molecule: } C_v = \underset{\text{(transational)}}{\frac{3}{2}R} + \underset{\text{(rotational)}}{\frac{3}{2}R} + \underset{\text{(vibrational)}}{(3n\text{-}6)R} \qquad (6\text{-}B\text{-}5)$$

The contribution of U_{other} to C_V has been neglected, since its temperature dependence is generally unknown. U_{other} would arise primarily from electronic energies in the molecule.

Thus, we see that the value of the heat capacity at constant volume depends on the various translational, rotational, and vibrational degrees of freedom of the substance. Each translational and rotational degree of freedom contributes an amount 1/2R, while each vibrational degree of freedom contributes an amount R. The equipartition principle thus predicts that the heat capacity is a constant, independent of temperature. As will be seen, this is not true over the entire temperature region, but the equipartition principle does predict the upper temperature limit of the heat capacity.

As examples, we will consider the gases Ar, Ne, H_2, N_2, Cl_2, CO_2, and H_2O. The equipartition principle would predict the following heat capacities:

Degrees of Freedom	Number of Nuclear Modes	Heat Capacity Contribution
Ar, Ne	3 translational modes	3/2 R
$3n = 3$	total	$\frac{3}{2}$ R ~ 3 cal/K-mol
H_2, N_2, Cl_2	3 translational modes	3/2R
$3n = 6$	2 rotational modes	R
	1 vibrational mode	R
	total	$\frac{7}{2}$ R ~ 7 cal/K-mol
CO_2	3 translational modes	3/2 R
$3n = 9$	2 rotational modes	R
	4 vibrational modes	4 R
	total	$\frac{13}{2}$ R ~ 13 cal/K-mol
H_2O	3 translational modes	3/2 R
$3n = 9$	3 rotational modes	3/2 R
	3 vibrational modes	3 R
	total	$\frac{12}{2}$ R ~ 12 cal/K-mol

The graph of heat capacity as a function of temperature is given in Figure 6-B-5. It is obvious from the curves that the heat capacities are not constant over the temperature range, except for the monatomic gases, He, Ne, Ar, and Hg. For the inert gases, not only is the heat capacity constant, but the equipartition principle properly predicts the magnitude of the heat capacity. For the other molecules, the heat capacity always increases with temperature, but is always lower than the prediction of the equipartition principle. If a high enough temperature were attained, the heat capacity would approach the equipartition principle limit.

It should be noted that the diatomics H_2, N_2, and O_2 all have C_V values greater than 3 cal/K-mole at -100 °C, while at 0°C, they are all close to 5 cal/K-mol. This would be the limit predicted by the equipartition principle, if only the translational and rotational degrees of freedom contributed to the heat capacity. This is quite obviously the situation, since at 100 °C the C_V values are not far from the predicted ~ 5 cal/K-mol. As the temperature is increased, the vibrational mode does begin to make a contribution. The curve for Cl_2 attains a value of 6.40 cal/K-mole. The Cl_2 bond is the weakest of the diatomic molecules given in Figure 6-B-5, and, as will be shown later, this is an important factor as to whether the vibrational mode can become active, that is, contribute to the heat capacity.

The triatomics cannot be investigated at low temperatures, since they will condense to a liquid (H_2O) and a solid (CO_2). At the initial temperature indicated in Figure 6-B-5, the heat capacity exceeds that predicted by just translational and rotational degrees of freedom. Apparently there are some vibrational modes that are contributing, even at the lower temperatures. The ν_3 vibration for H_2O, and the $\nu_3 = \nu_4$ vibrations for CO_2, occur at very low frequencies or energies, and apparently contribute to C_V, even at lower temperatures. In neither case does the heat capacity reach that predicted by the equipartition principle. The reason for the various contributions to C_V from the translational, rotational, and vibrational degrees of freedom at different temperatures, will be better understood as we go along. At this point, we can say that the equipartition principle predicts the upper limit for C_V, which is eventually approached at higher temperatures.

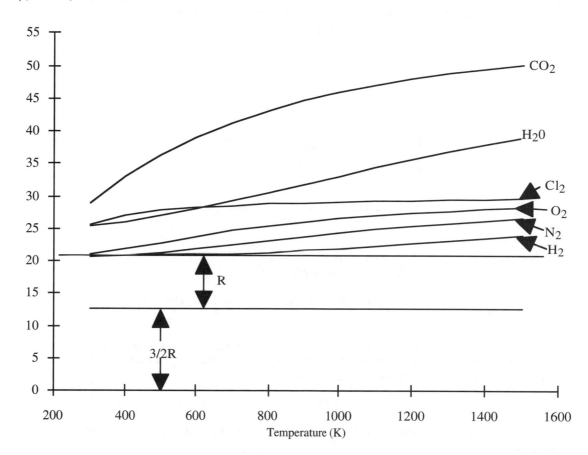

Figure 6-B-5. Heat Capacity Curves for Monatomic, Diatomic, Linear and Nonlinear Triatomic Molecules

Problem 6-B-2. Sketch heat capacity (C_V) versus T curves for (a) Xe, (b) HCl, (c) O_3, and (d) acetylene, from 200-1,000 K. Assume complete activation of the translational and rotational degrees of freedom at 200 K. Indicate the limiting value of C_V at high temperatures. Answer: (a) $C_V = (3/2)R$ = constant (b) changes from $C_V = (5/2)R$ to $(7/2)R$ (c) changes from $C_V = 3R$ to $6R$ (d) changes from $C_V = (5/2)R$ to $(19/2)R$.

C. Experimental Results Leading to Quantum Mechanics

Prior to the development of quantum mechanics, there were some essential scientific developments which suggest the foundation of quantum mechanics. Since most of these developments are discussed in the first year chemistry course, they will only be outlined for review at this time, with only the significant aspects emphasized in the following subsections.

Planck's Quantum Postulate

The wavelength distribution of the energy radiating from a so-called blackbody was experimentally known at the turn of the 20th century. The distribution is indicated by the solid line in Figure 6-C-1. The classical electromagnetic theory of radiation developed at that time was capable of predicting the low energy (high wavelength) region of the spectrum, as shown by the broken line in Figure 6-C-1. However, the theory was incapable of predicting the high energy (low wavelength) region of the spectra, called the ultraviolet. Hence, this disagreement between theory and experiment was referred to as the "ultraviolet catastrophe." Specifically, the ultraviolet region of the spectrum is that region of wavelengths shorter than the violet or blue region of the visible spectrum.

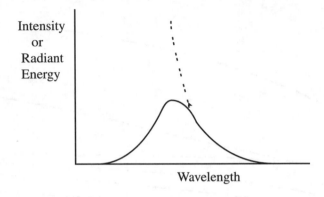

Figure 6-C-1. Distribution of Radiation from a Blackbody, T = Constant. Dashed Line
Represents the Ultraviolet Catastrophe

Planck was able to theoretically account for the experimental blackbody distribution by introducing the following postulates:

1. Electromagnetic radiation is composed of "bundles of energy" called *quanta*, or *photons*, which carry a discrete amount of energy. The quantity of energy is related to the frequency of the radiation by the simple expression

$$\varepsilon = h\nu \qquad \text{(6-C-1)}$$

    where ν = frequency of radiation, h is a constant, and ε is the energy of the quantum or photon.

2. The energy of an oscillator, such as a vibrating molecule, cannot have a continuous array of values, but can have only certain unique values given by the relationship

$$E = nh\nu \qquad \text{(6-C-2)}$$

    where $n = 1, 2, 3...$, and ν = frequency of the oscillator. Thus, the energy of the oscillator, E, must be an integral multiple of $h\nu$.

These two postulates are essential to the derivation of the distribution of blackbody radiation, but as the reader will see, they are also of fundamental importance to quantum theory.

The proportionality factor (h) is referred to as **Planck's constant** and is commonly represented in the units 6.626×10^{-27} erg-sec. The frequency is generally given in $\sec^{-1}$ and hence, ε has the units of ergs. The erg = 1 $g\text{-}cm^2/sec^2$ and is the basic energy unit in the cgs system of units: centimeter, gram, second. Alternatively, there is the MKS system of units where the basic energy unit is the joule(J) = 1 kg-m^2/sec^2. Planck's constant in the MKS system would be h = 6.627 x 10^{-34} J-sec. MKS stands for meter, kilogram, second. We will express our calculations in both systems of units so that you will be familiar with both systems as you may find them in the literature.

Also, the frequency of the light can be expressed in different terms, such as wavelength (λ), and frequency in wavenumber (cm^{-1} or kaysers). The relationship between these expressions is given by

$$\bar{\nu} = \frac{1}{\lambda} = \frac{\nu}{c} \qquad (6\text{-C-}3)$$

where c = velocity of light = 2.998 x 10^{10} cm/sec = 2.998 x 10^8 m/sec, ν, is the frequency (sec^{-1}), $\bar{\nu}$ is the wave number (cm^{-1}), and λ is the wavelength. λ is commonly expressed in angstrom units (A$^{\circ}$) or nanometers (nm) where 1 A$^{\circ}$ = 1 x 10^{-8} cm and 1 nm = 1 x 10^{-9} m.

Problem 6-C-1. Visible light falls within the range, 4,000 - 7,000 A. Calculate this range in (a) $\bar{\nu}$, (b) ν, (c) energy in erg/molecule, (d) kJ/mol, and (e) electron volts/molecule. Answers: (a) 25,000 - 14,200 cm^{-1} (b) 7.5 x 10^{14} - 4.29 x 10^{14} sec^{-1} (c) 4.96 x 10^{-12} - 2.84 x 10^{-12} erg/molecule (d) 298.7 - 170.7 kJ/mol (e) 3.1 - 1.77 eV/molecule.

The Photoelectric Effect

The **photoelectric efffect** is the emission of electrons from a metal surface, as a result of exposure to electromagnetic radiation. The classical electromagnetic theory was unable to explain the following experimental results:

1. Electrons are not emitted from the metal surface unless the frequency of the radiation exceeds some specific value (ν_o) which is characteristic of the metal used. If the emitted electrons were collected and the current measured, a graph of current (I) versus frequency would be represented by a discontinuous curve, as shown in Figure 6-C-2. The magnitude of the current (I) would remain constant above ν_o, providing

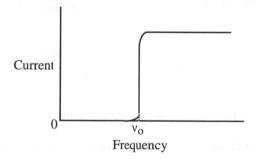

Figure 6-C-2. Photoelectric Current as a Function of Frequency

the intensity of light were constant at the different frequencies. The classical wave concept of light (oscillating electromagnetic fields) would predict quite different results. The energy carried by a light wave must be considered to be "spread out" over the entire region of the oscillation. Thus, when light waves (radiation) impinge on a metal surface, the energy would be more or less evenly distributed to all the electrons within the path of the wavefront. The result would be that each electron in the path of the radiation would receive a small amount of energy from each wave. If the intensity of the radiation were increased sufficiently, the electrons would finally acquire the amount of energy necessary to leave the surface of the metal. Thus, according to the wave theory of light, the emission of electrons would be a function of the total amount of energy absorbed, and therefore would be a function of the intensity of the radiation and the time of exposure of the metal surface to the radiation. However, no part of the classical wave theory suggests that there should be a dependence on the frequency of the radiation. Contrary to the experimental result of a threshold frequency, classical theory predicts that any frequency of radiation could eventually bring about the photoelectric effect.

2.   The energy of the emitted electrons increases with an increase in the frequency of the radiation above the characteristic frequency (v_0) . Again, this could not be explained by the classical theory.

Albert Einstein explained the photoelectric phenomenon by using Planck's postulate that radiation arrived at the surface as quanta of energy, hv. An electron at the surface received energy as a result of a collision with the photon. Thus, all the energy of one photon was acquired by a single electron. Unless the v exceeded v_0, there was insufficient energy in the photon for the electron emission process to occur. If the frequency exceeded v_0 the excess energy was retained by the electron after escaping the surface. Hence, the relationship

$$E_e = hv - hv_0$$

where E_e = energy of the electron. Figure 6-C-3 illustrates how Planck's constant can be evaluated from the linear graph of E_e- versus v with a slope = h. The term hv_0 is called the work function of the metal, and is the minimum energy required to remove an electron from the metal surface.

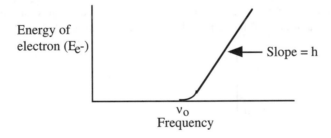

Figure 6-C-3. Photoelectron Energy as a Function of Frequency

The significance of these results are: (1) support of the particle concept of radiation (quantum postulate), and (2) the unique frequency v_0, which suggests a unique quantized event, namely, a collision between a photon (hv) and an electron.

Atomic Spectra

Both the absorption and the emission of light by atoms and ions in the gas phase are observed experimentally. The relative amount of light absorbed or emitted at each wavelength or frequency is known as the **absorption or emission spectrum** of the element. Since light contains energy, the absorption of light corresponds to an increase in the atom's energy. Before any substance can emit light energy, it must be excited (allowed to absorb energy). This is usually acomplished by heating to a high temperature, or by electrical discharge. The uniqueness of this experimental result is the fact that only certain frequencies or wavelengths of light are absorbed or emitted. Moreover, the particular frequencies of light absorbed or emitted by a particular element are unique to that element. An example of a "line" spectrum is shown in Figure 6-C-4.

The classical understanding of light and energy cannot possibly explain, or even attempt to predict these unusual observations. However, in view of the photon theory of light, we can at least see the significance of these observations. Since light of different frequencies corresponds to photons of different energies, given by equation (6-C-1), the fact that only certain frequencies are absorbed or emitted by the atom means that only certain energies can be absorbed or emitted by the atom. This suggests that atoms, like Planck's oscillators, can have only certain energies. This restriction of only certain energies is completely outside the framework of classical theory.

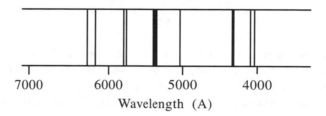

Figure 6-C-4. Atomic Line Spectrum

D. Bohr Theory of the Atom

Niels Bohr attempted to explain the atomic emission spectra, but was successful only after introducing a radically new concept of quantization. His theory was amazingly exact in describing the spectra of the hydrogen atom, or any hydrogen-like ion consisting of the positively charged nucleus plus a single electron. However, the theory was incapable of accurately describing more complex systems.

Bohr made the following basic assumptions in his model for systems consisting of a +Z charged nucleus and a single negatively charged electron:

1. There is a force of attraction between the +Z nucleus and the electron consisting of the classical coulombic attraction,

$$F = -\frac{Ze^2}{r^2} \qquad \text{(6-D-1)}$$

   where e is the charge of an electron in electrostatic units, and r is the distance between the electron and the nucleus in cm.

2. The electron moves in circular orbits at a speed and distance such that the classical centrifugal force from such motion balances the coulombic attractive force,

$$\frac{mv^2}{r} = \frac{Ze^2}{r^2} \qquad \text{(6-D-2)}$$

3. Bohr arbitrarily assumed that the orbital angular momentum was quantized. It could take on values which were only multiples of the quantity $h/2\pi$.

$$mvr = n\left(\frac{h}{2\pi}\right) \qquad n = 1, 2, \cdots \qquad \text{(6-D-3)}$$

   where mvr = orbital angular momentum.

4. Emission of light can occur only when the electron goes from a higher energy, E_2, to a lower energy, E_1, the frequency of the emission given by Planck's postulate.

$$E_2 - E_1 = \varepsilon_{photon} = h\nu = hc\,\bar{\nu} \qquad (6\text{-}D\text{-}4)$$

The radically new concept which was introduced was the quantization of the orbital angular momentum in the third assumption.

By employing these assumptions, Bohr was able to derive the following expression for the energy of the atom, E_n.

$$E_n = -\frac{2\pi^2 m e^4 Z^2}{h^2}\frac{1}{n^2} \qquad n = 1, 2, 3, \ldots \qquad (6\text{-}D\text{-}5)$$

where m = the mass of electron in grams, e = the charge on the electron in statcoulombs or esu, and n is the quantum number from the third postulate. Thus, the quantum postulate leads to the experimentally observed fact that the atom can have only certain unique energies, given by equation (6-D-5). The permitted values of the energy are referred to as **energy levels**.

In equation (6-D-5) only Z = atomic number and n can change and they can be factored out from the other constants.

$$E_n = -\left(\frac{2\pi^2 m e^4}{h^2}\right)\frac{Z^2}{n^2} \qquad n = 1, 2, 3, \ldots \qquad (6\text{-}D\text{-}6)$$

The term in parentheses is common to all hydrogenic systems.

According to the fourth postulate, equation (6-D-4) gives the frequency $\bar{\nu}$ of the photon of light, either emitted or absorbed when the atom changes from one energy level to another. Let us represent the upper (U) energy level by E_U and the lower (L) energy level by E_L. The quantum numbers corresponding to these states are n_U and n_L, respectively. The energy of the photon associated with this transition in energy levels is given by

$$\varepsilon = E_U - E_L = -\left(\frac{2\pi^2 m e^4}{h^2}\right)\frac{Z^2}{n_U^2} + \left(\frac{2\pi^2 m e^4}{h^2}\right)\frac{Z^2}{n_L^2} \qquad (6\text{-}D\text{-}7)$$

We can factor out the Z^2 and the term in parentheses, giving

$$\varepsilon = hc\,\bar{\nu} = \left(\frac{2\pi^2 m e^4}{h^2}\right) Z^2 \left(\frac{1}{n_L^2} - \frac{1}{n_U^2}\right).$$

Solving for the frequency of the photon in wave numbers associated with the transition

$$\bar{\nu} = \left(\frac{2\pi^2 m e^4}{ch^3}\right) Z^2 \left(\frac{1}{n_L^2} - \frac{1}{n_U^2}\right) \qquad (6\text{-}D\text{-}8)$$

Equation (6-D-8) can represent the frequencies of a series of spectral lines if the lower energy state, n_L, is held constant and the upper energy state, n_U, is increased. These are called Rydberg Series, being named after one of the original investigators in this area. The constants in the parentheses are appropriately defined as the Rydberg constant for the hydrogen atom, R_H, where $Z^2 = 1^2 = 1$.

$$R_H = \left(\frac{2\pi^2 me^4}{ch^3} \right) \qquad (6\text{-}D\text{-}9)$$

The Rydberg Constant for the hydrogen atom has been evaluated to be 109,677 cm^{-1}. Equation (6-D-8) can now be written in terms of R_H giving

$$\bar{\nu} = R_H \, Z^2 \left(\frac{1}{n_L^2} - \frac{1}{n_U^2} \right) \qquad (6\text{-}D\text{-}10)$$

The value for $\bar{\nu}$ in equation (6-D-10) is always positive and it will be understood that the photon is absorbed when going from $n_L \longrightarrow n_U$ and that the photon is emitted when going from $n_U \longrightarrow n_L$. Since n_L and n_U can have only integer values, only certain values of the frequency are possible. In this way, the existence of unique energy levels was accounted for. When the mass of the electron was replaced by the reduced mass of the system, the agreement with experiment was found to seven significant figures, an amazing result, indeed. Seldom can a theory be tested to such a degree, and even more seldom can a theory account for the results with such accuracy.

Problem 6-D-1.
(a) Calculate the energy of the transition from $n_L = 1$ to $n_U = 2$ for Li^{2+} and express the result in eV/molecule and kJ/mol. In what region of the spectrum would this transition occur?
(b) The ionization potential (IP) is the energy required to completely remove the electron, that is, $n_L = 1$ to $n_U = \infty$. Calculate the IP for Li^{2+} in eV. Answers: (a) 91.4 eV/molecule and 8,811 kJ/mol, vacuum ultraviolet (b) 122 eV.

E. Quantum Mechanics

Wave Nature of Particles

We have seen from the Planck postulate, supported by the distribution of blackbody radiation, the photoelectric effect, and the atomic spectra, that light displays particle (quantum or photon) behavior, as well as wave character. Louis de Broglie made an extremely important extension of this concept. He reasoned that particles of finite rest mass must also have a wave nature. De Broglie proposed the relationship: the product of momentum (mv) and wavelength of a particle is equal to Planck's constant.

$$(\lambda)(mv) = h$$

or

$$\lambda = \frac{h}{mv} \qquad (6\text{-}E\text{-}1)$$

This postulate was later confirmed experimentally by electron and neutron diffraction experiments. Both of these techniques play an important role in the study of the structure of matter. The wave character of finite particles is of pivotal importance to the development of quantum mechanics. For particles of a large mass and low velocity, the wavelength becomes exceedingly small and the wave nature is essentially negligible. Hence, such bodies obey the laws of classical mechanics. There is a very uniform conversion from small, high velocity objects, such as electrons, whose relatively long wavelengths must be taken into account, to large, low velocity objects, which obey the laws of classical mechanics.

Problem 6-E-1.

(a) Calculate the wavelength of a 100 eV electron. Compare this wavelength with typical internuclear separations. Hint: Kinetic Energy = $1/2mv^2$.

(b) Calculate the wavelength of a baseball (1/2 lb) traveling at a speed of 100 miles/hr. Can one account for a team's low batting averages with the de Broglie wave nature of the baseball? Answer: (a) 1.23 A° (b) 6.5 x 10^{-25} A°.

Schrödinger Equation

On the basis of the wave nature of particles, the subject of quantum mechanics was developed, independently, by Schrödinger and Heisenberg. Although the two approaches first appeared to be different, they were later shown to be equivalent. For systems which involve particles of small mass (for example, electrons), it is essential to use quantum mechanics. For larger masses, such as macroscopic bodies, classical mechanics is satisfactory. As we will see shortly, a quantum mechanical treatment of the translational motion of macroscopic particles automatically conforms to the classical result.

Our development of quantum mechanics will be far from complete or rigorous. However, it is hoped that the treatment presented here, specifically, the solution of the particle-in-a-box problem, will show the significance of the quantum mechanical description in contrast to the classical description. Only the solutions to the quantum mechanical analysis of the other molecular motions will be given. The knowledge obtained from the particle-in-a-box analysis will make it possible for the reader to properly interpret and use the quantum mechanical solutions for other problems.

Quantum mechanics is generally presented as a concise and complete set of postulates or laws. These are applied, in place of certain principles of classical mechanics, to a variety of problems. The quantum mechanical solutions to these problems provide information concerning such aspects of the system under consideration as the energy, momentum, angular momentum, and so forth. Since our interest is primarily the energies of systems and the associated spatial distribution of particles, we will concern ourselves with only a portion of quantum mechanics. The results which we need can be obtained from an equation known as the **Schrödinger equation**. This equation is a *wave* equation, since particles, such as electrons, are known to exhibit wave character. Hence, the solutions are amplitude functions commonly called wave functions, which do not have the physical significance that the solutions to problems in classical mechanics have.

The time-independent Schrödinger equation can be written simply in operator notation as

$$\hat{H}\psi = E\psi \tag{6-E-2}$$

where $\hat{H}$ is an operator, called the *Hamiltonian operator*, which will be explained in detail shortly. Equation (6-E-2) is a type known as an eigenvalue equation, where the solution involves finding ψ, the *eigenfunction*, in terms of the coordinates of the problem. The signficance of the eigenfunction is that the operation on it by $\hat{H}$, gives back the eigenfunction (ψ), mutliplied by a constant, called the corresponding *eigenvalue*. The constant in equation (6-E-2) is the energy (E) of the system being considered. In general, we will obtain a series of solutions (ψ_i), which correspond to a series of energies or eigenvalues (E_i).

The mathematical use of the term operator may not be familiar to everyone; however, even the reader with the most elementary math background has used them. An operator is a symbol that represents some kind of operation; for example, one of the simplest 2 x, which simply means "multiply by 2." We could use this operator to operate on any type of function, such as

$$(2 \text{ x })z = 2z$$
$$(2 \text{ x })y = 2y$$
$$(2 \text{ x })e^x = 2e^x$$
$$(2 \text{ x })f(z) = 2f(z)$$

Operators can also be of the differential type, such as taking a derivative with respect to x, written (d/dx). Again, the reader is familiar with this type of operation from the discussion in the Appendix, Section B.

The Hamiltonian operator of equation (6-E-2) is of the differential and multiplier type. $\hat{H}$ is defined as

$$\hat{H} = \left(-\frac{\hbar^2}{2m} \nabla^2 + V \right) \qquad (6\text{-}E\text{-}3)$$

where $\hbar = h/2\pi$ and h = Planck's constant, m = mass of the moving particle, V = potential energy for the particle. ∇^2 is called the *Laplacian operator*, which is a differential operator involving second derivatives.

$$\nabla^2 = \frac{\partial^2}{\partial x^2} + \frac{\partial^2}{\partial y^2} + \frac{\partial^2}{\partial z^2} \qquad (6\text{-}E\text{-}4)$$

The x, y, z coordinates in (6-E-4) refer to the coordinates of the moving particle. The "2" in (6-E-4) means second derivative, or "take the derivative of the derivative." If there is more than one moving particle, then subscripts must be used, and the term

$$-\frac{\hbar^2}{2m} \nabla^2$$

must be replaced by a summation

$$\sum_i -\frac{\hbar^2}{2m_i} \nabla_i^2$$

In our problems only a single particle will be free to move, so equation (6-E-3) can be used in its present form. Equation (6-E-3) can be substituted into equation (6-E-2) to give

$$\left(-\frac{\hbar^2}{2m} \nabla^2 + V \right) \psi = E \psi$$

$$-\frac{\hbar^2}{2m} \nabla^2 \psi + (V\text{-}E) \psi = 0 \qquad (6\text{-}E\text{-}5)$$

Equation (6-E-5) is called a second order partial differential equation, since it involves second derivatives. The solution of partial differential equations is generally covered in a higher mathematics course; thus, quite often, we will merely accept the solution as provided. Frequently, the specific differential equation which results from the application of the general equation to a particular problem cannot be solved exactly. In these cases, approximation methods must be employed to find approximate solutions. This is one of the biggest stumbling blocks in the quantum mechanical solution of problems.

No attempt is made to attach physical significance to ψ itself; rather, the square of the amplitude ψ^2 is considered to have physical meaning. Without any further comment then, we state that ψ^2 is proportional to the probablity of finding the particle in a small element of volume in the vicinity of the position x, y, z. ψ, and hence ψ^2, is a function of the spatial coordinates x, y, z, and therefore, the probablity per unit volume will also be a function of position in space, x, y, z. (Probability per unit volume $\alpha \psi^2$, where $\psi = f(x, y, z)$.)

The solutions (ψ) which we seek for (6-E-5) must be proper (physically meaningful) solutions. There are frequently many solutions to a differential equation, however, these may not be applicable to our problem, as the example to be presented shortly should make clear. The proper solutions, which we seek, must have the following properties:

(a) ψ must be single-valued for all values of x, y, z;
(b) ψ must be finite for all values of x, y, z;
(c) ψ must be continuous for all values of x, y, z;
(d) ψ must satisfy the boundary conditions of the problem.

Solutions which satisfy these four requirements are often referred to as "well-behaved functions."

Problem 6-E-2
(a) Carry out the operation (x) ln on the function e^x, where ln refers to the natural logarithm.

(b) Carry out the operation $\left(\dfrac{d}{dx}\right)\left(\dfrac{d}{dx}\right) = \dfrac{d^2}{dx^2}$ on e^{2x}. Answer: (a) x^2 (b) $4\,e^{2x}$.

Problem 6-E-3

(a) Show that e^{2x} is an eigenfunction of the eigenvalue equation:

$$\frac{d^2 \psi}{dx^2} = a\,\psi.$$

(b) What is the eigenvalue? Answer: (a) $\dfrac{d^2\psi}{dx^2} = 4\,e^{2x}$ (b) 4.

F. Particle in a One-Dimensional Box

As discussed earlier, the motion of atoms in a molecule can be separated into:

(a) the motion of the center of mass (translational motion),
(b) motion at fixed internuclear distance about the center of the mass (rotational motion),
(c) small displacements about the equilibrium internuclear separations, preserving the position of the center of mass (vibrational motion).

In this section, we will be concerned only with the translational motion of the center of mass of a molecule. The total mass of the molecule will be represented by m.

In order to treat the motion of a particle confined to a specific portion of space, or "particle in a box," using the principles of quantum mechanics, we must decide how to represent the walls of the container mathematically. This is accomplished by requiring that the potential energy of the particle goes to infinity at the boundary or wall. Recalling that potential energy merely reflects the existence of a force, an infinite potential at the boundary thus insists that an infinite retarding force exists at that position, and hence, successfully represents an impenetrable wall. If the particle is an ideal gas molecule, it must be free of intermolecular forces; hence, within the container, the potential, V, must be zero (force-free condition). For the purpose of mathematical simplicity, we will first consider this problem in *one dimension*. A graphical representation of the one-dimensional box is given in the following figure.

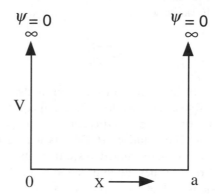

Although this simplified situation certainly cannot represent a molecule moving about in a three-dimensional container, it will be instructive with regard to the analysis and the interpretation of the results. Physically, this situation could represent a molecule confined to move on a straight line between certain limits. We will represent the length of our one-dimensional box by a, and we will use the variable x to represent the position in this one dimension. The three-dimensional solution to the particle in a box will be presented later.

The previously stated conditions can be expressed very precisely in mathematical terms, where V is the potential of the particle.

$$\begin{cases} x \leqq 0 \\ V = \infty \end{cases} \quad \begin{cases} x \geqq a \\ V = \infty \end{cases} \quad \begin{cases} a > x > 0 \\ V = 0 \end{cases} \tag{6-F-1}$$

The Schrödinger Equation, (6-E-5), written specifically for the motion of the particle in a one-dimensional box, is

$$\left(-\frac{\hbar^2}{2m} \nabla^2 + 0 \right) \psi = E \, \psi$$

or

$$-\frac{\hbar^2}{2m} \frac{d^2 \psi}{dx^2} = E \, \psi \tag{6-F-2}$$

The general solution (eigenfunction) to (6-F-2) is

$$\psi = A \sin Cx + B \cos Cx \tag{6-F-3}$$

where A, B, and C are constants which depend on the specific conditions of the problem. The function given in equation (6-F-3) can be shown to be a solution, an eigenfunction, to (6-F-2), by performing the operation $\overline{\overline{H}} = (-\hbar^2/2m)(d^2/dx^2)$ on it, and observing that it gives ψ times a constant.

$$\frac{d \psi}{dx} = AC \cos Cx - BC \sin Cx$$

$$\frac{d^2 \psi}{dx^2} = - AC^2 \sin Cx - BC^2 \cos Cx$$

$$\frac{d^2 \psi}{dx^2} = - C^2 (A \sin Cx + B \cos Cx)$$

$$\frac{d^2 \psi}{dx^2} = - C^2 \, \psi \tag{6-F-4}$$

Comparing (6-F-4) with (6-F-2), we see that

$$E = C^2 \frac{\hbar^2}{2m} \tag{6-F-5}$$

After we evaluate C, we can obtain a value or expression for the energy, E.

The general solution, (6-F-3), satisfies the differential equation in (6-F-4). However, it must be checked to see if it satisfies the specifications for "proper solutions," which we set forth in the previous section. The solution, (6-F-5), is indeed finite, single-valued, and continuous; however, it must also be examined to see if it satisfies the boundary conditions of this problem. We must first consider the implication so the limiting value of the potential, $V = \infty$. Since, as previously stated, this represents an impenetrable wall or boundary through which the particle cannot pass, we reason that ψ^2, the probabilty density, equals zero, and hence, ψ must be zero for $x \geq a$ and $x \leq 0$. A finite nonzero value of ψ^2 would imply that there is some chance of the particle getting past the boundary.

This exact conclusion, concerning the value of ψ at $x = 0$ and $x = a$, follows naturally from the mathematics of the problem. Consider the Schrödinger wave equation (6-E-5), when $V = \infty$ and $\nabla^2 = d^2/dx^2$.

$$-\frac{\hbar^2}{2m} \frac{d^2 \psi}{dx^2} + (\infty - E) \psi = 0 \tag{6-F-6}$$

For finite values of E, the quantity $(\infty - E)$ is also ∞. Thus, in order to satisfy equation (6-F-6), the solution ψ must be found, which, when differentiated twice, will yield a value of ∞ times the solution. No such solution can be found, which also satisfies the properties for proper solutions presented earlier. We thus conclude that there should be no solution ($\psi = 0$) in the region where $V = \infty$. The boundary conditions now become:

$$\begin{cases} x \leq 0 \\ \psi = 0 \end{cases} \begin{cases} x \geq a \\ \psi = 0 \end{cases} \tag{6-F-7}$$

We must now examine the general solution given in equation (6-F-3), to see if it satisfies these boundary conditions. When $x = 0$, ψ must be zero; that is,

$$0 = A \sin (C \bullet 0) + B \cos (C \bullet 0) \tag{6-F-8}$$

In order for (6-F-8) to be satisfied, B must be zero. This is obvious, since $\sin 0 = 0$, and $\cos 0 = 1$, or

$$0 = A \bullet 0 + B \bullet 1$$

or

$$B = 0$$

The general solution of (5-F-3) must then become

$$\psi = A \sin Cx \tag{6-F-9}$$

in order to satisfy the wave equation and the boundary conditions simultaneously.

Finally, we must satisfy the boundary conditions, $\psi = 0$, when $x = a$; that is,

$$0 = A \sin (C \bullet a) \tag{6-F-10}$$

Equation (6-F-10) will be satisfied, if $A = 0$, or if $\sin C \bullet a = 0$. However, it would be unsatisfactory for A to be zero, since then ψ would be identically zero for all values of x.

In other words, $\psi = 0$ would be no solution at all. Thus, we must accept the alternate possibility,

$$\sin (C \bullet a) = 0$$

which will occur whenever the $(C \bullet a)$ product is a multiple of $\pi = 180^\circ$, since

$$\sin n\pi = 0 \qquad\qquad n = 1, 2, 3, \ldots$$

Therefore, our condition is

$$C \bullet a = n\pi$$

or

$$C = \frac{n\pi}{a} \qquad\qquad\qquad (6\text{-}F\text{-}11)$$

Substituting (6-F-11) into our previous solution, our eigenfuctions are

$$\psi = A \sin \frac{n\pi}{a} x \qquad\qquad n = 1, 2, 3, \ldots \qquad (6\text{-}F\text{-}12)$$

Also, an explicit expression for the eigenvalues (energies) can be obtained by substituting (6-F-11) into the relationship given in (6-F-5).

$$E = n^2 \frac{\pi^2 \hbar^2}{2ma^2} = n^2 \left(\frac{h^2}{8ma^2} \right) \qquad n = 1, 2, 3, \ldots \qquad (6\text{-}F\text{-}13)$$

Thus, we arrive at the extremely important conclusion that a particle confined in a box can have only certain discrete values for its total energy. These energies are given by equation (6-F-13) and correspond to the various values of n, 1, 2, 3, ..., known as *quantum numbers*. Corresponding to each possible value of the energy, there is a unique function, equation (6-F-12), since ψ also depends on the quantum number, n. It should be pointed out that quantization occurs in this case as a direct consequence of the application of the principles of quantum mechanics, rather than as an arbitrary postulate.

Our solution is now complete, except for the evaluation of the constant A in (6-F-12). This is obtained by "normalizing" the wave function. It was previously stated that the physical significance of the eigenfunction was ψ^2, and its relationship to the probability of finding the particle in a small element of volume in the vicinity of x, y, z. The process of *normalization* is a process whereby ψ is scaled up or down, so that the total probability of finding a particle, somewhere in all of space, is unity. If A is left alone, or given any arbitrary value, then ψ^2 will be a relative measure of probability.

For the one-dimensional particle in a box problem, normalization can be expressed mathematically as

$$\int_0^a \psi^2 dx = 1 \qquad\qquad (6\text{-}F\text{-}14)$$

where the "element of volume" in this problem is dx, since there is only one dimension. All space in this problem is $0 \rightarrow a$, since the particle cannot escape a wall of infinite potential. Substituting (6-F-12) into (6-F-14),

$$\int_0^a A^2 \sin^2\left(\frac{n\pi}{a}\right) x \, dx = 1$$

$$A^2 \int_0^a \sin^2\left(\frac{n\pi}{a}\right) x \, dx = 1 \qquad (6\text{-F-}15)$$

From the integral tables,

$$\int \sin^2 bx \, dx = \frac{1}{2} x - \frac{1}{4b} \sin 2bx \qquad (6\text{-F-}16)$$

Hence, setting $b = (n\pi/a)$, equation (6-F-15) becomes

$$A^2 \left[\frac{1}{2} x - \frac{1}{4} \frac{a}{n\pi} \sin 2 \frac{n\pi}{a} x \right]_0^a = 1$$

$$A^2 \left[\frac{1}{2} a - \frac{1}{4} \frac{a}{n\pi} \sin 2 \frac{n\pi}{a} a - \frac{1}{2} \cdot 0 - \frac{1}{4} \frac{a}{n\pi} \sin 2 \frac{n\pi}{a} \cdot 0 \right] = 1$$

$$A^2 \left[\frac{a}{2} - 0 - 0 - 0 \right] = 1$$

A is found to be

$$A = \left(\frac{2}{a}\right)^{1/2}$$

which we can substitute into equation (6-F-12) to get the final, normalized solution.

$$\psi = \left(\frac{2}{a}\right)^{1/2} \sin \frac{n\pi}{a} x \qquad n = 1, 2, 3 \ldots \qquad (6\text{-F-}17)$$

We now ask, "What is the meaning of these results? What is the significance of the quantum mechanical solution, in contrast to the classical solution of the problem?" The significance of the eigenfunctions (6-F-17) and the eigenvalues (6-F-13) is essentially twofold:

1. The *energy* of the particle can have only *discrete* values. In other words, the particle can have only certain energies, and not all possible values. This is because the energy is a function of n, which can have only integer values. In classical mechanics, there is no such restriction, and the energies are said to be continuous. These discrete energies are known as *energy levels*.

2. The position of the particle is represented by a probability distribution (ψ^2), which is different for each value of the energy. The probability distribution gives the probability per unit distance of finding the particle at a certain position, because the exact position cannot be known. In classical mechanics, there is no such restriction, and the precise position of an object can always be specified.

The results of the one-dimensional particle in a box are represented graphically in Figure 6-F-1. The horizontal lines on the graph correspond to the first four energy levels.

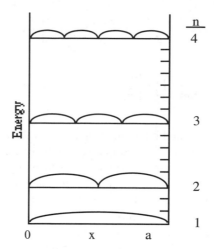

Figure 6-F-1. Graph of ψ^2 as a Function of x for the Lowest Four Energy Levels

Obviously, the energy levels are not equally spaced, since the energy depends on the *square* of n. At higher energy levels, the separation between levels increases. The curve drawn at each energy level represents the function ψ^2 (square of (6-F-17)). It should be noted that ψ^2 goes to zero at certain values of x between 0 and a. These points are called **nodes**—points where there is zero probability of finding the particle. It should also be noted that the number of nodes increases as n increases, in fact, for this problem, the number of nodes is simply n - 1. The existence of nodes in wave functions is a general occurrence in quantum mechanical solutions. However, generally, more than one dimension is involved in a problem, and as a result, more than one type of quantum number arises. In such cases, the relationship between the number of nodes and the quantum numbers will be more complicated. Futhermore, for problems in three dimensions, the nodes are nodal surfaces; that is, surfaces on which the probability of finding the particle is zero.

Problem 6-F-1. Calculate the 5 lowest energy levels for an electron in a one-dimensional box of length 8.4Å. ($1Å = 10^{-8}$ cm).
Answer: 8.5×10^{-13} erg = 8.5×10^{-20} J, 3.4×10^{-12} erg = 3.4×10^{-19} J, 7.7×10^{-12} erg = 7.7×10^{-19} J, 1.37×10^{-11} erg = 1.37×10^{-18} J, 2.14×10^{-11} erg = 2.14×10^{-18} J.

G. Particle in Two- or Three-Dimensional Box (Translational Energy)

Two-Dimensional Particle in a Box

The two-dimensional particle in the box would have a Hamiltonian operator similar to the one-dimensional particle in the box except that the Laplacian operator would be

$$\nabla^2 = \frac{\partial^2}{\partial x^2} + \frac{\partial^2}{\partial y^2}$$

The Schrödinger equation would then be

$$-\frac{\hbar^2}{2m}\left(\frac{\partial^2 \psi}{\partial x^2} + \frac{\partial^2 \psi}{\partial y^2}\right) = E\psi \qquad (6\text{-}G\text{-}1)$$

The solution to equation (6-G-1) involves a technique called separation of variables which breaks the equation down to the solution of the one-dimensional particle in the box in the x and y directions. We will not go through this derivation, but the resulting solutions are:

$$\psi = \left(\frac{4}{ab}\right)^{1/2} \sin \frac{n_x \pi}{a} x \; \sin \frac{n_y \pi}{b} y \qquad \text{(6-G-2)}$$

$$E = \frac{h^2}{8m}\left(\frac{n_x^2}{a^2} + \frac{n_y^2}{b^2}\right) \qquad n_x, n_y, = 1, 2, \cdots \qquad \text{(6-G-3)}$$

where n_x, n_y = 1, 2, 3, ..., and a and b are the dimensions of the two-dimensional box in the x and y directions. Note that there are two quantum numbers, n_x and n_y, which may take on any positive whole numbers. In this case the values taken by n_x and n_y are independent of each other; i.e., there are no restrictions on the combination of n_x and n_y values. As we will see later, this is not always true and frequently the value of one quantum number will restrict the value of the other quantum number.

Note in equation (6-G-2) that the wave function is a simple product of two one-dimensional equations, like (6-F-17). Therefore, the properties of the one-dimensional particle-in-the-box wave function apply here, if one of the variables x or y is held constant. However, in order to represent ψ or ψ^2 for the two-dimensional particle-in-the-box problem we must show how ψ or ψ^2 varies when both x and y vary simultaneously.

Equation (6-G-2) could be represented by a three-dimensional surface of ψ versus x and y. When the surface is somewhat complicated with several maxima and minima, it is difficult to draw the surface and equally difficult to interpret the surface. An alternative is to show the magnitude of ψ^2 (probability density) by the density of dots in a two-dimensional graph of y versus x. Each eigenfunction would be represented by a different probability density diagram.

For simplicity we will assume the two-dimensional box is a square; i.e., a = b. The wave functions and corresponding energies are then given by

$$\psi = \frac{2}{a} \sin \frac{n_x \pi}{a} x \; \sin \frac{n_y \pi}{a} y \qquad \text{(6-G-4)}$$

$$E = \frac{h^2}{8ma^2}\left(n_x^2 + n_y^2\right) \qquad \text{(6-G-5)}$$

The energy level diagram is shown in Figure 6-G-1 for various combinations of n_x and n_y. Note that when the box is square, certain combinations of n_x and n_y can give the same

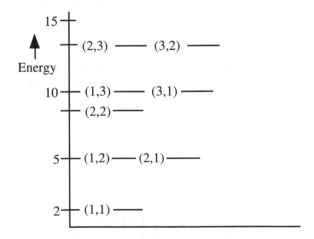

Figure 6-G-1. Energy Level Diagram for a Particle in a Two-Dimensional Box of Equal Sides. Numbers in Parentheses Represent the Values of n_x and n_y, Respectively

energy. For example, the combination $n_x = 1$ and $n_y = 2$ gives the same energy as the combination $n_x = 2$ and $n_y = 1$. Each combination of quantum numbers corresponds to a different wave function, equation (6-G-4). Different solutions, eigenfunctions, which happen to correspond to the same value of the energy, are said to be *degenerate*. When there are two energy levels that are degenerate, as in the example above, we refer to the levels as being two-fold or doubly degenerate.

The probability density diagrams for some of the wave functions for the Particle in a Two-Dimensional Box of equal sides are shown in Figure 6-G-2. Note in Figure 6-G-2 that the nodes are dashed lines in a two-dimensional representation of wave functions in contrast to points in the One-Dimensional Particle-in-the-Box wave functions, shown in Figure 6-F-1. Again the number of nodes for a given coordinate is $n - 1$. The wave function with $n_x = 1$, $n_y = 2$ has a node at $y = a/2$ since

$$\sin \frac{n_y \pi}{a} \ y = \sin \frac{2\pi}{a} \ \frac{a}{2} \ = \sin \pi = 0$$

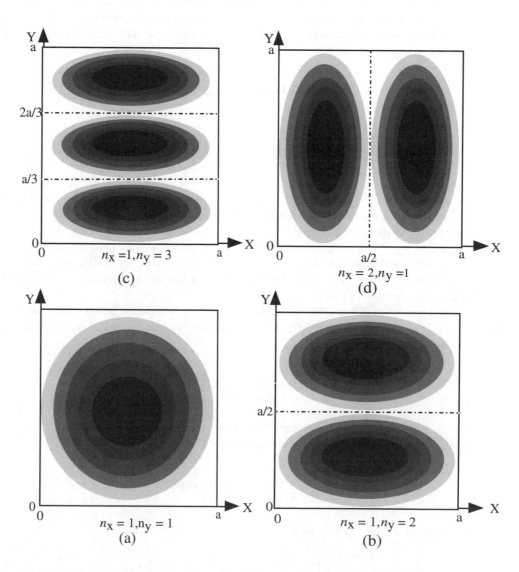

Figure 6-G-2. Probability Density Diagrams for a Particle in a Two-Dimensional Box

This is shown by the dashed line in Figure 6-G-2(b). Similarly, the wave function with $n_x = 2$, $n_y = 1$ has a single node at x = a/2, as shown in Figure 6-G-2(d). The wave function with $n_x = 1$, $n_y = 3$ would have two nodes along the y-axis at a/3 and 2a/3, as shown in Figure 6-G-2(c). By simply determining the nodes in the x and y directions for a given wave function, the probability density diagram can be described in a qualitative manner.

Problem 6-G-1. Sketch the probability density diagram for the wave function from the Particle in a Two-Dimensional Box (equal sides) with quantum numbers $n_x = 2$, $n_y = 3$.

Three-Dimensional Particle in the Box

The problem of the particle in a three-dimensional box can also be solved readily. If we have an actual particle in a box (for example, a gas molecule in a container), then the three-dimensional solution would necessarily apply, since physically we require three coordinates to describe the position of the particle. As with the Two-Dimensional Particle in a Box, we will not formally derive the three-dimensional solution, but only consider the result. The similarities between the one-dimensional, two-dimensional, and three-dimensional solutions are readily noted. The eigenfunctions and eigenvalues for the particle in a three-dimensional box are

$$\psi = \left(\frac{8}{abc}\right)^{1/2} \sin \frac{n_x \pi}{a} x \sin \frac{n_y \pi}{b} y \sin \frac{n_z \pi}{c} z \qquad (6\text{-}G\text{-}6)$$

$$E = \varepsilon_{trans} = \frac{h^2}{8m}\left(\frac{n_x^2}{a^2} + \frac{n_y^2}{b^2} + \frac{n_z^2}{c^2}\right) \qquad n_x, n_y, n_z = 1, 2, \cdots \qquad (6\text{-}G\text{-}7)$$

where a, b, c are the sides of the box in the x, y, z directions, respectively. Three quantum numbers, n_x, n_y, n_z, which arise, are obviously related to the x, y, z portions of the wave function ψ. As in the case for the two-dimensional particle in the box, the values of n_x, n_y, n_z must be whole or integral numbers, but the values each can take on are independent; that is, any combination is allowed. This is not always true for the quantum mechanical solution to problems. The energy for the particle, given by equation (6-G-7), is dependent on all three quantum numbers.

If the sides of the box are all equal, (a = b = c), a situation arises where some of the energy levels are equal for different combinations of n_x, n_y, n_z. If a = b = c, then E becomes

$$E = \frac{h^2}{8ma^2}\left(n_x^2 + n_y^2 + n_z^2\right) \qquad (6\text{-}G\text{-}8)$$

The energy levels can be designated by subscripts which refer to the n_x, n_y, n_z values characteristic of that level. For example, for $n_x = 1$, $n_y = 2$, $n_z = 1$,

$$E_{1,2,1} = \frac{h^2}{8ma^2} (1^2 + 2^2 + 1^2) = 6\left(\frac{h^2}{8ma^2}\right)$$

Similarly, for $n_x = 1$, $n_y = 1$, $n_z = 2$,

$$E_{1,1,2} = \frac{h^2}{8ma^2} (1^2 + 1^2 + 2^2) = 6\left(\frac{h^2}{8ma^2}\right) = E_{1,2,1} \text{ etc.}$$

Possible energy levels, along with their degeneracies, are given in Figure 6-G-1.

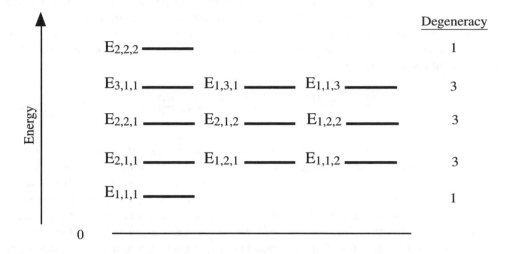

Figure 6-G-3. Energy Levels for a Particle in a Three-Dimensional Box of Equal Sides

The degeneracy is indicated to the right of the energy level. The terminology frequently used can be illustrated by the example, $E = 6(h^2/8ma^2)$, which is said to be *threefold* or *triply degenerate*. As we will see in quantum mechanical solutions of other problems, degeneracies arise quite frequently, and many times can be related directly to the quantum numbers.

The solution to the three-dimensional particle-in-a-box problem can be applied to the ideal gas, since an ideal gas is free of intermolecular forces or potentials. When the particle is a molecule or atom, the energy levels are extremely closely spaced, compared to the value of kT, where k = Boltzmann constant = gas constant/molecule; i.e.,

$$k = \frac{R}{N} = \frac{8.314 \times 10^7 \text{ergs/K-mol}}{6.023 \times 10^{23} \text{molecules/mol}} = 1.3804 \times 10^{-16} \text{ erg/K-molecule}$$
$$= 1.3804 \times 10^{-23} \text{ J/K-molecule}$$

The reader will recall that 1/2RT is the classical value for the average energy in one degree of freedom/mol which is equivalent to 1/2kT in one degree of freedom/molecule. As we will see later, when the energy levels are more closely spaced that kT, a continuum of states or energies can be assumed. This situation approximates the classical result where the molecules are free to have any energy.

Problem 6-G-2. Calculate the three lowest different eigenvalues (translational motion) for H_2, and compare them with kT at 300 K. Assume the container is a cube, whose sides are each 1 mm. Recall that 1 erg = 1 g-cm^2/sec^2. Answer: 4.9 x 10^{-28} erg, = 4.9 x 10^{-35} J, 9.8 x 10^{-28} erg = 9.8 x 10^{-35} J, 14.7 x 10^{-28} erg = 14.7 x 10^{-35} J, kT = 4.14 x 10^{-14} erg = 4.14 x 10^{-21} J. Eigenvalues for the translational motion of H_2 are much lower than kT.

It is difficult to describe the probability density for the wave functions for the Three-Dimensional Particle in the Box since we have four variables: x, y, z, ψ^2. A true dot diagram, where the density of dots represents ψ^2, loses its perspective in three dimensions since the location of a dot in space cannot be perceived. The alternative is to assume the dots define some object with a "fuzzy" surface, but this is not very quantitative. The alternative is to show a two-dimensional probability density diagram of the projection of the orbital on some plane. Viewing the orbital on three perpendicular planes gives a reasonable representation of the shape. As an example we will consider the probability density diagram for the wave function with the quantum numbers $n_x = 1$, $n_y = 3$, $n_z = 2$. We will assume equal sides of the box in which the particle resides; i.e., the box is a cube. The wave function would then be

$$\psi = \left(\frac{8}{a^3}\right)^{1/2} \sin \frac{n_x \pi}{a} x \sin \frac{n_y \pi}{a} y \sin \frac{n_z \pi}{a} z \qquad (6\text{-G-}9)$$

The number of nodal planes perpendicular to the axes is: x-axis 0, y-axis 2, z-axis 1. A three dimensional diagram showing these nodal surfaces is given in Figure 6-G-4. If one were to use dots in this diagram to represent probability density, it is obvious that it would be impossible to view in three dimensions. In Figure 6-G-5 we represent the projections of the dot diagrams onto the three pairs of axes. If the cube in Figure 6-G-4 were rotated about the z-axis counter clockwise by 90°, we would observe the projection on the x-z plane, as shown in Figure 6-G-5(a). The nodal plane perpendicular to the Z-axis appears as a line in Figure 6-G-5(a). As viewed from this perspective, the probability density would appear in just two regions. If the cube in Figure 6-G-4 were viewed from the front along the x-axis, we would observe the projection in the y-z plane, as shown in Figure 6-G-5(b). In the projection we observe the three nodel planes perpendicular to the y-axis and the single nodal plane perpendicular to the z-axis. Consequently, the probability density is separated into six different regions. Finally viewing the cube from above along the z-axis we observe the projection on the x-y plane, as shown in Figure 6-G-5(c). Only the three nodal planes perpendicular to the y-axis are observed and the probability density is restricted to three regions.

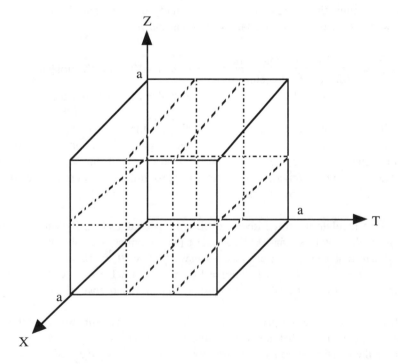

Figure 6-G-4. Nodal Planes for the Wave Function for the Three-Dimensional Particle in the Box. $n_x = 1$, $n_y = 3$, $n_z = 2$

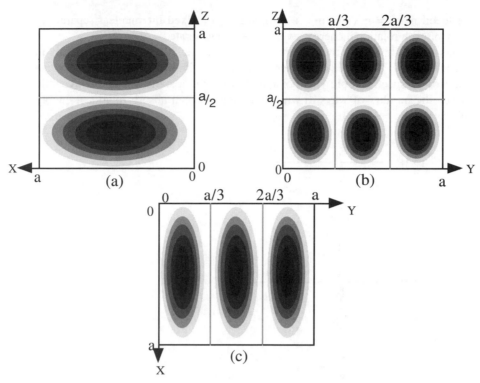

Figure 6-G-5. Projection of Probability Density on Three Planes for the Wave Function Described in Figure 6-G-4

Problem 6-G-3. For the wave function from the Three-Dimensional Particle in the Box (equal sides) with $n_x = 2$, $n_y = 2$, $n_z = 3$, show the projection of the probability density on the x-z, y-z, and x-y planes.

H. Rigid Rotor (Rotational Energies)

As discussed previously, molecules also have energy due to rotational motions. Since molecules possess a moment of inertia, a certain kinetic energy can be associated with rotation about the center of mass. These rotational motions can be investigated through quantum mechanics, using the Schrödinger equation (6-E-5), as in the case of the one-dimensional particle in a box. In this case, V = 0, since there is no force or potential restricting the motion of a free molecule (collisions and intermolecular attractions are neglected). The solution of this problem is conveniently carried out in the angular coordinates of a spherical or polar coordinate system. The radial coordinate is unnecessary, since we assume a fixed internuclear distance. Hence, the name, *rigid rotor*.

For a diatomic molecule, there is only one moment of inertia, and the orientation of the single internuclear axis can be represented by two angles, as shown previously in Figure 6-B-2. The eigenvalues which result from the quantum mechanical solution are

$$\mathbf{E_{rot}} = \mathbf{J(J+1)} \; \frac{\hbar^2}{2\mathbf{I}} = \mathbf{J(J+1)} \; \mathbf{B} \qquad\qquad J = 0, 1, 2 \;\cdots\cdots \qquad (6\text{-}H\text{-}1)$$

where I = moment of inertia, and J = the rotational quantum number, B = constant characteristic of the molecule = $\hbar^2/2I$. Again, we see that the energies are quantized. The moment of inertia is

$$\mathbf{I} = \mathbf{\mu r_{AB}^2} \qquad\qquad\qquad (6\text{-}H\text{-}2)$$

where μ = reduced mass = $m_A m_B/(m_A + m_B)$, and r_{AB} = fixed internuclear separation. The degeneracy of each rotational energy level is $2J + 1$ degenerate.

The various energy levels for the first three values of J are shown diagrammatically in Figure 6-H-1. Obviously, the degeneracy increases rapidly with increasing energy. This is of importance from a spectroscopic viewpoint, and also from a statistical thermodynamic standpoint, since the number of ways in which the molecule can absorb rotational energy increases with increasing energy.

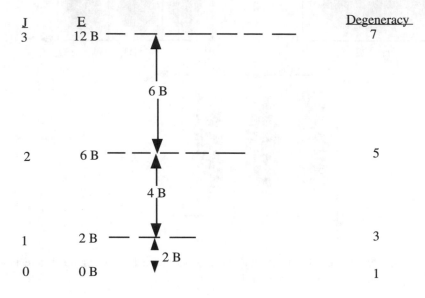

Figure 6-H-1. Energy Levels for a Linear Rigid Rotator. $B = \hbar^2/2I$

A molecule which exists in a particular rotational energy level can absorb a quantity of energy equal to the difference between its energy and the next higher energy, and thus, can be changed to the next higher energy level. This process is referred to as a rotational transition, and requires exactly the proper amount or quantity of energy before it can occur. In general, when a quantity of energy is gained or lost so that a molecule changes from one energy level to another, we say that a transition has occurred. There are certain rules which govern transitions, and consequently, not all possible transitions, as determined by the position of the levels, can actually occur. If a molecule encounters a photon of such frequency that its energy corresponds to the difference between the two energy levels of the molecule, then the molecule may absorb the photon and be raised in energy to the next higher level. Thus, energy transitions may occur as a result of the molecule absorbing electromagnetic radiation of the proper frequency. The transitions between rotational levels can be observed spectroscopically in the *microwave region* of the *electromagnetic spectrum*. The energy differences between rotational levels are in the range of the energies of photons whose frequencies lie in the microwave region. The transitions which can occur (allowed transitions) can be predicted from the quantum mechanical solution. This *selection rule*, as it is called, is

$$\Delta J = \pm 1$$

The transitions which are allowed for absorption are indicated by the arrows in Figure 6-H-1. Note that as the quantum number of the lower state increases, the energy of the transition increases. In fact, the successive transitions differ by $2(\hbar^2/2I)$ units, and the absorption lines are observed to be equidistant from each other.

In order for a molecule to absorb the electromagnetic radiation and undergo these transitions between rotational energy levels, it is necessary that the molecule have a dipole moment. The reason for this is simply that the rotating polar ends of the molecule can interact with the oscillating electric field associated with the electromagnetic radiation. Consequently, homonuclear molecules, such as O_2 and N_2, will not absorb microwave radiation since they do not have dipole moments. The same applies to non-polar molecules in general such as the linear CO_2 and C_2H_2. On the other hand, polar molecules such as HCl, HBr, and H_2O absorb microwave radiation since they have dipole moments.

Example 6-H-1. Calculate the rotational energy level for J = 3 for I_2 from equations (6-H-1) and (6-H-2). The internuclear separation is 2.662Å and the atomic mass of I is 127.

$$I = \frac{m_I^2}{2m_I}\, r_{I_2}^2 = \frac{127g/mol}{2}(2.662 \times 10^{-8}\,cm)^2/(6.023 \times 10^{23}\,atoms/mol)$$

$$E_{rot} = J(J+1)\frac{\hbar^2}{2I} = \frac{3(3+1)}{4\pi^2}\frac{(6.62 \times 10^{-27})^2\,erg^2\,sec^2}{\dfrac{127(2.662 \times 10^{-8}\,cm)^2}{6.023 \times 10^{23}}}$$

$$E_{rot} = 8.92 \times 10^{-17}\,erg/molecule = 8.92 \times 10^{-24}\,J/molecule$$

Problem 6-H-1. Calculate the 4 lowest different rotational energy levels for HI. Assume an internuclear separation of 1.604Å. Show that the 3 lowest energy transitions would be equally spaced, and give the separation in $\bar{v}$ (cm^{-1}). Compare the rotational energy level separation with the translational energy separation for H_2 in Problem 6-G-2. Also, compare the rotational separation with kT at 300 K. Answer: 0, 2.6 x 10^{-15} erg = 2.6 x 10^{-22} J, 7.9 x 10^{-15} erg = 7.9 x 10^{-22} J, 15.8 x 10^{-15} erg = 15.8 x 10^{-22} J, spacing = 13.2 cm^{-1}, rotational energy levels are greater than translational, on the order of kT.

Problem 6-H-2. Order the following molecules: O_2, HI, N_2, HCl in ascending order of their rotational energy level spacing. Answer: O_2, N_2, HI, HCl.

Problem 6-H-3. The lowest frequency, in the microwave region for HI, is 13.08 cm^{-1}. Calculate the internuclear distance in HI. The atomic mass for iodine is 127 g/mol. Answer: 1.604 A^o.

The problem of rotations of polyatomic molecules is more formidable than that of the diatomic molecules, but, nevertheless, can be handled by quantum mechanics. The complications arise from the fact that nonlinear molecules have moments of inertia in more than one direction. In general, a nonlinear polyatomic molecule will have three moments, I_x, I_y, and I_z. Several different categories exist, depending on whether $I_x = I_y = I_z$, $I_x = I_y \neq I_z$, or $I_x \neq I_y \neq I_z$. Each of these cases constitues a separate problem with its own solution. These more involved rotational situations will not be considered at this time.

I. Harmonic Oscillator (Vibrational Energies)

The vibrational motion of a diatomic molecule can be quite well approximated by a harmonic oscillator. The motion of a harmonic oscillator is like the oscillatory motion of a ball on a spring. The designation, harmonic oscillator, specifically requires that the mass, undergoing the motion, experience a restoring force, which is *directly proportional to the displacement* from the equilibrium position. In the case of a diatomic molecule, the displacement (x) is then

$$x = r - r_e, \tag{6-I-1}$$

where r is the internuclear separation at any time, and r_e is the equilibrium internuclear separation.

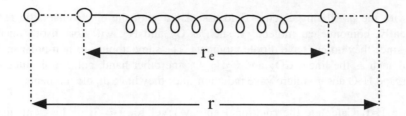

Note that the *magnitude* of the restoring force is independent of the sign of x. In other words, the displacement could be compression as well as extension, and the force would be the same magnitude, if the displacement were the same. Mathematically, the harmonic oscillator is represented by the equation

$$f = -kx \qquad (6\text{-}I\text{-}2)$$

where f represents the restoring force and k is a proportionality constant called the *force constant*. The negative sign signifies that the direction of the restoring force is in the opposite direction of the displacement.

According to the principles of physics, the potential is defined (in one dimension) by the relation

$$-\frac{dV}{dx} = f \qquad (6\text{-}I\text{-}3)$$

Substituting (6-I-2) into (6-I-3), and using the boundary condition V = 0 when x = 0, we obtain the potential energy of a harmonic oscillator.

$$V = \frac{1}{2}kx^2 \qquad (6\text{-}I\text{-}4)$$

If we plot V versus x or r, the parabolic-shaped curve, shown by the solid line in Figure 6-I-1, is obtained. This curve illustrates how the potential energy increases as r either in-

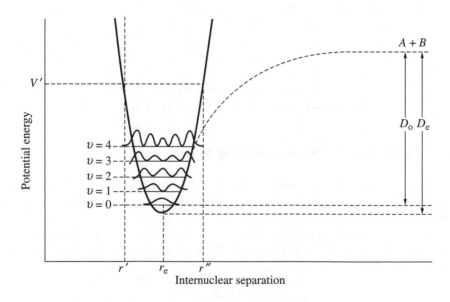

Figure 6-I-1. Vibrational Energy Levels for a Harmonic Oscillator

creases or decreases from the equilibrium position, r_e. Suppose the limits of the oscillation are r', equal to the minimum separation distance, and r'', equal to the maximum separation distance. When the oscillation reaches its minimum or maximum position, r' or r'', the motion ceases for an instant and all of the energy of the oscillator is potential energy. As the oscillator begins to move back toward the equilibrium position, the potential energy decreases, while the kinetic energy of motion increases. The total energy of the oscillator, which is, of course, the sum of the kinetic and potential energies, remains constant at all times. This is represented by the horizontal line corresponding to V'. When the oscillator reaches the equilibrium position, the potential energy has decreased to zero (x = 0), and the kinetic energy is equal to the total energy of the oscillator, which has the value V'. The broken-line curve in Figure 6-I-1 is representative of the potential energy curve for a diatomic molecule. At low energies, the harmonic oscillator is a good approximation to the potential energy curve for vibration.

The quantum mechanical solution to the harmonic oscillator is obtained by substituting (6-I-4) into the Schrödinger equation (6-E-5). Again, only certain values of the total energy are possible, and these are given by

$$\mathbf{E_{vib}} = \left(v + \frac{1}{2} \right) \mathbf{h}\,\boldsymbol{\nu} \qquad\qquad v = 0, 1, 2 \cdots \qquad (6\text{-}I\text{-}5)$$

where v is a vibrational quantum number, which can have only integer values $(0, 1, 2, \cdots)$, ν is the frequency of vibration, and it can be calculated from the classical relationship,

$$\nu = \frac{1}{2\pi} \sqrt{\frac{k}{\mu}} \qquad (6\text{-}I\text{-}6)$$

where k = force constant. The total energy of the oscillator for the four lowest quantum levels, v = 0, 1, 2, 3, is shown by the horizontal lines in Figure 6-I-1. The values of r which correspond to the intersection of a horizontal line (energy level) with the potential energy curve, represent the limits of the oscillation when the oscillator has that particular total energy. Each vibrational energy level is singly degenerate. This result occurs because the harmonic oscillator is a one-dimensional problem.

The harmonic oscillator eigenfunctions are somewhat complicated to express mathematically, so only a graphical representation will be given. These are plotted as ψ^2 in Figure 6-I-1 for the various values of the quantum number, v. Note that with increasing v, there is a corresponding increase in nodal points. It should also be noted that the energy spacings are equal in this case. The selection rule for the absorption of electromagnetic radiation is $\Delta v = \pm 1$.

In order for a molecule to absorb infrared radiation by excitation of vibrational energy levels, the vibration involved must experience a <u>change</u> in dipole moment during the vibration. For a diatomic molecule which has no dipole moment, such as with a homonuclear diatomic, the stretching vibration is symmetric and no dipole moment will be induced. Consequently, a diatomic molecule with zero dipole moment cannot absorb infrared radiation. On the otherhand, a diatomic molecule with a dipole moment, a heteronuclear diatomic molecule, will undergo a <u>change</u> in dipole moment upon excitation to the next vibrational level and will absorb infrared radiation corresponding to that vibrational energy transition.

For a polyatomic molecule it may be more difficult to deduce whether a vibrational excitation will cause a change in dipole moment. However, for a simple molecule, such as a triatomic the task is not so formidable. Let us consider the 4 vibrations of CO_2 shown in Figure 6-B-3. The symmetrical stretch (ν_1) shows equal displacement of the oxygen atoms and obviously would not create a dipole moment. On the other hand, the unsymmetrical stretch (ν_2) has unequal displacement of the oxygen atoms and consequently there is a temporary dipole moment created and obviously there is a <u>change</u> in dipole moment. The unsymmetrical stretch (ν_2) would absorb infrared radiation in the course of excitation from v = 0 to v = 1. The remaining two vibrations (v_3 and v_4) show a bending motion in which the molecule becomes non-linear during the course of the vibration. Again a dipole moment is created when the molecule is non-linear and these vibrations cause a <u>change</u> in dipole moment. Consequently, these vibrational modes will absorb infrared radiation in undergoing a transition between vibrational energy levels.

In the cgs-system of units the force is expressed in dynes where 1 dyne = 1 g-cm/sec^2 and obviously 1 dyne-cm = 1 erg. The force constant, k, would be expressed as 1 dyne/cm = 1 g/sec^2. In the MKS-system of units the force is expressed in Newtons (N) where 1 N = 1kg-m/sec^2 and 1 N-m = 1 J. The force constant would then be expressed as 1 N/m = 1 kg/sec^2.

Example 6-I-1. The force constant for HI is 3.2 x 10^5 dynes/cm = 3.2 x 10^2 N/m. The fundamental frequency can be calculated from equation (6-I-6). The atomic mass of I = 128 g/mol.

$$v = \frac{1}{2\pi} \sqrt{\frac{3.2 \times 10^5 \text{dynes/cm}}{\frac{(1)(128)g}{1+128} \Big/ 6.023 \times 10^{23}}}$$

$$v = \frac{1}{2\pi} \sqrt{\frac{3.2 \times 10^2 \text{ N/m}}{\frac{(1)(128)\text{kg}}{1+128} \Big/ 6.023 \times 10^{23}}} = 6.93 \times 10^{13} \text{ sec}^{-1}$$

Problem 6-I-1.
(a) Calculate the two lowest vibrational energy levels for H_2. The force constant is 5.69 x 10^5 dynes/cm.
(b) What is the energy separation between these two lowest vibrational energy levels? Compare with kT at 300 K, with the translational energy of H_2 (Problem 6-G-2), and with the rotational energy of HI (Problem 6-H-1). Calculate the frequency of the vibrational transition (cm^{-1}) between these two lowest levels.
Answer: (a) 4.4 x 10^{-13} erg = 4.4 x 10^{-20} J, 13.1 x 10^{-13} erg = 13.1 x 10^{-20} J (b) $\Delta\varepsilon_{vib}$ = 8.7 x 10^{-13} erg >> kT = 4.14 x 10^{-17} erg ~ $\Delta\varepsilon_{rot}$ >> $\Delta\varepsilon_{trans}$; 4396 cm^{-1} .

Problem 6-I-2. Which of the following molecules would you expect to absorb infrared radiation by undergoing transitions between vibrational levels: HI, O_2, N_2, H_2O, N_2O (linear), NO_2 (non-linear)? Answer: HI, H_2O, N_2O, NO_2.

Problem 6-I-3. Order the following molecules according to their force constants in ascending order: O_2, HI, N_2, HCl. Use vibrational frequencies from Appendix D. Answer: HI < HCl < O_2 < N_2.

Problem 6-I-4. Sketch the potential energy curve as a function of internuclear separation for F_2. Use data in the Appendices. Express the potential energy in terms of kcal/mol. Assume the harmonic oscillator approximation at small displacements but show the dissociation limit at large internuclear separation. Show all calculations.

If the diatomic molecule is represented by the general formula AB, the bond dissociation energy is the energy for the reaction

AB = A + B

This is related to the potential energy curve, shown in Figure 6-I-1, by the distance D_0. The energy difference, from the minimum of the potential energy curve to the dissociated state A + B, is represented by D_e, the spectroscopic dissociation energy. The difference between D_0 and D_e, 1/2hv, is known as the zero point energy.

$$D_e = D_0 + \frac{1}{2} \, hv \qquad (6\text{-}I\text{-}7)$$

where v is the vibrational frequency.

For polyatomic molecules, the vibrational motion involves the simultaneous motion of more than two atoms, as described in Section B of this chapter. In these so-called "normal" modes of vibration, the displacement of all the atoms occurs with the same frequency. It would require a multi-dimensional surface to represent diagrammatically the potential energy as a function of internuclear separation. For example, the symmetrical and unsymmetrical modes for a linear triatomic could be represented in three dimensions, as shown in Figure 6-I-2 for CO_2. The coordinates are internuclear separations, r_1 and r_2.

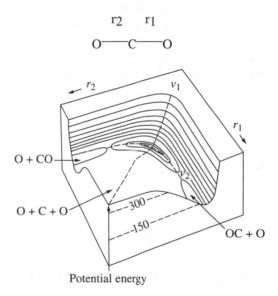

Figure 6-I-2. Potential Energy Diagram for the Linear Triatomic Molecule CO_2. The Contours Indicate Constant Values of the Potential Energy Expressed in kcal/mole

The magnitude of the potential energy in kcal is the vertical axis. The broken lines in Figure 6-I-2 show the paths followed during the vibration for the two modes being represented.

If the polyatomic molecule undergoes small internuclear displacements, the potential function can be approximated by the harmonic oscillator expression (6-I-4), where the displacement x corresponds to a so-called normal coordinate, which is a combination of several of the Cartesian coordinates of the various atoms. With this approximation, an analogous expression for the eigenvalues can be derived from quantum mechanics, as given in (6-I-5). In this case, however, there is a different expression for each normal vibrational mode with its characteristic frequency.

Problem 6-I-5. Calculate the spacing in the energy levels (v = 0 → v = 1) for the four vibrational frequencies for CO_2, as described in Figure 6-B-3. Compare these with kT at 600°C. Does this result explain roughly why the heat capacity of CO_2 does not approach the upper temperature limit predicted by the equipartition principle? $\bar{v}_1 = 1340$ cm^{-1}, $\bar{v}_2$

= 2349.3 cm^{-1}, $\bar{v}_3 = \bar{v} = 667.3$ cm^{-1}. Answer: $\Delta E_1 = 26.6 \times 10^{-14}$ erg $= 26.6 \times 10^{-21}$ J, $\Delta E_2 = 46.7 \times 10^{-14}$ erg $= 46.7 \times 10^{-21}$ J, $\Delta E_3 = \Delta E_4 = 13.3 \times 10^{-14}$ erg $= 13.3 \times 10^{-21}$ J, $\Delta E_3 = \Delta E_4 \sim kT = 12.05 \times 10^{-14}$ erg $= 12.05 \times 10^{-21}$ J.

Despite the fact that a normal vibration mode for a polyatomic molecule involves the displacement of nearly all the atoms in the molecule; frequently, some vibrational modes center primarily about a given bond in the molecule. For example, v_1 for H_2O in Figure 6-B-4 involves a stretch and compression of the O-H bond. For this reason, the frequency of this normal mode is referred to as the O-H *stretching frequency*. Amazingly, normal modes in other polyatomic molecules, such as CH_3OH, C_2H_5OH, have at least one vibrational frequency in this same range, known as the O-H stretching frequency region. Characteristic frequencies occur for other groups, such as C-H, N-H, C-Cl, C=O, and C-O. For this reason, infrared (IR) absorption spectra are an excellent means for the identification of molecular structure. The vibrational energy level spacing is generally such that the fundamental transition (v = 0 to v = 1) lies in the infrared region. Thus, the vibrational frequencies of a molecule can be determined from infrared absorption measurements.

In addition to the vibrational modes which are characteristic of some specific group in the molecule, there are also vibrational modes which involve the entire molecule, called *group vibrations*. A good example of this type of vibrational mode is in benzene, which has some 3(12) - 6 = 30 vibrational modes. Some of these which involve the entire molecule are shown as follows:

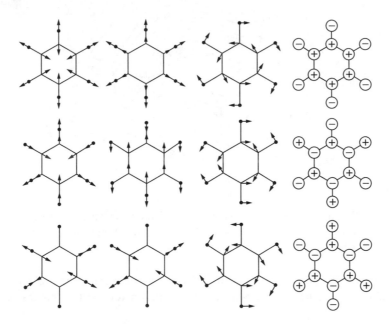

Figure 6-I-3. Some Normal Vibrational Modes of Benzene

J. Summary

The following are the basic concepts that were covered in this chapter:

1. A molecule has 3n nuclear motions or degrees of freedom (df) where n = number of atoms in the molecule. The nuclear motions can be represented as:
   (a) 3 translational modes which describe the motion of the center of mass.
   (b) 2 rotational modes for a linear molecule.
       3 rotational modes for a non-linear molecule.
       Rotational modes describe the rotation of the molecule about the center of mass.
   (c) (3n - 5) vibrational modes for a linear molecule.

(3n - 6) vibrational modes for a non-linear molecule.
Vibrational modes describe oscillatory displacements of the nuclei such that the center of mass does not change.

2. Each of the nuclear motions contributes the following maximum values to the internal energy, U, and heat capacity, C_V for each degree of freedom (df):
   (a) 1/2 RT to U for each translational df; 1/2 R to C_V for each translational df.
   (b) 1/2 RT to U for each rotational df; 1/2 R to C_V for each rotational df.
   (c) RT to U for each vibrational df; R to C_V for each vibrational df.

3. The frequency (ν) of a photon is related to the energy (Joule) and the wavelength, λ, by

$$\varepsilon = h\nu = h\frac{c}{\lambda}$$

where h = Planck's constant = 6.626×10^{-27} erg-sec = 6.626×10^{-34} Joule-sec and c = speed of light.

4. Bohr orbital energies for a hydrogenic species are given by

$$E_n = -\frac{2\pi^2 m e^4 Z^2}{h^2}\frac{1}{n^2}$$

where Z = nuclear charge, m = mass of electron, e = charge on the electron, n = quantum number = whole number.

5. Transitions occur between Bohr orbital energy levels to cause absorption or emission of radiation of frequency, ν.

$$\nu = \frac{\Delta E}{h} = \frac{E_2 - E_1}{h}$$

6. A particle of mass , m, possesses a wave character with a de Broglie wavelength, λ, given by

$$\lambda = \frac{h}{mv}$$

7. The quantum mechanical solutions for a particle in a one-dimensional box are

$$\psi = \left(\frac{2}{a}\right)^{1/2}\sin\frac{n\pi}{a}x \qquad \text{(eigenfunction)}$$

$$E = n^2\left(\frac{h^2}{8ma^2}\right) \qquad \text{(eigenvalue)}$$

where n = translational quantum number = whole number.

8. The rotational energy levels for a rigid rotor are given by

$$E_{rot} = J(J+1)\frac{\hbar^2}{2I}$$

where I = moment of enertia = μr_{AB}^2 , μ = reduced mass = $(m_A m_B)/(m_A + m_B)$, r_{AB} = internuclear distance, J = rotational quantum number = whole number, $\hbar$ = h/2π.

9. The vibrational energy levels for a harmonic oscillator are given by

$$E_{vib} = \left(v + \frac{1}{2}\right) h\nu$$

where v = vibrational quantum number = whole number, and ν = fundamental vibrational frequency. ν is related to the force constant, k, and reduced mass, μ, by

$$\nu = \frac{1}{2\pi} \sqrt{\frac{k}{\mu}}$$

Exercises

1. How many translational, rotational, and vibrational degrees of freedom are there in the molecules, I_2, N_2O (linear), H_2S (bent), CH_4 (tetrahedral)? Sketch out the heat capacity curves as a function of temperature for I_2, N_2O, and H_2S. Answer: I_2: 3 trans, 2 rot, 1 vib; N_2O: 3 trans, 2 rot, 4 vib, etc.

2. A molecule must have a permanent dipole moment in order to absorb radiation in the microwave region by transition between rotational energy levels.
   (a) Which of the following molecules will show a rotational spectra? HF, ICl, I_2, Kr, CO_2 (linear), SO_2 (bent), BeF_2 (linear), HCN (linear), CCl_4, $CHCl_3$, H_2.
   (b) Radar utilizes frequencies in the microwave region. Which of the following atmospheric gases would you expect to interfere by absorption of the radar signal? N_2, O_2, CO_2, H_2O, He? Answer: (a) HF, ICl, SO_2, HCN, $CHCl_3$ (b) H_2O.

3. Using the force constant of 9.7×10^5 dyne/cm = 9.7×10^2 N/m and the reduced mass of 1.58×10^{-24} g = 1.58×10^{-27} kg for HF:
   (a) Plot the potential energy versus internuclear distance curve near the equilibrium distance of 0.92 Å.
   (b) Add the allowed vibrational energies as horizontal lines to the diagram. Sketch in the actual potential energy curve for large displacements from the equilibrium distance.
   (c) Add the rotational levels for the zero vibrational state ($v = 0$).

4. $H^{35}Cl$ absorbs infrared radiation at 2560 cm^{-1}. If the force constant in $D^{35}Cl$ is the same as that for HCl, predict the frequency at which DCl would absorb infrared radiation. Would the frequency of absorption be much different in $H^{37}Cl$ compared to $H^{35}Cl$? (^{35}Cl and ^{37}Cl designate the isotopes of mass numbers 35 and 37, respectively; D is the hydrogen isotope of mass 2, ^2H.) Answer: $\bar{\nu}_{D35Cl} = 1{,}834.7$ cm^{-1}; No.

5. (a) Calculate the number of translational, rotational, and vibrational degrees of freedom for the linear molecule COS.
   (b) Calculate the internal energy that arises from translational, rotational, and vibrational motion in COS.
   (c) Sketch the heat capacity, C_V, as a function of temperature for COS, showing the limits that you would expect in going from 300 K to 1,000 K.
   (d) Would you expect COS to absorb radiation in the infrared region? Briefly explain.
   (e) Would you expect COS to absorb radiation in the microwave region? Briefly explain. Answer: (a) 3 trans, 2 rot, 4 vib (b) 3/2 RT, 2/2 RT, 4 RT (d) yes (e) yes.

6. Calculate the second ionization potential for helium; the energy for the process

$$He^+ \longrightarrow He^{2+} + e^-$$

Answer : 54.0 eV.

7. (a) The hydroxyl radical, OH, has an absorption in the infrared region at a wave length of 2,677.2 nm = 26,772A$^\circ$. Calculate the force constant of OH.
   (b) Graph the potential energy as a function of internuclear distance for OH and show the vibrational energy levels on the diagram. The bond distance in OH is 0.971A$^\circ$. Answer : (a) 7.74 x 10^5 dynes/cm = 7.74 x 10^2 N/m.

8. NO$_2$ could have a linear or bent structure. Using microwave or infrared absorption spectra, explain how you could differentiate between these two structures.
   Answer: linear: no microwave abs., 2 IR abs. bands; non-linear: microwave abs., 3 IR abs. bands.

9. Calculate the wavelength of the photon given off when Li^{2+} undergoes an electronic transition from n = 5 to n = 3. Answer: 143 nm = 1430 A$^\circ$.

10. H^{79}Br has an infrared absorption at 2849.7 cm^{-1} which results from the transition from the zero vibrational level (v = 0) to the first vibrational energy level (v = 1). Calculate the infrared absorption frequency for H^{81}Br assuming the force constant for H^{81}Br is the same as that for H^{79}Br. Answer : 2,849.3 cm^{-1}.

11. Which of the following molecules would have the greatest separation of rotational energy levels: H$_2$, HBr, Br$_2$? The atomic weight of Br is 80.0 g/mol. The bond distances are: r$_{H-H}$ = 0.741 A$^\circ$, r$_{H-Br}$ = 1.414A$^\circ$, r$_{Br-Br}$ = 2.290A$^\circ$. <u>Show your calculations</u>. Answer: H$_2$.

12. Which of the following electronic transitions will have the lowest energy?

$$H: n = 1 \text{ to } n = 3$$
$$He^+: n = 2 \text{ to } n = 4$$
$$Li^{2+}: n = 3 \text{ to } n = 8$$

<u>Show your calculations</u>. Answer: He$^+$: n = 2 to n = 4.

13. The fundamental vibrational frequency for H$_2$ is 4395.2 cm^{-1}. Assuming the force constant is independent of the isotope, calculate the fundamental vibrational frequencies for HD and D$_2$. D = $_1^2$H with atomic mass = 2.00 Answer:
    $\bar{v}_{HD}$=3,806.4cm^{-1},$\bar{v}_{D_2}$ =3,107.9cm^{-1}.

14. Nitrous oxide, N$_2$O, has the structure NNO. How could you determine whether this molecule has a linear or a bent structure?

linear   bent

(Hint: Consider the possible modes of motion which each of these two structures would have and how these can be detected spectroscopically.) Answer: The linear structure would have a microwave absorption which is identical to that for a diatomic molecule. The non-linear structure would have a more complicated spectra due to three different moments of inertia.

15. The lowest frequency observed in the microwave region for $^{12}C^{16}O$ is 3.86248 cm^{-1}. Calculate the internuclear distance for $^{12}C^{16}O$. Answer: 1.13 A$^{\circ}$.

16. Briefly describe how quantization of energy levels occurs in quantum mechanics.

17. The change in standard enthalpy (ΔH^o) was determined at 515 K for the reaction

$$I_{(g)} + CH_3COI_{(g)} = CH_3CO_{(g)} + I_{2(g)} \quad \Delta H^o_{515} = 61.1 \text{ kJ/mol.}$$

Estimate the ΔC_P for this reaction and calculate ΔH^o_{298}. Assume the C_P for CH_3COI and CH_3CO are equal. Can you support this argument? Assume two situations:
(a) I_2 vibrational mode is completely active,
(b) I_2 vibrational mode is inactive.
Answer: CH_3COI and CH_3CO have numerous vibrational modes and the C_P's should be large and similiar in magnitude. (a) $\Delta C_P = 16.7$ J/K-mol, $\Delta H^o_{298} = 57.5$ kJ (b) $\Delta C_P = 8.3$ J/K-mol, $\Delta H^o_{298} = 59.3$ kJ.

18. Consider the gas phase reaction,

$$H_{2(g)} + I_{2(g)} = 2HI_{(g)}$$

Use appropriate data from the Appendix, Section D to make the necessary calculations in the following questions:
(a) Calculate and compare the rotational energy levels for H_2, I_2, and HI. Compare these with kT at 500 K. In order to simplify your calculations, recall Example 6-B-1 and Problem 6-H-1.
(b) Calculate and compare the vibrational energy levels for H_2, I_2, and HI. Compare these with kT at 500 K. Recall Example 6-I-1 and Problem 6-I-1.
(c) Sketch the heat capacity curves as a function of temperature, which you might expect for H_2, I_2, and HI. Point out the distinguishing features and differences between the curves. For specific C_P values, use Table 4, Appendix D.
(d) Calculate ΔH^o_{298}, using the heats of formation in the Appendix.
(e) Estimating ΔC_P from the result of (c), estimate ΔH^o_{500} for the reaction.

   Answer: (a) kT > H_2 > HI > I_2 (b) H_2 > HI > kT > I_2 (c) C_{V,I_2} ~ 26.8 J/K-mol, $C_{V,HI}$ ~ 20.9 J/K-mol (d) $\Delta H^o_{298} = -9.4$ kJ (e) $\Delta H^o_{298} = -11.1$ kJ.

19. The one-dimensional particle-in-a-box solution has several important applications to physical problems. One of these is the so-called "free electron model" for conjugated molecules. Specifically, the type of molecule referred to can be represented by the general formula

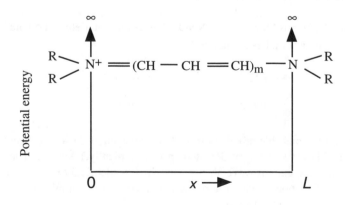

The potential, which the electron in a π orbital sees in the conjugated chain (CH-CH=CH)$_m$, is assumed to be zero (hence, the name free electron model). At the nitrogen atoms it is assumed that the potential suddenly becomes infinite. Hence, the quantum mechanical solution is the one-dimensional particle in a box. The length of the chain or box can be approximated by 1.4 Å/bond. In other words,

$$L = 2.8 + 2.8m = 2.8(m + 1)$$

For every double bond, there are 2 electrons which can occupy the various energy levels (Pauli principle). For example, let us take m = 2. In this case, there are 6π electrons which must be put into molecular energy levels. For m = 2, L = 8.4 Å. The energy levels can be calculated using the results of the particle-in-a-box. The lowest 5 energy levels are represented by E_1, E_2, ..., E_5. If we place 2 electrons in each level, we will obtain the following electronic configuration (lowest energy or ground state):

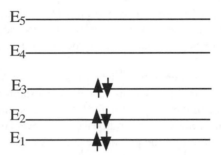

If we wanted to calculate the wavelength for the lowest energy transition as shown in the diagram, we would use

$$\Delta E = E_4 - E_3 = h\nu = \frac{hc}{\lambda}$$

(a) Calculate the lowest energy transition for the molecules m = 2, m = 3, m = 4. (Recall that the number of electrons in a molecule is 2(m + 1).)

(b) Plot λ for the lowest energy transition versus m, noting the variation of λ with m. The free electron theory correctly predicts that λ will continually increase with increasing m. Experimentally, there is an increase in λ with increasing m; however, for large m, the wavelength of the lowest energy transition approaches a limiting value. A more sophisticated theory is required to explain this effect. Answer: (a) m = 2, Δ E = 3.74 eV; m = 3, ΔE = 2.7 eV; m = 4, Δ E= 2.1 eV (b) m = 2, λ = 3,310 Å°; m = 3, λ = 4 ,580 Å°; m = 4, λ = 5,830 Å°.

20. For the three-dimensional particle in the box describe the probability density diagram for a particle in the energy level with quantum numbers n_x = 1, n_y = 2, n_z = 3. Show projections on the XY, YZ, and XZ planes. Assume the box has equal sides, i.e., a cube. Answer: XY planc has one node along Y, XZ plane has one node along Y and 2 along Z, XZ plane has 3 nodes along Z.

21. (a) Calculate the lowest energy rotational transition in HF. Express the energy in eV.
(b) Calculate the frequency of this transition in wavenumbers ($\bar{\nu}$) which should occur in the microwave region. Answer: (a) 0.0052 eV (b) 42.2 cm^{-1}.

22. Formaldehyde, CH_2O, has a planar structure

(a) Calculate the number of translational, rotational, and vibrational modes for CH_2O.

(b) Assume CH_2O is an ideal gas and $C_P = C_V + R$. Estimate the Cp at T= 298K and at a high temperature, say 1500K.

(c) Compare your results from Part (b) with those in Table 4 of the Appendices.

(d) Would you expect CH_2O to absorb microwave radiation? Briefly explain.

(e) Would you expect CH_2O to absorb infrared radiation? Briefly explain.
Answer: (a) 3 trans, 3 rot, 6vib (b) 298K $C_P \sim$ 34 J/K, 1500K 84 J/K (c) 298K actual C_P = 35.4 J/K, 1500 K actual C_P = 71.1 J/K (d) Yes, since CH_2O has a dipole moment. (e) Yes, all six vibrational modes should absorb in the IR since the dipole moment will change during the vibrations.

23. (a) The vibrational frequency for $^{79}Br^{81}Br$ is 323.2 cm^{-1}. Calculate the vibrational frequency for $^{79}Br^{79}Br$, assuming the force constant is the same as in $^{79}Br^{81}Br$.

(b) Sketch the potential energy curve for $^{79}Br^{81}Br$, assuming the vibration is a harmonic oscillator. Show the vibrational energy levels on this diagram. Show at least one calculation of the potential energy. Answer: (a) 325.3 cm^{-1} (b) x = 0.1 A$^{\circ}$, V = 1.23 x 10^{-13} erg = 1.23 x 10^{-20} J.

24. The fundamental vibrational frequency for OH is 3,735.2 cm^{-1}. Assuming the force constant for OD is the same as that for OH, calculate the fundamental vibrational frequency for OD. $D = {}_1^2 H$ with an atomic mass of 2.000, $H = {}_1^1 H$ with an atomic mass of 1.000 Answer: 2,717.8 cm^{-1}.

25. A diatomic molecule (AB) has a microwave absorption spectrum consisting of equally spaced absorption lines on a graph of absorption versus $\bar{v}$ (cm^{-1}). The energy difference between these lines can be represented by

$$\Delta\varepsilon = hc\Delta\bar{v}$$

Derive the relationship between $\Delta\bar{v}$ and the internuclear distance, r_{AB}. Answer: $r_{AB} = (\hbar^2/h\,c\,\mu\Delta\bar{v})^{1/2}$.

7

ATOMIC AND MOLECULAR STRUCTURE

In 1998, for the first time, the Nobel Prize in Chemistry was awarded to two scientists working in the field of Theoretical Chemistry. These recipients are John A. Pople, professor at Northwestern University, and Walter Kohn, professor at the University of California—Santa Barbara. These scientists along with many others have developed theoretical calculations to the point that they can be used to predict chemical behavior. This, along with the rapid development of personal computers with greater speed and memory makes theoretical calculations available to the experimental chemist. One application of this is the prediction of the stability of aromatic compounds due to resonance energy. This is especially true for large molecules containing many benzene-like rings in a structure such as

a compound called circumanthracene. Because these molecules are so stable they are readily formed in soot from diesel engines. There is also evidence that they may be formed in extraterrestrial nebulae that are formed at extremely high temperatures. Theoretical calculations are very valuable in predicting the stability of these large aromatic molecules. For a further understanding of the importance of resonance to aromatic compounds, the reader is referred to a paper by Jun-ichi Aihara in the March issue of Scientific American, *1992.*

The wave nature of particles was discussed in Chapter 6, and the principles of quantum mechanics were introduced and applied to several elementary problems. It was emphasized in Chapter 6 that a correct description of the motions and energies of microscopic particles of matter requires a quantum mechanical analysis, rather than an analysis based on the principles of classical mechanics. With this in mind, it might seem that all problems concerned with the energies, motions, and structural characteristics of the electrons and nuclei of atoms, and molecules could be solved by carefully applying the principles of quantum mechanics. Unfortunately, this is not the case. In most instances, once the problem has been formulated in terms of the quantum mechanical requirements for the system, the resulting equations are not analytically solvable. However, this is not always the case, as we shall see shortly. Moreover, even when an exact solution to the problem is not possible, approximate solutions, which provide much insight into the nature of the system, can often be found.

The natural starting point for any discussion concerning atomic and molecular structure is with the simplest possible atomic system, the hydrogen atom. A quantum mechanical analysis of the hydrogen atom can be performed analytically.

A. Hydrogen Atomic Orbitals

Formulation of the Problem

The hydrogen atom is known to consist of one proton, the nucleus, and one electron with a mass 1/1836 of the mass of the proton. In order to formulate the problem in quantum mechanical terms, the Schrödinger equation (6-E-2), $\hat{H}\psi = E\psi$, must be written specifically for the hydrogen atom system. Since the Hamiltonian operator, $\hat{H}$, involves partial derivatives with respect to three spatial coordinates and a potential energy term which is also a function of position coordinates, a suitable coordinate system should be chosen. For simplicity, it is best to choose the coordinate system so that its center is coincident with the nucleus of the atom. The position of the electron is then denoted by either the rectangular coordinates, x,y, and z, or the spherical coordinates r, θ and ϕ. The coordinate system is shown in Figure 7-A-1, where the proton nucleus is denoted by the dot at the center of the coordinate system, and the electron position is represented by the dot at the coordinates (x, y, z) or (r, θ, ϕ).

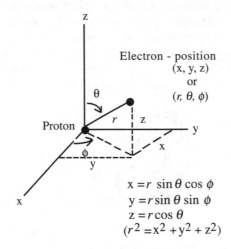

$$x = r \sin\theta \cos\phi$$
$$y = r \sin\theta \sin\phi$$
$$z = r \cos\theta$$
$$(r^2 = x^2 + y^2 + z^2)$$

Figure 7-A-1. Hydrogen Atomic System, Showing Rectangular and Spherical Coordinates

The potential energy of the system is due to the coulombic attraction between the negative electron and the positive nucleus, and can be expressed in electrostatic units.

$$V = -\frac{e^2}{r} \tag{7-A-1}$$

where r is the distance between the proton and the electron, in centimeters, and e is the absolute value of the charge on the electron in statcoulombs.

The Hamiltonian operator, equation (6-E-3),

$$\hat{H} = -\frac{\hbar^2}{2m}\nabla^2 - V$$

can then be expressed for the hydrogen atom as

$$\hat{H} = -\frac{\hbar^2}{2m}\nabla^2 - \frac{e^2}{r} \tag{7-A-2}$$

or

$$\hat{H} = -\frac{\hbar^2}{2m}\left(\frac{\partial^2}{\partial x^2} + \frac{\partial^2}{\partial y^2} + \frac{\partial^2}{\partial z^2}\right) - \frac{e^2}{r}$$

where m equals the mass of the electron. For a precise development, the origin of the coordinate system should coincide with the center of mass of the system, and the reduced mass should appear in the Hamiltonian operator. However, because of the large difference in the masses of the electron and the proton, the reduced mass, μ, is very nearly equal to the mass of the electron, m. The Schrödinger equation for the system becomes

$$\left(-\frac{\hbar 2}{2m} \nabla^2 - \frac{e^2}{r} \right) \psi = E \psi \qquad (7\text{-A-}3)$$

Equation (7-A-3) contains the spherical coordinate r and implicitly, the second partial derivatives with respect to the rectangular coordinates, x, y, and z. The solution of the problem is less difficult if equation (7-A-3) is expressed in spherical coordinates, rather than in rectangular coordinates. This is possible since the Laplacian operator, ∇^2, can be expressed in terms of spherical coordinates and partial derivatives of spherical coordinates.

Expressing ∇^2 in spherical coordinates, multiplying both sides of equation (7-A-3) by $-2m/(\hbar)^2$, and collecting all terms on the left side of the equals sign, we obtain

$$\frac{1}{r^2} \frac{\partial}{\partial r} \left(r^2 \frac{\partial \psi}{\partial r} \right) + \frac{1}{r^2 \sin^2 \theta} \frac{\partial^2 \psi}{\partial \phi^2} + \frac{1}{r^2 \sin \theta} \frac{\partial}{\partial \theta} \left(\sin \theta \frac{\partial \psi}{\partial \theta} \right) + \frac{2m}{\hbar^2} \left(E + \frac{e^2}{r} \right) \psi = 0 \qquad (7\text{-A-}4)$$

This is the **Schrödinger Equation for the Hydrogen Atom in Spherical Coordinates.** Though formidable in its appearance, equation (7-A-4) can be solved by the method of solution of partial differential equations known as "separation of variables".

Problem 7-A-1. Write the Schrödinger equation for He$^+$ in the form of equation (7-A-3).

Answer : $\left(-\dfrac{\hbar 2}{2m} \nabla^2 - \dfrac{2e^2}{r} \right) \psi = E \psi.$

The Eigenfunctions (Wave Functions)

The general solutions of equation (7-A-4) have the form of a product of three functions, each of which involves only one of the coordinates.

$$\psi = R(r)^\Theta (\theta)^\Phi (\phi) \qquad (7\text{-A-}5)$$

In addition to being dependent on the spatial coordinates, r, θ, and ϕ, Ψ is also a function of three parameters, n, ℓ, and m. For each different set of values of n, ℓ, and m, a different solution or eigenfuntion, ψ, is obtained. The eigenfunctions are often referred to as wave functions.

In order that the solutions (7-A-5) be physically meaningful, as discussed in Chapter 6, the parameters n, ℓ, and m_ℓ can have only integer values. Moreover, even the integer values are restricted in this instance by certain relationships between n, ℓ, and m_ℓ). The allowed values of n, ℓ, and m_ℓ are given in (7-A-6).

$$n = 1, 2, 3 \cdots$$
$$\ell = 0, 1, \cdots (n\text{-}1) \qquad (7\text{-A-}6)$$
$$m_\ell = -\ell, -(\ell - 1), \dots 0, \dots (\ell - 1), \ell$$

Except for the restricting relationships between n, ℓ, and m_ℓ, the situation is analogous to the solution of the particle-in-a-three-dimensional-box problem discussed in Chapter 5. The fact that n, ℓ, and m_ℓ can have only integer values arises naturally in the solution of the problem, as a result of the boundary conditions and the specific requirements inherent in the problem. Thus, by analogy, n, ℓ, and m_ℓ are referred to as quantum numbers; n is known as the principal quantum number, ℓ is known as the angular momentum or azimuthal quantum number, and m_ℓ is called the magnetic quantum number. For a specific

n, ℓ, and m_ℓ, a wave function ($\psi_{n,\ \ell,\ m_\ell}$) is defined, and a corresponding eigenvalue ($E_{n,\ \ell,\ m_\ell}$) can be obtained by operating on the wave function with the Hamiltonian operator.

   Table (7-A-1) shows the various combinations of the values of the three quantum numbers up through $n = 4$.

n	1	2		3			4			
ℓ	0	0	1	0	1	2	0	1	2	3
m_ℓ	0	0	-1,0,1	0	-1,0,1	-2,-1,0,1,2	0	-1,0,1	-2,-1,0,1,2	-3,-2,-1,0,1,2,3
orbital designation	1s	2s	2p	3s	3p	3d	4s	4p	4d	4f
number of orbitals	one	one	three	one	three	five	one	three	five	seven

Table 7-A-1. Hydrogenic Atomic Orbitals

Each combination differs from the others in the value of at least one of the quantum numbers, defining a unique wave function, and hence a unique atomic orbital. However, orbitals having the same values of n and ℓ, but different values of m_ℓ, are given the same orbital designation. In the orbital designation, the numerical (integer) prefix indicates the value of the principal quantum number, and the letters s, p, d, and f stand for ℓ equals 0, 1, 2, and 3, respectively.

   Each wave function describes a possible state of the atom. As stated in Chapter 6, ψ^2 is a measure of the probability per unit volume of finding the electron at a particular position in space, denoted by the coordinates of ψ^2. The corresponding eigenvalue is the energy of the system in the state described by the particular wave function. It is not surprising to discover that the hydrogen atom system is quantized; that is, only certain discrete values of the energy of the system corresponding to specific states of the system are possible. It is worth reemphasizing the fact that the quantum numbers, which lead to the concept of discrete energy states of the system, arise naturally by requiring physically meaningful solutions to the problem. The reader is referred again to the discussion of "well-behaved solutions" given in Chapter 6.

   The solutions to the hydrogen atom problem discussed here constitute many possible states of the system, each described by a unique wave function defined by the specific values of the three quantum numbers. However, an important feature of these solutions is that the corresponding energies are not all unique—sets of the wave functions are found to give the same value of the energy of the system (identical eigenvalues). Sets of wave functions which correspond to the same energy are said to represent degenerate states of the system. For the specific case of the hydrogen atom, all of the wave functions having the same principal quantum number correspond to the same eigenvalue. Hence, the energy of the hydrogen atom, in any one of its possible states, depends only on the value of the principal quantum number of the state. The degeneracy of a particular energy state of the hydrogen atom is thus equal to the total number of wave functions having the principal quantum number value which corresponds to the particular energy state. The number of wave functions which can have the same value of the principal quantum number, n, is given by n^2, and hence,

State Degeneracy = n^2

   The eigenvalues or energies of the hydrogen atomic orbitals is given by the formula

$$E_n = -\ \frac{2\pi^2 m e^4}{h^2}\ \frac{1}{n^2} \qquad (n = 1, 2, 3, \cdots) \qquad\qquad (7\text{-}A\text{-}7)$$

where n = principal quantum number. This equation for energies is identical to that obtained from the Bohr theory, equation (6-D-5), where Z = atomic number = 1 for hydrogen.

We can represent the orbital energy diagram knowing the E_n for each of the degenerate states where the degeneracy is n^2, as shown in Figure 7-A-2. The calculation of the frequency, $\bar{\nu}$ change between energy levels is identical to that carried out previously in Chapter 6 using equation (6-D-11). Spectral series such as the following have been observed

Lyman Series	Balmer Series	Paschen Series
$\Delta = -1$	$\Delta = -1$	$\Delta = \pm 1$
2p -----> 1s		
3p -----> 1s	3p -----> 2s or 3d -----> 2p	n_2 -----> n_1
4p -----> 1s	4p -----> 2s or 4d -----> 2p	4 ------> 3
5p -----> 1s	5p -----> 2s or 5d -----> 2p	5 ------> 3
.	.	.
.	.	.
.	.	.
∞p -----> 1s	∞p -----> 2s or ∞d -----> 2p	∞ ------> 3

There is a selection rule that arises from the quantum mechanical solution that restricts the change in ℓ

$$\Delta \ell = \pm 1$$

in order for the transition to occur through absorption or emission of electromagnetic radiation. We will not go into this theory since it involves mathematical concepts in excess of that expected of the reader of this text.

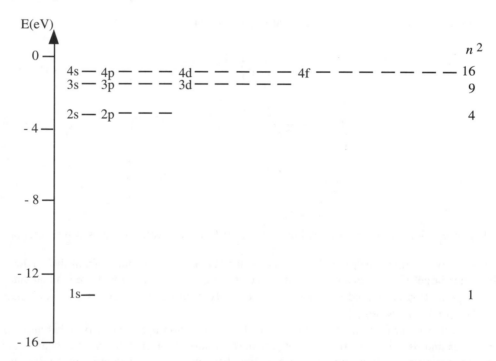

Figure 7-A-2. Energy Level Diagram for Atomic Hydrogen

Problem 7-A-2. How many wave functions are there with $n = 4$, and what are the values of ℓ and m_ℓ for each of them? Answer: 16 wave functions; $\ell = 0$ and $m_\ell = 0$; $\ell = 1$ and $m_\ell = -1, 0, 1$; $\ell = 2$ and $m_\ell = -2, -1, 0, 1, 2$; $\ell = 3$ and $m_\ell = -3, -2, -1, 0, 1, 2, 3$.

Problem 7-A-3. Calculate the wavelength of the transition in hydrogen from a 4p to a 2s orbital. How would this compare to a transition from a 4s to a 2p orbital? Answer: 4,863 A° = 486.3 nm. It would be the same.

In order to have a better understanding of the wave functions for the hydrogen atom, we will examine the actual mathematical form for some of these. As we noted in Chapter 6, the equations for the energy levels for the hydrogen atom also apply to any hydrogenic system which consists of a single electron plus the positive nucleus: He^+, Li^{2+}, Be^{3+}, etc. However, for these hydrogenic systems the nuclear charge, which is the atomic number, is greater than one and this greatly affects the magnitude of the energy levels. In the following wave functions we will express them in terms of Z = atomic number. You will recall from Chapter 6 that the energy levels from the Bohr theory for a hydrogenic system are dependent upon a quantum number, n.

$$E_n = -\frac{2\pi^2 m e^4 Z^2}{h^2}\frac{1}{n^2} \qquad (n = 1, 2, 3, \cdots) \qquad (6\text{-}D\text{-}5)$$

In solving the Schrödinger equation for a hydrogenic system we obtain an identical solution for the eigenvalues, where n is now the principal quantum number.

Wave Functions for s Orbitals

For the s orbitals the $\Theta\,(\theta)\,\Phi\,(\phi)$ are constant and the wave functions for the s orbitals for a hydrogenic system depend only on the radial function

$$\Psi_{ns} = R(r).$$

For this reason the s orbitals are spherically symmetrical.

$$\Psi_{1s} = \frac{1}{\sqrt{\pi}}\left(\frac{Z}{a_o}\right)^{3/2} e^{-\sigma} \qquad (7\text{-}A\text{-}8)$$

$$\Psi_{2s} = \frac{1}{4\sqrt{2\pi}}\left(\frac{Z}{a_o}\right)^{3/2} (2-\sigma)\,e^{-\sigma/2} \qquad (7\text{-}A\text{-}9)$$

$$\Psi_{3s} = \frac{1}{81\sqrt{3\pi}}\left(\frac{Z}{a_o}\right)^{3/2} (27 - 18\sigma + 2\sigma^2)\,e^{-\sigma/3} \qquad (7\text{-}A\text{-}10)$$

where $\sigma = \dfrac{Z}{a_o} r$ and a_o = Bohr radius = 5.2 x 10^{-11} meter. Note that all s wave functions have an exponential term which diminishes in value as r, the distance from the nucleus, becomes large. For the 2s and 3s orbitals the drop off in Ψ at large r is more gradual since the exponent $-\sigma$ is divided by 2 and 3, respectively. Consequently, the 2s and 3s orbitals are larger than the 1s orbital

Another point to be noted in the s-orbital wave function is the quantity in parentheses in the 2s and 3s wave functions. These are polynomials of the first and second degree and there are roots to these polynomials where they become zero. For example, in the 2s wave function when $\sigma = 2$ the quantity $(2 - \sigma) = 0$. This is called a **node**, similar to the node observed for the particle in the box problem in Chapter 6. However, in this case the node is at a given distance $\sigma = 2$ and the nodal surface is a sphere with radius $\sigma = 2$. Since the quantity in parentheses for the 3s orbital is a second order polynomial (a quadratic), there are two roots which cause this quantity to go to zero. These can be found from the quadratic formula to be approximately $\sigma = 1.9$ and 7.1. In this case there are two nodal spheres at distances of $\sigma = 1.9$ and 7.1, corresponding to radii of $\left(\dfrac{a_o}{Z}\right)1.9$ and $\left(\dfrac{a_o}{Z}\right)7.1$.

In Figure 7-A-3 we show graphs of the wave function, ψ, for the s orbitals.

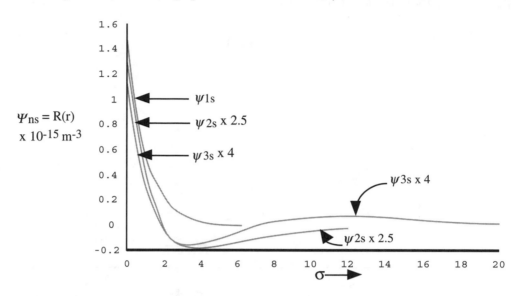

Figure 7-A-3. Wave Functions for the s Orbitals as a Function of r

Note the larger size for the 3s and 2s orbitals and the nodes ($\psi = 0$) for the 2s and 3s orbitals.

Wave Functions for p Orbitals

The wave functions for the $2p_z$ and $3p_z$ are

$$\psi_{2p_z} = \frac{1}{4\sqrt{2\pi}}\left(\frac{Z}{a_o}\right)^{3/2}\sigma\, e^{-\sigma/2}\cos\theta \tag{7-A-11}$$

$$\psi_{3p_z} = \frac{\sqrt{2}}{81\sqrt{\pi}}\left(\frac{Z}{a_o}\right)^{3/2}(6-\sigma)\,\sigma\, e^{-\sigma/3}\cos\theta \tag{7-A-12}$$

In this case the wave function consists of a radial portion, $R(r)$, and an angular portion, $\Theta(\theta)$. Since these p_z orbitals are directed along the z-axis, they are independent of the angle ϕ, as defined in Figure 7-A-1. On the other hand, the wave functions for the p_x and p_y orbitals contain functions of both θ and ϕ in order to describe their orientation along the x and y axes, respectively.

$$\psi_{2p_x} = \frac{1}{4\sqrt{2\pi}}\left(\frac{Z}{a_o}\right)^{3/2}\sigma\, e^{-\sigma/2}\sin\theta\cos\phi \tag{7-A-13}$$

$$\psi_{2p_y} = \frac{1}{4\sqrt{2\pi}}\left(\frac{Z}{a_o}\right)^{3/2}\sigma\, e^{-\sigma/2}\sin\theta\sin\phi \tag{7-A-14}$$

Since the wave functions for the p orbitals contain both radial and angular functions, graphical representation of these orbitals is more difficult. For simplicity we consider the two functions separately so their properties can be examined independently. The combined effect of both functions on the shape of a p orbital will be presented in the following section.

We can graph the radial portion of the $2p_z$ and $3p_z$ orbitals as we did previously for the s orbitals. These are shown in Figure 7-A-4.

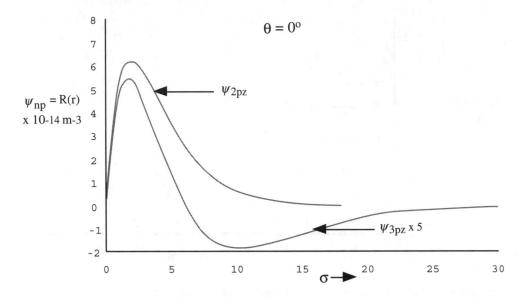

Figure 7-A-4. Graph of the Radial Portion of the Wave Functions Versus r for the $2p_z$ and $3p_z$ Orbitals

These $R(r)$ functions for the $2p_z$ and $3p_z$ orbitals differ from the s-orbitals in Figure 7-A-2 in two respects. First note that both the ψ_{2p_z} and ψ_{3p_z} have a σ in front of the exponential term. This causes the wave function to become zero at $\sigma = 0$ (the nucleus), whereas the s-orbitals have a significant value for ψ at $r = 0$. This will become important in Section 7-C when we consider the effect that other electrons can have on these orbital energies in the case of Many - Electron - Atoms. Secondly, note that the $2p_z$ wave function does not have any nodes whereas the $3p_z$ wave function has one radial node at $\sigma = 6$. For the same principal quantum number the number of radial nodes for the p orbitals is one less than that for the s orbitals. A general formula can be derived which relates the number of radial nodes to the quantum numbers n and ℓ.

$$\text{\# radial nodes} = n - \ell - 1 \qquad (7\text{-A-15})$$

This will be very useful in determining the shapes of other orbitals, as described in Section 7-B.

The angular portion of the ψ_{2p_z} and ψ_{3p_z} orbitals consists of simply $\cos\theta$. This $\Theta(\theta)$ = $\cos\theta$ is the same for all p_z orbitals, regardless of the principal quantum number. This is also true for the p_x and p_y orbitals in that the angular functions are the same regardless of the principal quantum number, n. This is also true for d and f orbitals. For example if we have a $d_{x^2-y^2}$ orbital, the angular portion of the $d_{x^2-y^2}$ wave function is the same for the $3d_{x^2-y^2}$, $4d_{x^2-y^2}$, $5d_{x^2-y^2}$, ···· orbitals. Only the radial portion changes as n increases.

The $\Theta(\theta) = \cos\theta$ function can be graphed as a function of θ, the angle from the z-axis. Note that where $\theta = 0^\circ$, $\cos\theta$ becomes a maximum value of 1.00 and that when $\theta = 90^\circ$, $\cos\theta$ becomes zero. All other values of θ between 0° and 90° give $\cos\theta$ values intermediate between 1.00 and zero. It is customary to graph the magnitude of $\cos\theta$ as a radial distance from the nucleus as a function of θ. The graph would have the following shape

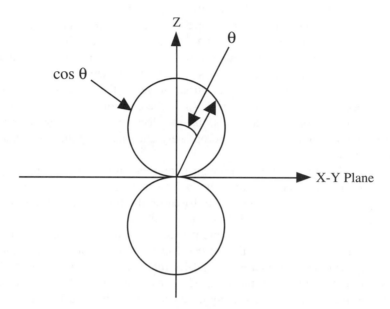

This type of graph emphasizes the influence that the angular portion of ψ has upon the directionality of the orbital, in this case, the directionality of the p_z orbital.

The representation of ψ in three dimensions showing the effect of all variables r, θ, ϕ obviously is very difficult or impossible. Alternatively, we will describe the orbital shapes by graphs of ψ^2 in terms of probability density diagrams, as discussed in Section 7-C.

Problem 7-A-4. Sketch ψ^2 for the 1s, 2s, and 3s hydrogen atomic orbitals as a function of r.

Problem 7-A-5. Show that ψ, given in equation (7-A-8), is a solution to the Schrödinger equation for the hydrogen atom given in equation (7-A-4). Assume $Z = 1$ and evaluate the eigenvalue. The eigenvalue should be identical to the energy in equation (6-D-5) for $n = 1$. Hint: Since ψ is independent of θ and ϕ, the derivatives with respect to θ and ϕ in equation (7-A-4) are zero.

Problem 7-A-6. Account for the nodal properties for the $3p_x$ wave function given by

$$\psi_{3p_x} = \frac{\sqrt{2}}{81\sqrt{\pi}} \left(\frac{z}{a_o}\right)^{3/2} (6 - \sigma)\, \sigma e^{-\sigma/3} \sin\theta \cos\phi$$

Answer: radial node occurs at $(6 - \sigma) = 0$ or $\sigma = 6$; planar node perpendicular to the x-axis occurs since $\phi = 90^\circ$ in the YZ plane passing through the origin and $\cos 90^\circ = 0$.

B. Graphical Representation of Hydrogenic Wave Functions

The Orbital Concept

In the previous section we showed graphical representations of the s and p orbitals in terms of the separate radial and angular portions. In this section we wish to show a three-dimensional representation of the hydrogenic wave functions.

In general, ψ and ψ^2 are functions of the three space coordinates and, consequently, have precise values at any point in space, defined by the values of the coordinates r, θ, ϕ. The wave function ψ provides us with two things: the energy of the state described by the wave function, and the probability distribution for the electron, ψ^2, when in that state.

If ψ has been normalized, the value of ψ^2 for any particular values of the coordinates is equal to the probability per unit volume of finding the electron at those coordinates. The results of the quantum mechanical analysis do not provide any information whatsoever regarding an exact path or orbit for the electron. However, the wave functions do provide information concerning in which regions of space the electron is most likely to be found; that is, how much more or less likely it is, that the electron be in one region of space than another. The wave functions, or, more appropriately, the squares of the wave functions, are referred to as **orbitals**.

A somewhat different and quite popular interpretation of the meaning of ψ^2 is that often referred to as the electron charge cloud. The rationale for this attitude toward the meaning of the atomic orbitals is generally based on the following logic. Since the electron, according to the postulates of quantum mechanics, has a wave nature, it can be thought of as "smeared" out in space around the nucleus of the atom. This idea is, in a way, consistent with the fact that the position of the electron as a particle of matter can never be known precisely. The physical picture of the electron is then a nebulous cloud of negative charge surrounding the nucleus of the atom. The density of the cloud is considered to be an exact analogy to the probability of finding the electron at a given position per unit volume of space. Hence, the density of the charge cloud, which varies as a function of the position in space, is identified with ψ^2.

Shapes of Hydrogenic Orbitals

Since it is impossible to describe exactly where the electron is, we must resort to some other type of description. Of course, the wave function or orbitals themselves contain, in mathematical form, all the information concerning the system which is possible for us to know. However, it is desirable to illustrate the orbitals graphically, so that at a glance we can obtain an overall view of the situation described by any one of the orbitals. In the case of some of the orbitals, such as the s-type orbitals, the probability distribution is totally symmetrical about the nucleus. For all other types of orbitals, p, d, f, and so forth, there exist surfaces containing the nucleus such that the value of ψ^2 is equal to zero at all points on these surfaces. Such surfaces of zero probability for the electron are known as nodal surfaces.

A number of methods have been employed for graphically illustrating the atomic orbitals, none of which is entirely satisfactory. The optimal graphic representation of an orbital would be to construct closed surfaces in three dimensions, so that the value of ψ^2 would be constant for every point on any given surface, and each successive surface would contain an increasing percentage of the total probability of finding the electron within the enclosed region of space.

Figure 7-B-1 illustrates three of the more common methods of graphically representing the hydrogen atomic orbitals. Figure 7-B-1 includes graphical representations for the 1s, 2s, $2p_x$, $2p_y$, and $2p_z$ atomic orbitals. The type of orbital representation shown in Figure 7-B-1 (a) is known as a dot diagram. The two-dimensional dot diagrams represent the value of ψ^2 in a plane containing the nucleus, similar to a cross section of a three-dimensional object. The more closely the dots are spaced in any region, the larger is the value of ψ^2 in that region, and the greater the probability of finding the electron in that region. It can also be said that the denser the dots, the greater is the electron density. Nodal planes or surfaces which intersect the planes of the dot diagrams are indicated by solid lines or closed curves, respectively, on the diagrams.

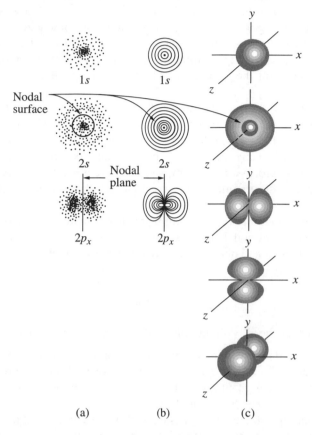

Nodal surface

Nodal plane

(a)   (b)   (c)

Figure 7-B-1. Atomic orbital representations. (a) Two-dimensional dot diagrams; (b) two-dimensional contour diagrams; (c) three-dimensional surface contours (90% probability of the electron being within the enclosed space)

Two-dimensional contour diagrams for the orbitals are shown in Figure 7-B-1 (b). The contour diagrams, similar to the dot diagrams, provide a representation of the change in the value of ψ^2 in a plane containing the nucleus. Each contour corresponds to a constant value of ψ^2. In three dimensions, the contours would become closed surfaces. Each successively larger contour surface would correspond to a greater probability for finding the electron within the region of space enclosed by the imaginary surface. A contour surface is a surface where the density of the dots in a dot diagram would be constant.

Figure 7-B-1 (c) illustrates the 90% contour surfaces in three-dimensional space. These three-dimensional diagrams are similar to the "shapes of the orbitals" found in most general and organic chemistry texts with which the reader is probably familiar. The three-dimensional surfaces are constructed in such a manner that the wave function is equal to the same value at each and every point on any of the closed surfaces, and each surface contour encloses a portion of space which corresponds to 90% of the total probability for finding the electron.

One can see from Figure 7-B-1 that the 1s orbital is symmetrical about the nucleus, with no nodal planes or surfaces. The value of the wave function is greatest at the nucleus, and decreases with increasing distance from the nucleus. The 2s orbital is similar to the 1s orbital, except in regard to how the value of the wave function changes with distance from the nucleus. As shown previously in Figure 7-A-2, ψ^2 for the 2s orbital decreases as the distance from the nucleus increases, until it becomes equal to zero at a given distance from the nucleus ($\sigma = 2$). Thus, the 2s orbital contains a spherical nodal surface . As the distance from the nucleus continues to increase, the value of ψ^2 increases from zero to a maximum value, and then decreases again.

The three 2p orbitals each contain a nodal plane. The $2p_x$ orbital is equal to zero at every point in the yz plane, the $2p_y$ orbital is equal to zero at every point in the xz plane, and the $2p_z$ orbital is equal to zero at every point in the xy plane. These nodal planes arise from the angular portion $\Theta(\theta)\Phi(\phi)$ of the wave function becoming zero. The symmetry of the 2p orbitals can be seen from the diagrams in Figure 7-B-1 (c). The $2p_x$ orbital is symmetrical about the x axis, the $2p_y$ orbital is symmetrical about the y axis, and the $2p_z$ orbital is symmetrical about the z axis. The value of each of the 2p orbitals is zero at the nucleus.

The way in which the value of a 2p wave function changes with distance from the nucleus on either side of the nodal plane can best be seen from the contour diagram for the $2p_x$ orbital in Figure 7-B-1(b). The largest values of ψ^2 of the $2p_x$ orbitals are at points on the x axis, as compared to any other radial line extending out from the nucleus. The value of ψ^2 increases from zero to a maximum, with distance along the x axis on either side of the yz (nodal) plane. After the maximum, ψ^2 decreases with increasing distance along the x axis.

The shapes of the atomic orbitals can be readily visualized by noting the nodal surfaces. The total number of nodal surfaces is equal to $(n - 1)$. This total number of nodes is distributed between radial and planar nodes according to the simple rules:

$$\text{\# radial nodes} = n - \ell - 1$$
$$\text{\# planar nodes} = \ell$$

In a few cases, such as the d_{z^2} orbital, one of the planar nodes is actually a conical surface, but most of the time they are planar surfaces. We show in Figure 7-B-2 the hydrogenic atomic orbitals categorized by the number of nodes of each type. Across the top of the table we give the principal quantum number, n , and $n - 1 =$ total \# nodal surfaces. Along the vertical axis we designate the ℓ = \# planar nodes.

The 1s, 2s, and 2p orbitals are consistent with those represented in Figure 7-B-1. For $n = 3$ we have $(n - 1) = 2$ nodal surfaces. For the 3s orbital $\ell = 0$, so there are no nodal planes. The 2 nodal surfaces must therefore be nodal spheres, as shown by the two dashed, nodal circles shown in Figure 7-B-2.

The 3p orbital ($n = 3$, $\ell = 1$) has

$$n - 1 = 3\text{-}1 = 2 \text{ nodal surfaces}$$
$$n - \ell - 1 = 3\text{-}1\text{-}1 = 1 \text{ nodal sphere}$$
$$\ell = 1 \text{ nodal plane}$$

The probability density diagram for the 3p orbital in Figure 7-B-2 shows the single nodal plane and a single spherical node. In a qualitative manner probability density is indicated between the nodal surfaces.

The 3d orbital ($n = 3$, $\ell = 2$) has 2 nodal surfaces. Since

$$n - \ell - 1 = 3\text{-}2\text{-}1 = 0 \text{ spherical nodes}$$
$$\ell = 2 \text{ planar nodes}$$

The probability density diagram in Figure 7-B-2 shows the 3d orbital characteristic of the d_{xy}, d_{xz}, d_{yz}, and $d_{x^2-y^2}$ orbitals.

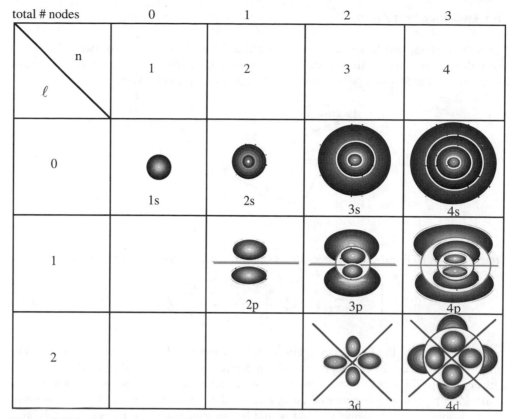

total # nodes	0	1	2	3
ℓ \\ n	1	2	3	4
0	1s	2s	3s	4s
1		2p	3p	4p
2			3d	4d

Figure 7-B-2. Hydrogenic Orbitals Classified by the Nodal Surfaces

For $n = 4$ we have the 4s, 4p, 4d orbitals. A summary of the nodal surfaces is

$$n - 1 = 4 - 1 = 3 \text{ nodal surfaces}$$

	4s	4p	4d
# radial nodes = $n - \ell - 1$	$4 - 0 - 1 = 3$	$4 - 1 - 1 = 2$	$4 - 2 - 1 = 1$
# planar nodes = ℓ	0	1	2

The probability density diagrams shown in Figure 7-B-2 for $n = 4$ are self-explanatory after the appropriate radial and planar nodes are drawn.

Problem 7-B-1. Sketch the probabilty density diagram for a 5d orbital showing all nodal surfaces. Answer: two radial nodes and two planar nodes.

C. Many-Electron Atoms

In principle, an atom containing more than one electron can be treated in the same manner as was the one-electron hydrogen atom. The Schrödinger wave equation must be written for the entire system, being certain to include all potential energy terms corresponding to both the attractive forces between the nucleus and electrons and the repulsive forces between pairs of electrons. In addition, the Schrödinger equation must contain a Laplacian operator for each of the electrons present. The simplest example of a many-electron atom is the helium atom.

The Two-Electron Atom

The helium atom, which contains two electrons and a nucleus made up of two protons and two neutrons, is pictured diagramatically in Figure 7-C-1. With reference to the diagram in Figure 7-C-1, the Schrödinger equation for the helium atom can be written immediately.

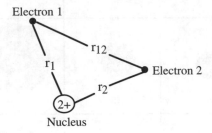

Figure 7-C-1. Helium Atom (Three-dimensional Coordinate System Not Shown)

$$\left[-\frac{\hbar^2}{2m}\left(\nabla_1^2 + \nabla_2^2\right) - \frac{2e^2}{r_1} - \frac{2e^2}{r_2} + \frac{e^2}{r_{12}} \right] \psi = E\psi \tag{7-C-1}$$

∇_1^2 and ∇_2^2 are the Laplacian operators for electron 1 and electron 2, respectively. The terms $-2e^2/r_1$ and $-2e^2/r_2$ represent the potential energy terms due to the attraction between the positive (+2) nucleus and electron 1 and electron 2, respectively, whereas the term e^2/r_{12} represents the potential energy due to the repulsion between the two electrons. The solutions, ψ, of equation (7-C-1) would be two electron wave functions involving a total of six spatial coordinates each, three pertaining to each of the electrons in the system.

The only difficulty with this analysis of the helium atom is that analytic solutions to equation (7-C-1) cannot be found. Furthermore, the problem is equally unsolvable for all other atoms having more than one electron. In order to gain any insight into atoms larger than the hydrogen atom, some additional reasoning must be brought to bear on the problem.

There exists a number of approximate techniques which can be applied to the solution of the helium atom problem, as well as to atoms of larger atomic number. One general method is to assume complicated approximate wave functions involving many undetermined parameters, and then vary the values of these parameters until a wave function is obtained which gives the lowest possible energy. The general form of the wave function is altered until the lowest possible energy is obtained. It is assumed that the best wave function which can be obtained by any approximation technique is the one which corresponds to the lowest possible energy.

Another method often employed is to write the Schrödinger equation for only one of the electrons. However, in order to express the portion of the potential energy of the single electron due to the repulsion between it and the other electrons present, approximate wave functions must be assumed for the other electrons. This is done, and the one-electron approximate solutions which are obtained are considered to be better than the wave functions originally assumed for the other electrons. Consequently, the process is repeated, employing the approximate solutions of the first analysis, instead of the originally assumed wave functions, to express the repulsive potential energy terms of the system. The same procedure is repeated again and again (by computer), until the energies given by the approximate solutions cease to change appreciably, hence the name *self-consistent field* solutions. These final approximate wave functions are taken to be the best approximations of the actual solutions to the problem. In the case of helium, very precise energies can be obtained using extremely complicated mathematical expressions.

The results of numerous extensive approximate calculations for many-electron atoms have indicated that the correct solutions, if known, would probably have angular dependencies of the hydrogen atom wave functions (orbitals). Logic also suggests that the potential field of a many-electron atom should be spherically symmetrical about the nucleus. Hence, the angular dependencies of the one-electron wave functions would be expected to be similar to those for the hydrogen atom wave function. The spatial extent of the wave functions would be expected to differ for different atoms. Also, the energies for many-electron atoms will depend on the angular momentum quantum number, as well as on the principle quantum number. This is a result of the electronic repulsion between electrons.

Problem 7-C-1. (a) Write the complete Schrödinger equation for the lithium atom. (b) How many spatial coordinates would the actual correct solutions involve? Answer:

$$\text{(a) } \left[-\left[-\frac{\hbar^2}{2m}\left(\nabla_1^2 + \nabla_2^2 + \nabla_3^2 \right) - \frac{3e^2}{r_1} - \frac{3e^2}{r_2} - \frac{3e^2}{r_3} + \frac{e^2}{r_{12}} + \frac{e^2}{r_{23}} + \frac{e^2}{r_{13}} \right] \psi = E \, \psi \text{ (b) } 9$$

Hydrogenic Wave Functions

The simplest method for describing many-electron atoms is to assume one-electron wave functions of the same general form as the hydrogen wave functions, modified for the increased nuclear charge of any particular atom. Such wave functions are known as *hydrogenic wave functions. Each of the hydrogenic wave functions represents the coordinates of a single electron, and the wave function or orbital is said to contain the electron.*

It is worth noting, at least qualitatively, that the increased nuclear charge tends to contract each orbital. This contraction is quite easily determined for a single electron in an orbital influenced only by the +Z nucleus, that is, if the effects due to the presence of other electrons are ignored. For example, the average radius of a 1s orbital for helium (Z = +2) would be one-fourth that of the 1s orbital for hydrogen (Z = +1). However, the electrons present in inner orbitals have a definite effect on the interaction of the nucleus and any particular electron in an outer orbital. "Inner electrons" tend to *shield* or *screen the nucleus* from electrons further from the nucleus (outer electrons). In other words, the probability distribution of a 2s electron will be contracted less, if there are other electrons in the 1s orbital, than if the 1s electrons were not present. Thus, the contraction of any hydrogenic orbital will depend on both the nuclear charge, Z, and on the other electrons present.

The spatial extent of hydrogenic orbitals is not related directly to the nuclear charge, but rather to a defined quantity known as the *effective nuclear charge*. The effective nuclear charge is equal to the actual nuclear charge minus a screening constant. The value of the screening constant for an electron in a particular orbital is dependent on the inner electrons relative to it, and on the types of orbitals occupied by the inner electrons. We will discuss the effect that shielding has on the energies of the atomic orbitals later in the following two sections.

D. Aufbau Principle—Electronic Configurations of Atoms

Electron Spin

As a consequence of several experimental results dating back to the Stern-Gerlach experiments of 1922 and 1924, the electron must be assumed to have an additional property called *electron spin*. Furthermore, it is necessary to assume that the spin coordinate is quantized. The spin quantum numbers (m_s) can have only two possible values, 1/2 or -1/2. The possible spins are, therefore, of equal magnitude, but opposite in direction.

The spin of the electron is described by a spin wave function, represented by either α or β, containing $m_s = +1/2$ and $m_s = -1/2$, respectively. The total wave function for a single electron, ψ, is thus equal to the product of the spatial wave function (ψ_{n,ℓ,m_ℓ}) and a spin wave function, α or β.

$$\psi = \psi_{n,l,m}, \alpha \text{ or } \psi = \psi_{n,l,m}, \beta \quad \textbf{Total one-electron wave functions} \quad (7\text{-}D\text{-}1)$$

Thus, any electron is characterized by *four* quantum numbers, n, l, m_ℓ, and m_s.

The Pauli Principle

One additional postulate is necessary before it is possible to describe the electrons in a many-electron atom, that is, before it becomes possible to elucidate the electronic structure of atoms. The postulate which cannot be proven nor derived from first principles is known as the **Pauli Principle**. In its simplest form, the Pauli Principle states that no two electrons in the same atom can have the same values for all four quantum numbers. The direct consequence of the Pauli Principle is that no two electrons in the same atom can have the same total wave function. If two electrons have identical values for three of the quantum numbers, they must have different values for the fourth.

If the spatial portion of the one-electron wave functions for two electrons in an atom is the same, the total wave functions can be different only if the spin quantum numbers of the two electrons are different. Since only two values for the spin quantum number are possible, this places a restriction of a maximum of two electrons in an atom which can have the same spatial wave function. On this basis, it is said that *an atomic orbital can contain a maximum of two electrons.*

Electron Configuration of Atoms

The electronic configuration of an atom refers to a description of the electrons in the atom according to the wave function which best describes each of them. The popular way of expressing this is to simply identify the orbitals which are occupied by the electrons in the atom. The electron configuration can be determined by placing the electrons one by one into the hydrogenic atomic orbitals, such that each successive electron is placed in the lowest possible energy orbital without disobeying the Pauli Principle. This method of discerning the electron configuration is known as the **Aufbau** (building up) **Principle.**

The chemistry student has been introduced to the idea of an electron configuration of an atom in an introductory chemistry course. For a given element, the number of electrons occupying each type of orbital is designated by a superscript. For example:

He: $1s^2$ (both electrons in the 1s orbital with opposite spins);
B: $1s^2, 2s^2, 2p^1$ (the fifth electron goes into one of the three 2p orbitals);
Cl: $1s^2, 2s^2, 2p^6, 3s^2, 3p^5$ (the 17 electrons are used before the 3p orbitals are completely filled).

In these examples, the electronic configuration is for the lowest energy state of the atom, referred to as the *ground electronic state*. Electronic configurations can also be written for "excited" states of the atom, where an electron has been promoted to a higher energy, unfilled atomic orbital. The electron configuration of the ground state is that obtained by application of the Aufbau Principle.

It was previously mentioned that the energies of the hydrogenic orbitals for many electron atoms, unlike the case for the hydrogen atom, depend on the orbital angular momentum quantum number, ℓ. This splitting of the orbital degeneracies results from the electron-electron interactions in the many-electron atoms. The energy of each type orbital is different, and moreover, the energy order of the orbitals is not always the same as their order according to the principal quantum number.

In order to rationalize this splitting of the orbital degeneracies we will consider the 2s and 2p orbitals. As stated previously, helium has 2 electrons and according to the Aufbau

Principle we would put these into the 1s orbital. However, when we add a third electron, as for the Li atom, we have a choice of adding the electron to the 2s orbital or the 2p orbital. In the H atom these two orbitals have the same energy. However, the presence of the 2 electrons in the 1s orbital splits this degeneracy, causing the 2s orbital to have a lower energy (be more stable) than the 2p orbital. This can be rationalized by the fact that the two 1s electrons shield an electron in a 2p orbital more than an electron in a 2s orbital, thus making the electron in the 2p orbital have a lower effective nuclear charge than an electron in a 2s orbital. The energy of an orbital is related to the square of the effective nuclear charge, and a lower effective nuclear charge would cause the orbital to have a higher energy (be less stable). Consequently, the 2s and 2p orbitals split with the 2s orbital being lower in energy than the 2p orbital. We show this in the following diagram for the Li-atom.

Figure 7-D-1. Orbital Energy Diagram for the Li Atom

The reason for the greater effective nuclear charge for an electron in a 2s orbital compared to a 2p orbital can be understood in terms of the spatial characteristics of the orbitals. In Figure 7-B-1 we showed the shapes of the H-atomic orbitals in terms of their probability density diagrams. Note that the 2s orbital has a significant probability density at the nucleus in the vicinity of the 1s orbital. Consequently, when the electron in the 2s orbital is in this vicinity, it experiences a sizeable portion of the nuclear charge. On the otherhand, the 2p orbital has a planar node passing through the nucleus and the probability of finding an electron in a 2p orbital at the nucleus is zero. Consequently, since an electron in a 2s orbital has a greater probability in the vicinity of the nucleus than an electron in a 2p orbital, its effective nuclear charge is greater and its energy is lower (more stable).

A more quantitative representation of the probability density near the nucleus can be obtained by observing a graph of the radial portion of the wave function as shown in Figures 7-A-2 and 7-A-3. Note that the wave function for the 2s orbital has a finite value at $r = 0$ (Figure 7-A-2) whereas the ψ_{2p_z} has a value of zero at $r = 0$. This increased probability of finding the electron near the nucleus is sometimes referred to as the <u>penetration</u> of the electron; i.e., an electron in a 2s orbital is said to be more <u>penetrating</u> than an electron in a 2p orbital. Orbitals with greater <u>penetration</u> have lower energies (more stable).

The relative energies of the atomic orbitals are highly dependent upon the effective nuclear charge which the electron in a given orbital experiences. This, of course, is dependent on the electrons in the other occupied orbitals. The occupied atomic orbitals vary with increasing atomic number, and therefore, the relative order of the energies of the orbitals changes. The variation of the energy order for the atomic orbitals as a function of atomic number is shown in Figure 7-D-2.

Having had an introductory course in chemistry, the student should be familiar with the filling order of the atomic orbitals; that is, the order in which electrons are placed in successive atomic orbitals in atoms of increasing atomic number. The filling order is necessarily dependent on the way in which the energies of the orbitals change with atomic number. Hence, the well-known filling order rule is:

$$1s^2, 2s^2, 2p^6, 3s^2, 3p^6, 4s^2, 3d^{10}, 4p^6, 5s^2, 4d^{10}, 5p^6, 6s^2,$$
$$4f^{14}, 5d^{10}, 6p^6, 7s^2, 5f^{14}, 6d^{10}.....$$

$$(7\text{-}D\text{-}2)$$

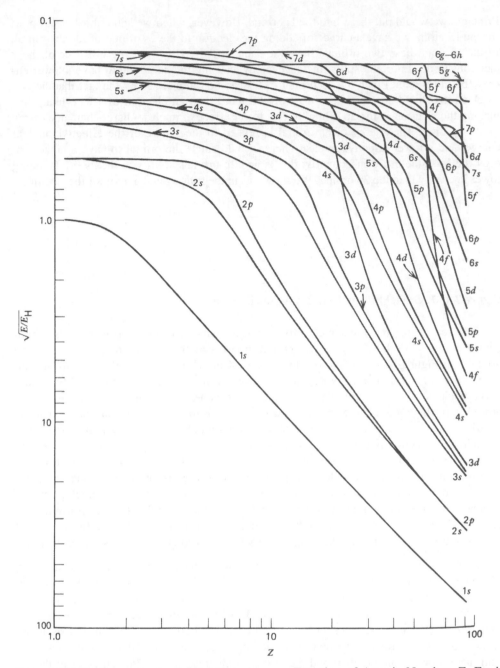

Figure 7-D-2. Approximate Orbital Energies as a Function of Atomic Number, Z. E_H is the Energy of the Ground State Hydrogen Atom. (From R. Latter, *Phys. Rev.* **1955**, *99*, 510.)

This order is most readily remembered by referring to the *Periodic Table* as shown in Figure 7-D-3. The principal quantum number, n, designates the period. The type of orbital being filled is designated within the region of the periodic table as shown. The order of filling is found by proceeding from left to right across the successive periods of the table, which is, of course, the order of increasing atomic number. The principal quantum number of the d orbitals is one less than the principal quantum number for that period as indicated by (n - 1). Similarly, the principal quantum number of the f orbitals is two less than the principal quantum number of that period as shown by (n - 2). The elements in which f orbitals are being filled are inserted at the locations indicated by the asterisks.

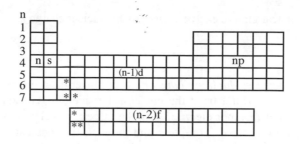

Figure 7-D-3. Filling Order to Obtain Electron Configurations of the Elements in the Periodic Table

In conclusion, two important aspects of the electron configuration of atoms must be mentioned. The first is that the order of filling of atomic orbitals, as the name suggests, is only the filling order, and *not* the actual order within the atom. The second has to do with the general manner in which the description of the electronic configuration is approached. The expression, "placing electrons in orbitals," is merely a convenient manner of deducing the electronic configuration. An orbital refers only to a possible wave function for the atom, and not, in any sense, to a thing which exists irrespective of an electron. Placing the electrons in orbitals means that each successive electron is described by the next possible wave function corresponding to the lowest energy for that electron.

Problem 7-D-1. Write the electron configurations for the elements phosphorous, cobalt, xenon, and samarium according to (a) the filling order of the orbitals (b) orbital energy as shown in Figure 7-D-2.
Answer: (a)P:$1s^2 2s^2 2p^6 3s^2 3p^3$;Co:$1s^2 2s^2 2p^6 3s^2 3p^6 4s^2 3d^7$Xe:$1s^2 2s^2 2p^6 3s^2 3p^6 4s^2 3d^{10} 4p^6 5s^2 4d^{10} 5p^6$;Sm:$1s^2 2s^2 2p^6 3s^2 3p^6 4s^2 3d^{10} 4p^6 5s^2 4d^{10} 5p^6 6s^2 4f^5$(b) P:$1s^2 2s^2 2p^6 3s^2 3p^3$;Co:$1s^2 2s^2 2p^6 3s^2 3p^6 3d^6 4s^2$;Xe:$1s^2 2s^2 2p^6 3s^2 3p^6 3d^{10} 4s^2 4p^6 4d^{10} 5s^2 5p^6$;Sm:$1s^2 2s^2 2p^6 3s^2 3p^6 3d^{10} 4s^2 4p^6 4d^{10} 5s^2 5p^6 4f^5 6s^2$.

E. Analysis of He and Li Emission Spectra—Evaluation of Effective Nuclear Charge

Transition Energies

In section 7-A we discussed the spectral transitions that are observed for the excited H-atom. The mathematical equation which describes the transitions is based upon the equation for the eigenvalues for the H-atom equation (7-A-7). This is the same equation as that for a hydrogenic system, equation (6-D-5), where Z = 1, the atomic number for the H-atom. In the following discussion we will assume the energy of the atom can be expressed in a similar formula except that the atomic number Z will be replaced by the <u>effective nuclear charge</u>, Z_{eff}. The equation for the energy for the electron in the nth orbital is then

$$E_i = -\frac{2\pi^2 me^4 (Z_{eff})^2}{h^2} \frac{1}{n^2} \tag{7-E-1}$$

where *n* is now the principal quantum number for the nth orbital and all other terms are as defined previously. The effective nuclear charge on an electron will be dependent not only on the nature of the atomic orbital in which it resides but also upon the atomic number and electronic configuration of the atom.

The effective nuclear charge can be related to the nuclear charge Z, by

$$Z_{eff} = Z - \sigma$$

where σ is referred to as the *shielding constant*. If the inner electrons completely shield an electron in an outermost orbital from the nucleus, the shielding constant for this electron would equal the number of inner electrons and the effective nuclear charge would be one. Of course the *shielding* is never complete and the effective nuclear charge is generally greater than one. Nevertheless, comparing σ with the number of inner electrons reveals the effectiveness of the inner electrons to shield the electron from the +Z nuclear charge.

A spectral transition would be represented by the difference in two energy terms of the type shown in equation (7-E-1). This is similar to our previous equation (6-D-7) <u>except</u> that now we assume the Z_{eff} may be different between the upper and lower states. As before, we will let the subscript U refer to the upper state and the subscript L to the lower state.

$$e = hc\ \bar{v} = E_U - E_L = - \frac{2\pi^2 me^4 Z_U^2}{h^2} \frac{1}{n_U^2} + \frac{2\pi^2 me^4 Z_L^2}{h^2} \frac{1}{n_L^2} \qquad (7\text{-}E\text{-}2)$$

Solving for $\bar{v}$

$$\bar{v} = - \frac{2\pi^2 me^4 Z_U^2}{h^3 c} \frac{1}{n_U^2} + \frac{2\pi^2 me^4 Z_L^2}{h^3 c} \frac{1}{n_L^2} \qquad (7\text{-}E\text{-}3)$$

Equation (7-E-3) can be simplified, as we did in Chapter 6, by using the Rydberg Constant for hydrogen, R_H

$$R_H = \frac{2\pi^2 me^4}{h^3 c} = 109,677 \text{ cm}^{-1} \qquad (7\text{-}E\text{-}4)$$

Equation (7-E-3) can then be simplified

$$\bar{v} = - R_H Z_U^2 \frac{1}{n_U^2} + R_H Z_L^2 \frac{1}{n_L^2} \qquad (7\text{-}E\text{-}5)$$

As we will see shortly, the emission spectra can be categorized by a series where the upper state increases in principal quantum number, whereas the lower state remains constant. Equation (7-E-5) then becomes a linear equation with the variables $\bar{v}$ and $1/n_U^2$.

Vector Model For He

In order to interpret the helium atomic spectrum, we must first present the vector model for the various atomic states for helium. The helium excited states are a good introduction to the vector model since it is the simplest two electron system. In excited states of He one electron remains in the 1s orbital and since it has $\ell = 0$, the coupling with the orbital angular momentum of the excited electron is simple. Here the spin-orbit coupling is weak and we couple the orbital angular momentum and spin vectors first before spin orbit coupling. We use the conventional nomenclature n = principal quantum number (q.n.), ℓ = orbital angular momentum q.n., and s = spin q.n. . The rules for Russell-Saunders coupling are

$$L = |\ell_1 + \ell_2|, |\ell_1 + \ell_2 - 1|, \ldots |\ell_1 - \ell_2|$$

L	0	1	2	3
Term Symbol	S	P	D	F

$$S = |s_1 + s_2|, |s_1 + s_2 - 1|, \cdots |s_1 - s_2|$$

$2S + 1$ is the pre-superscript

$$J = |L + S|, |L + S - 1|, \ldots |L - S|$$

$S < J$ #J levels $= 2S + 1$
$S > J$ #J levels $= 2L + 1$

Applying the rules to helium we have the following states where $n \geq 2$.

Table 7-E-1. Helium Atomic States Based on the Russell-Saunders Approximation to the Vector Model

Electronic Configuration	ℓ_1	ℓ_2	s_1	s_2	L	S	State 2S +1	J	Designation
$1s^2$	0	0	1/2	-1/2	0	0	1	0	1^1S_0
$1s^1 ns^1$	0	0	1/2	1/2	0	0	1	0	n^1S_0
					0	1	3	1	n^3S_1
$1s^1 np^1$	0	1	1/2	1/2	1	0	1	1	n^1P_1
					1	1	3	2,1,0	$n^3P_{2,1,0}$
$1s^1 nd^1$	0	2	1/2	1/2	2	0	1	2	n^1D_2
					2	1	3	3,2,1	$n^3D_{3,2,1}$

Most of the observed spectra involve only these states. Note that the excited states are either singlet or triplet states and on the energy diagram we separate these two manifolds of states. The selection rules for an atomic transition are

$$\Delta L = \pm 1$$
$$\Delta S = 0$$

The energy level diagram for helium is shown in Figure 7-E-1. For simplicity the J values are left off the state designation. The helium emission spectrum contains transitions between both singlet and triplet states. However, as a rule the transitions between triplet states have on the order of three times greater intensity than those between singlet states. The Rydberg Series analysis is most readily carried out using the more intense triplet transitions. The transitions between the triplet states occurs in the ultraviolet-visible spectral region and this is convenient from an experimental standpoint since air is transparent to these frequencies. From the ionization limit of 24.47 eV down to the 2^3S state (19.82 eV) we can place a lower limit on the wavelength of these triplet transitions: $\lambda > 266$ nm.

In Figure 7-E-1 three Rydberg series are identified as

$$n^3D \text{--------} > 2^3P$$
$$n^3P \text{---------} > 2^3S$$
$$n^3S \text{---------} > 2^3P$$

We will analyze these series shortly.

In Figure 7-E-1 the well-known resonance transition $2^1P \longrightarrow 1^1S$ at 58.44 nm is shown. This transition could be observed in the vacuum UV region but only if the pressure in the discharge region were low, since the radiation is self absorbed by the helium through the reverse transition $1^1S \longrightarrow 2^1P$. We would never observe this resonance radiation at a pressure of ~1 atm. The 2^1S state at 20.61 eV is also occupied but the

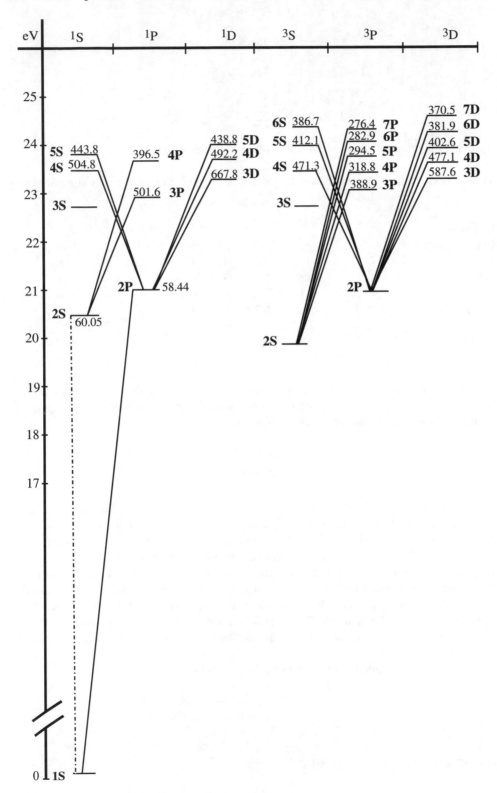

Figure 7-E-1. Energy Level Diagram for Helium

transition at 60.05 nm is forbidden ($\Delta L = 0$). This state would be long-lived and most certainly is involved in the formation of He_2. Transitions from excited n^1P to 1^1S would be allowed, but these would be of higher energy than the $2^1P \longrightarrow 1^1S$ transition and would have $\lambda < 58.44$ nm, farther into the vacuum UV.

Analysis of the He Emission Spectra

Previously we derived equation (7-E-5) which expresses the frequency of a transition assuming the energy terms can be expressed as Rydberg-like using Z_{eff} to account for the attractive force of the nucleus. This expression for $\bar{v}$ is convenient in the analysis of a spectral series in He where the lower state is the same for all members of the series. In Table 7-E-1 we list the wavelengths and frequencies of the transitions involved in the three previously mentioned series. This data has been taken from a reference book which has compiled very precise data on atomic transitions:

Table 7-E-1. Data for Correlation of $\bar{v}$ Versus $1/n^2$ for Atomic Helium Transitions

$n^3D \longrightarrow 2^3P$	n	$\lambda(nm)$	$\Delta\lambda$	$\bar{v}(cm^{-1})$	$\Delta\bar{v}$
	3	587.562	+0.020	17,018.48	-0.06
	4	447.148	-0.007	22,363.96	+0.33
	5	402.619	-0.003	24,837.38	+0.21
	6	381.961	+0.002	26,180.68	-0.14
	7	370.500	+0.006	26,990.55	-0.46
$n^3P \longrightarrow 2^3S$					
	3	388.865	+0.10	25,715.87	-6.
	4	318.774	-0.28	31,370.19	+27.
	5	294.510	-0.09	33,954.70	+10.
	6	282.907	+0.08	35,347.30	- 10.
	7	276.380	+0.21	36,182.07	- 27.
$n^3S \longrightarrow 2^3P$					
	4	471.314	+0.12	21,218.28	-5.3
	5	412.081	-0.44	24,267.07	+25.
	6	386.748	+0.32	25,856.63	- 21.

Note in each of the series there is a common lower state. Since the upper states have principal quantum numbers of 3 or greater, electrons in the higher atomic orbital of these states should be well shielded from the nuclear charge and Z_U should be approximately constant. Consequently, equation (7-E-5) becomes a linear equation, since the term representing the lower state is constant with $n_L = 2$ and Z_U is assumed constant for $n_U \geq 3$.

$$\bar{v} = - \left(R_H Z_U^2\right)\frac{1}{n_U^2} + \left(R_H Z_L^2 \frac{1}{4}\right) \tag{7-E-6}$$

The linear equation $y = m x + b$ is related to equation (6-E-6) by $y = \bar{v}$, $m = - (R_H Z_U^2)$, $x = 1/n_U^2$ and $b = (R_H Z_L^2 \, 1/4)$.

Problem 7-E-1. For the three series given in Table 7-E-1 graph the three linear functions defined by equation (7-E-6). From the slopes and intercepts evaluate the Z_U and Z_L for each of the series.

From the graphs in Problem 7-E-1 one can see that the data fit the linear function very precisely. In fact the deviations from the linear functions are so small that they cannot be evaluated graphically. For this reason a statistical analysis of the data has been performed which is called a "least squares analysis." We do not go into this type analysis in this text, but it is a somewhat common procedure in various aspects of science. A result of the statistical analysis is an estimate of error in the slope and intercept and a listing of the deviations of the data points from the linear function. The results of the statistical analysis of the three spectral series are given in Table 7-E-2.

Table 7-E-2. Least Squares Evaluation of Z_{eff} from the Rydberg Series for Atomic Helium Transitions. Z_U, Z_L = Effective Nuclear Charge for the Upper and Lower States, Respectively; σ_{Z_U}, σ_{Z_L} = Standard Deviations in Z_U, Z_L, Respectively

Transition	$m(cm^{-1})$	$b(cm^{-1})$	Z_U	$\sigma_{Z_U^-}$	Z_L	$\sigma_{Z_L^-}$
$n^3D \longrightarrow 2^3P$	-109,935.	29,234.6	1.00118	0.00003	1.0326	0.00001
$n^3P \longrightarrow 2^3S$	-115,619.	38,568.	1.027	0.002	1.1860	0.0006
$n^3S \longrightarrow 2^3P$	-134,171.	29,608.	1.106	0.01	1.039	0.002

Identifying the specific states involved, we find the following order of Z_{eff} and shielding constants, σ, for the various orbitals:

$$Z(n^3D) = 1.00118 \pm 0.00003 \qquad \sigma(n^3D) = .9988 \pm 0.00003$$
$$Z(n^3P) = 1.027 \pm 0.002 \qquad \sigma(n^3P) = .973 \pm 0.002$$
$$Z(2^3P) = 1.0326 \pm 0.00001 \qquad \sigma(2^3P) = .9674 \pm 0.00001$$
$$Z(2^3P) = 1.039 \pm 0.002 \qquad \sigma(2^3P) = .961 \pm 0.002$$
$$Z(n^3S) = 1.106 \pm 0.01 \qquad \sigma(n^3S) = .894 \pm 0.01$$
$$Z(2^3S) = 1.1860 \pm 0.0006 \qquad \sigma(2^3S) = .8140 \pm 0.0006$$

Note the small errors in Z_U and Z_L in Table 7-E-2. The errors for the series $n^3D \longrightarrow 2^3P$ are exceptionally small (0.001% and 0.003%). This means that the assumption $Z(n^3D)$ = constant is well supported. This is understandable since the effective nuclear charge for the n^3D states is close to unity which reflects complete shielding by the inner 1s electron. Since it is close to unity it must necessarily be close to a constant. The deviations in λ and $\bar{v}$ from the linear relationship are given in Table 7-E-1. Again note the exceedingly small deviations for the $n^3D \longrightarrow 2^3P$ series.

Examination of Tables 7-E-1 and 7-E-2 reveals that the errors associated with the $n^3P \longrightarrow 2^3S$ are greater than with the $n^3D \longrightarrow 2^3P$ series. Again this can be associated with the greater Z_{eff} (n^3P) and the, consequently, greater variation in Z_{eff} for the various n levels. The errors associated with the linear fit for the $n^3S \longrightarrow 2^3P$ series are the largest, but yet still satisfactory (<1%). Again the larger deviations can be associated with the greater value for Z_{eff} (n^3S) and the apparent variation with n.

The ordering of the Z_{eff} and shielding constants for the various orbitals should be noted. As expected from the orbital shapes and sizes it is not surprising that the s-orbitals have a higher Z_{eff} than p-orbitals since the probability of being near the nucleus is greater in an S-orbital. And as we would expect, the Z_{eff} for an orbital with a lower principal quantum number is greater than one with a higher principal quantum number. Thus

$$Z(2^3S) > Z(n^3S) \qquad n \geq 3$$
$$Z(2^3P) > Z(n^3p) \qquad n \geq 3$$

Since σ is calculated by subtracting Z_{eff} from a constant, the atomic number, the error in σ will be identical to the error in Z_{eff}. Based upon our previous discussion, the order of Z_{eff} and σ's follows what we would expect:

$$\sigma_D > \sigma_P > \sigma_S$$

The principal quantum number n has a less significant effect on the σ values than the nature of the atomic orbital; however, note that an electron in an atomic orbital with n>2 has a greater shielding constant than one where n = 2.

$$\sigma\,(n^3\text{P}) > \sigma\,(2^3\text{P}) \text{ and } \sigma\,(n^3\text{S}) > \sigma\,(2^3\text{S})$$

Also note the consistency in the two estimates of the $Z(2^3\text{P})$ and the $\sigma(2^3\text{P})$. The two estimates of each differ by only 0.6%.

A similar analysis can be performed on the atomic emission spectra from lithium. Bettelheim (Frederick A. Bettelheim, *Experimental Physical Chemistry*, W.B. Saunders Co., Philadelphia, 1971) gives the wavelengths of the transitions for three series, as shown in Table 7-E-3.

Table 7-E-3. Data for Correlation of $\bar{\nu}$ versus $1/n^2$ for Atomic Lithium Transitions. $\Delta\,\lambda$ and $\Delta\,\bar{\nu}$ Are the Residuals from the Least Squares Analysis

$n^2\text{D} \longrightarrow 2^2\text{P}$	n	λ(nm)	$\Delta\lambda$	$\bar{\nu}$(cm^{-1})	$\Delta\bar{\nu}$
	3	610.359	+0.001	16,383.8	-0.04
	4	460.286	-0.005	21,725.6	+0.24
	5	413.259	-0.002	24,197.9	+0.15
	6	391.532	+0.001	25,540.7	-0.07
	7	379.504	+0.005	26,350.2	-0.38
$n^2\text{P} \longrightarrow 2^2\text{S}$					
	3	323.266	+0.031	30,934.28	-3.0
	4	274.12	-0.010	36,480.38	+14.
	5	256.231	-0.008	39,027.29	+1.3
	6	247.506	+0.081	40,403.06	-13.2
$n^2\text{S} \longrightarrow 2^2\text{P}$					
	3	812.634	+1.1	12,305.66	-17.
	4	497.17	-6.5	20,113.84	+258.
	5	427.31	-1.1	23,402.21	+60.
	6	398.551	+2.3	25,090.89	-145.
	7	383.561	+4.5	26,071.47	-307.

The analysis can be carried out in the same manner as with the helium atomic spectra in that $\bar{\nu}$ should be linearly related to $1/n_U^2$, as given by equation (7-E-6). The analysis can be carried out using a graph of the data or a least squares analysis of the linear function. The latter has been performed in order to obtain greater accuracy and to obtain an estimate of error in Z_U and Z_L. The results are shown in Table 7-E-4.

Table 7-E-4. Least Squares Evaluation of Z_{eff} from the Rydberg Series for Atomic Lithium Transitions. Z_U, Z_L = Effective Nuclear Charge for the Upper and Lower States, Respectively; σ_{Z_U}, σ_{Z_L} = Standard Deviations in Z_U and Z_L, Respectively

Transition	m(cm^{-1})	b(cm^{-1})	Z_U	σ_{Z_U}	Z_L	σ_{Z_L}
nd $\longrightarrow$ 2p	-109,883.1	28,593.07	1.00094	0.00002	1.02117	0.000008
np $\longrightarrow$ 2s	-113,747.9	43,575.95	1.018	0.002	1.2607	0.0005
ns $\longrightarrow$ 2p	-154,963.7	29,540.94	1.19	0.02	1.038	0.008

Problem 7-E-2. For the three series given in Table 6-E-3, graph $\bar{\nu}$ versus $1/n_U^2$, as suggested by equation (7-E-6). Evaluate the Z_L and Z_U for each series. Compare your results with those given in Table 7-E-4 which were evaluated from a Least Squares Analysis.

The spectral emission data for the lithium atom fit equation (7-E-6) in much the same way as that for the helium atom. The lithium $n^2D \longrightarrow 2^2P$ series fit the linear relationship exceedingly well with deviations in $\bar{v}$ less than one cm^{-1}. The slope and intercept are evaluated to 6 significant figures. The lithium $n^2P \longrightarrow 2^2S$ series also fit the linear relationships well with deviations in $\bar{v}$ of ~ 20 cm^{-1}. The slopes and intercepts are evaluated to 4 significant figures. The lithium $n^2S \longrightarrow 2^2P$ series fits the linear equation less precisely, apparently due to the variation in the Z_{eff} for the ns orbitals. For this series the slope and intercept are evaluated to 3 significant figures.

From the slopes and intercepts in Table 7-E-3 the following Z_{eff} have been evaluated for the different orbitals. The error shown is the standard deviation from the least squares analysis.

$$Z (nd) = 1.00094 \pm 0.00002 \qquad \sigma (nd) = 1.99906 \pm 0.00002$$
$$Z (np) = 1.018 \pm 0.002 \qquad \sigma (np) = 1.982 \pm 0.002$$
$$Z (2p) = 1.021178 \pm 0.000008 \qquad \sigma (2p) = 1.978822 \pm 0.000008$$
$$Z (2p) = 1.038 \pm 0.008 \qquad \sigma (2p) = 1.962 \pm 0.008$$
$$Z (ns) = 1.19 \pm 0.02 \qquad \sigma (ns) = 1.81 \pm 0.02$$
$$Z (2s) = 1.2606 \pm 0.0005 \qquad \sigma (2s) = 1.7394 \pm 0.0005$$

It should be noted that the σ values for the above states follow the same order as those derived from the He-atom transitions, given previously. However, since there are two 1s electrons shielding the electrons in the orbitals with $n \geq 2$, the shielding constants are close to two in this case. Closer examination reveals that the $\sigma (ns)$ and $\sigma (2s)$ are considerably less than the maximum shielding of 2 from the two 1s electrons. This is consistent with the σ values from the He-atom transitions and can be rationalized by the greater penetration of the s-orbitals.

The shielding constants for the two 1s electrons in Li are approximately twice that for the single 1s electron in He, as one might expect. However, closer examination reveals that the shielding per 1s electron is greater in Li than He. This likely results from the fact that the 1s electrons in Li reside closer to the nucleus than in He due to the greater nuclear charge (Z = 3) for Li compared to (Z = 2) for He.

F. Approach to Molecular Bonding

A group of atoms united together in some manner, such that they have unique properties attributable to the whole, which translate together as a unit, undergo rotational and vibrational motions, and in general, exhibit a tendency to remain together, is known as a **molecule**. The motions and associated energies of molecules were discussed in Chapter 6. However, no mention has been made, thus far, concerning what holds the atoms together, except for an occasional reference to the common idea that the atoms in the molecule are "bonded" together. The term *molecular bond,* though familiar to even the beginning student of general chemistry, is quite nebulous and, in fact, meaningless without further clarification.

Simple logic insists that if a group of atoms unite together to form a stable molecule, the energy of the molecular system must be lower than that of the individual atoms, collectively. The obvious question is, what causes the decrease in the energy of the system? The only plausible answer to this question is the following: The energy of the system of atoms will depend on the various interactions of the electrons with each other and with the various nuclei involved in the system. Thus, the total energy, taking into account both the attractive and the repulsive terms, must exhibit a minimum for some particular geometry and nuclear separations of the atoms.

In order to analyze a molecular system correctly, we must, as in the case of individual atoms, resort to a quantum mechanical treatment of the problem. This will involve expressing the Schrödinger Wave Equation for the system and attempting to find physically meaningful solutions. In the case of molecules, however, we will encounter some new terms and features which do not pertain to atomic systems. For example, V, the potential energy portion of the Hamiltonian, will include terms representing the attraction of each electron to more than one nucleus, as well as the repulsion between the nuclei. In addition, the nuclei of molecular systems are in motion relative to each other, and, consequently, a strictly correct analysis must include Laplacian operators for the nuclei. There exists, however, an approximation known as the **Born-Oppenheimer Approximation**, which alleviates the quantum mechanical analysis of this particular difficulty.

In effect, the Born-Oppenheimer Approximation states that the motion of the nuclei in a molecule is much slower that the motion of the electrons. In other words, the electrons traverse many times throughout the molecular system during the course of a single vibration of the nuclei. Consequently, in analyzing the motion of the electrons, the nuclei can be considered fixed with respect to the center of mass. In theory, the resulting solutions will be varied in order to obtain the best possible wave functions and energies.

In particular, the result of the quantum mechanical analysis of a proposed molecular system would be a set of wave functions which describe the electrons relative to the entire system of nuclei, and the energies (eigenvalues) associated with each wave function. The Pauli Principle applies to molecules just as to atoms; hence, the total energy of the molecule could be found as a function of the geometry and internuclear distances. If a minimum in the energy of the system is of great enough magnitude so that the atoms remain together as a relatively stable system, the atoms are said to bond to one another and the molecule exists. If, on the other hand, no such minimum energy is possible for the system, then it is said that the atoms do not bond and the molecule does not exist.

In all cases but one, it is impossible, even when employing the Born-Oppenheimer Approximation, to find exact solutions to the Schödinger equation for molecular systems. However, several methods exist for finding approximate solutions for molecular systems. Moreover, the energies corresponding to approximate wave functions can be compared to experimentally determined values of the bond dissociation energies of molecules. Much information concerning molecular properties is to be gained from such procedures.

G. Molecular Orbitals of H_2^+

The simplest of all molecular systems is the H_2^+ molecular ion, which is composed of two hydrogen nuclei (two protons) and a single electron. Because it is so basic, it is appropriate to introduce the subject of molecular bonding with this example. Although an exact solution for the H_2^+ system is possible when the Born-Oppenheimer Approximation is employed, it will not be presented here. It will be more valuable for the reader to gain an understanding of certain approximate solutions to the problem, which can then be applied to other molecular systems. Exact solutions cannot be found for any other molecular system.

Formulation of the Problem

The H_2^+ molecular ion is shown in Figure 7-G-1

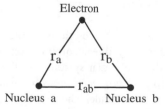

Figure 7-G-1. H_2^+ Molecular Ion

The two nuclei are designated "a" and "b," and the internuclear distance, r_{ab}, is considered to be fixed. The Hamiltonian for the system contains the potential energy terms $-e^2/r_a$, $-e^2/r_b$, and e^2/r_{ab}, representing the attractions between the electron and nucleus "a," the electron and nucleus "b," and the repulsion between the two nuclei, respectively, and also the Laplacian for the single electron, ∇^2. Hence, the Schrödinger wave equation for the system is

$$\left(-\frac{\hbar^2}{2m}\nabla^2 - \frac{e^2}{r_a} - \frac{e^2}{r_b} + \frac{e^2}{r_{ab}}\right)\psi = E\psi \qquad (7\text{-}G\text{-}1)$$

In equation 7-G-1, E is the electronic energy of the molecular ion and is certainly a function of the parameter r_{ab}. The electronic energy for a two-nuclei system is equivalent to the potential energy for vibrational motion of the two nuclei referred to in equations (6-I-3) and (6-I-4) and by the broken line in Figure 6-I-1 of Chapter 6. Equation (7-G-1) can be solved analytically using elliptical coordinates, as mentioned previously, and the electronic energies, as functions of the internuclear distance, can be determined.

Problem 7-G-1. Write the Schrödinger Equation for H_2.

Answer: $\left\{-\dfrac{\hbar^2}{2m}\left(\nabla_2^1 + \nabla_2^2\right) - \dfrac{e^2}{r_{a1}} - \dfrac{e^2}{r_{a2}} - \dfrac{e^2}{r_{b1}} - \dfrac{e^2}{r_{b2}} + \dfrac{e^2}{r_{ab}} + \dfrac{e^2}{r_{12}}\right\}\psi = E\psi$

Since approximate solutions must be employed for all other molecular systems, an approximate method of solution will be illustrated for the H_2^+ molecular ion. The approximate solutions are easily described physically, and, consequently, the application to other molecules, homonuclear diatomics in particular, will not be difficult to comprehend. The relationship between other diatomic systems and the H_2^+ molecular ion is quite analogous to the relationship between many-electron atoms and the hydrogen atom.

Approximate Methods of Solution

There are two principal approximate techniques used to describe molecular systems. These are the Valence Bond (VB) and the Molecular Orbital (MO) approximations. The former should be familiar to the reader, since it is generally discussed in introductory chemistry courses and used extensively in organic chemistry courses. The VB approximation is quite suitable for organic chemistry applications since it can conveniently describe the ground electronic states of molecules. Most organic reactions involve thermal energies, and hence, only the ground state energies of the molecules are important for understanding the energetics of these reactions. The MO approximation, on the other hand, can describe both the ground and excited energy states of molecules. Moreover, the MO approximation leads to a series of approximate spatial wave functions for molecules and inevitably, as its name implies, to the concept of molecular orbitals. The MO approach is more appropriate for our purposes and will be employed almost exclusively in the discussions which follow.

Molecular Orbital Approximation

The MO approximation assumes that the solutions to the Schrödinger equation for molecular systems can be represented by linear combinations of the hydrogenic wave functions of the atoms composing the molecular system. In other words, approximate molecular orbitals are constructed from *linear combinations of atomic orbitals* (LCAO) each located on a different atom in the molecule. The one-electron molecular orbitals are constructed from only the spatial portions of the atomic orbitals, and again, the spin coordinate must be introduced independently. The total one-electron molecular wave function for any electron is then the product of the spatial portion and the spin wave function. An immediate consequnce of the LCAO-MO method is that there are as many molecular orbitals possible as there are atomic orbitals for all the atoms present, collectively. Moreover, since the molecular orbitals are simple functions of the atomic orbitals, they can be described physically by the same methods used for describing the atomic orbitals.

Certain criteria exist which govern the proper choice of atomic orbitals to be used in constructing molecular orbitals. In general, atomic orbitals with comparable energies which have the same symmetry with respect to a line joining the nuclei should be used in constructing each molecular orbital. These criteria arise naturally from a careful mathematical development of the MO approximation. Atomic orbitals which do not meet the stated symmetry requirement cannot be combined to form meaningful molecular orbitals. Moreover, the more nearly equal the energies of the atomic orbitals which are combined, the more effective will be the combination in producing new orbitals which are truly molecular in their character. This can lead to more stabilization for the molecule, that is, to a lower energy for the system. An a priori principle taken to be true is that the approximate wave functions which result in the lowest energy for the molecular system, provide the most nearly correct description of the system.

In the case of H_2^+, the preceding criteria are satisfied exactly, if each molecular orbital is formed from the same type of atomic orbital, one from each of the two nuclei "a" and "b." This is also true for the case of any other homonuclear diatomic molecules. From each pair of atomic orbitals, it is possible to obtain two linear combinations in which each atomic orbital is represented equally. One of the molecular orbitals is obtained from the sum of the two atomic orbitals, and the other is obtained by taking the difference of the two atomic orbitals. The molecular orbital which results from the sum of the atomic orbitals will have a corresponding energy which is lower than the energy of either atomic orbital. Hence, it is called a *bonding molecular orbital*. The other molecular orbital, resulting from the difference of the two atomic orbitals, will have a corresponding energy which is higher than the energy of either atomic orbital and is referred to as an *anti-bonding molecular orbital*. An electron in a bonding molecule orbital adds to the stability of a molecule, while an electron in an anti-bonding molecular orbital decreases the stability of the molecule. We will not concern ourselves with the details of the mathematics, though it is not difficult, but we shall attempt to describe the results diagramatically.

Figure 7-G-2 illustrates some of the resulting molecular orbitals for the H_2^+ molecule ion and the atomic orbitals from which they are constructed. The orbitals are represented in Figure 7-G-2 in the same manner as the hydrogen atomic orbitals were represented in Figure 7-B-1 (c). The molecular orbitals are designated either σ or π, and a superscript * indicates anti-bonding molecular orbitals. The lines in Figure 7-G-2 indicate the atomic orbitals which are combined to form each molecular orbital. While s type atomic orbitals form only sigma type molecular orbitals, p type atomic orbitals can form both σ and π type molecular orbitals.

The σ type molecular orbitals do not have any nodal planes which contain the internuclear axis. The π type molecular orbitals each have one nodal plane which contains the internuclear axis. For example, the nodal plane for the π_{2p_y}, molecular orbital is perpendicular to the plane of the diagram and contains both nuclei. Other molecular orbitals, designated delta, δ, not shown in Figure 7-G-2, are also possible.

Delta type molecular orbitals can be constructed from d atomic orbitals and involve two nodal planes which contain the internuclear axis. There is an extremely important feature of the molecular orbitals which distinguishes the anti-bonding from the bonding

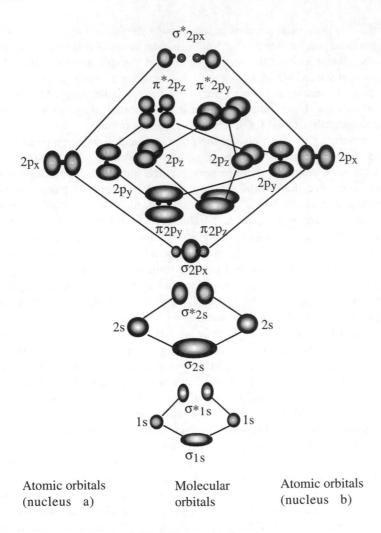

Figure 7-G-2. LCAO Molecular Orbital Representations

molecular orbitals. All anti-bonding molecular orbitals have a nodal plane *perpendicular* to the internuclear axis at a point between the nuclei. This leads to a low probability for the electron occupying an anti-bonding molecular orbital to be between the nuclei. On the other hand, no such perpendicular nodal plane exists in the case of bonding molecular orbitals, hence, there exists a great probability for the electron to be found between the nuclei.

An important concept, known as *overlap*, is concerned with the probability density for finding the electron between the two nuclei. From the diagrams in Figure 7-G-2, it can be seen that the bonding molecular orbitals have relatively large probabilities for the electron to be found between the two nuclei. This is a direct result of taking the sum of the atomic orbitals, and is a quite important factor in regard to the stability of the bond. The value of the probability for finding the electron between the nuclei depends on the extent of the overlap of the atomic orbitals between the nuclei. The extent of the overlap, in turn, depends on the particular atomic orbitals involved, their orientation, and on the distance of separation of the two nuclei. At first thought, it might seem that since the overlap increases as the nuclei move closer together, the stability due to a bonding molecular orbital should continually increase. This has the opposite effect on the stabilization of the molecule, since it increases the energy of the system. At some specific internuclear distance, a balance between these two opposing effects is achieved. The specific internuclear distance at which the balance occurs corresponds to the maximum stabilization, that is, to the minimum electronic energy.

When the single electron in H_2^+ occupies any one of the possible molecular orbitals shown in Figure 7-G-2, it represents a specific electronic energy state of the molecular ion. Each of these states has a given energy, E, which is a function of the internuclear separation, as discussed above. A plot of the energy versus the internuclear distance, r_{ab}, results in what is commonly known as the *potential energy curve*. Each electronic state will have its own characteristic potential energy curve, such as the potential energy curve for the ground electronic state which has already been mentioned in regard to the vibrational motions of molecules. Figure 7-G-3 shows a graph of the potential energy curve for the ground electronic state of the H_2^+ molecular ion which corresponds to the single electron of H_2^+ occupying the σ_{1s} molecular orbital. The minimum in the potential energy curve corresponds to the average internuclear distance. The molecular ion could never be expected to exist in a static state corresponding to the minimum in the potential energy curve, since it must always have a least one-half quantum of vibrational energy. The zero vibrational level was discussed in Chapter 6.

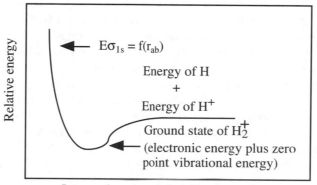

Figure 7-G-3. Potentional Energy for the Ground Electronic State H_2^+

H. Homonuclear Diatomic Molecules

In order to describe the bonding in homonuclear diatomic molecules, we employ the one-electron molecular orbitals from the H_2^+ system discussed in the previous section. The molecular orbitals will be smaller in size, of course, since Z has increased for both nuclei. As in the case of many-electron atoms, the contraction of the orbitals will depend on the effective nuclear charge in each case, rather than on Z itself. Nevertheless, the molecular orbitals will be more stable due to the higher atomic numbers. The bonding in homonuclear diatomic systems of higher atomic number, such as N_2 and O_2, can then be described in terms of a *molecular electronic configuration*, analogous to the electronic configuration for atoms. We will then see how these results can be used to predict certain properties of the molecules, which are consistent with experimental observations.

Electron Configuration and Bonding Scheme

The energy order for the one-electron LCAO molecular orbitals for homonuclear diatomic molecules varies as a function of the internuclear distance. However, in the cases which will be discussed here, the order with respect to increasing energy can be taken to be:

Energy Order for LCAO-MO's

$$\sigma_{1s} < \sigma_{1s}^* < \sigma_{2s} < \sigma_{2s}^* < \sigma_{2p_x} < \pi_{2p_y} = \pi_{2p_z} < \pi_{2p_y}^* = \pi_{2p_z}^* < \sigma_{2p_x}^* \qquad (7\text{-H-}1)$$

According to the Pauli Principle, which applies equally to molecules, each molecular orbital can contain a maximum of two electrons. In order to determine the electron configuration for any homonuclear diatomic molecule, the electrons from both atoms are placed, two by two, into the molecular orbitals according to the Aufbau Principle. The order of filling of the molecular orbitals is the same as the order of increasing energy given in (7-H-1). A pair of electrons in a single molecular orbital will have opposite spins, resulting in cancellation of the magnetic fields due to their spins. The two electrons are referred to as having *paired spins*, or, simply, as being *paired*.

One additional principle, known as **Hund's Rule,** is necessary before we will be able to determine the electron configurations of certain homonuclear diatomic molecules, and it will simply be stated without any justification or proof. When two molecular orbitals of equal energies, such as π_{2p_y} and $_{2p_y}$, are available, and only two electrons remain, each electron will occupy a different molecular orbital with their spins aligned. Such a pair of electrons, having the same value for their spin quantum numbers, are referred to as *unpaired electrons*. Unpaired electron spins result in a net spin magnetic field known as *paramagnetism*, a property which can be measured experimentally. A similar phenomenon occurs with atoms where there are atomic orbitals of equal energy.

Table 7-H-1 gives the electron configurations for a number of homonuclear diatomic molecules and molecular ions. The electron configurations are represented in the same manner as they are for the atoms, employing a superscript to indicate the number of electrons occupying a particular molecular orbital. For each example given in Table 7-H-1, the number of electrons occupying bonding molecular orbitals and the number occupying anti-bonding molecular orbitals are given. In addition, a quantity known as the **bond order** is given in each case. The bond order is defined as one-half of the number of electrons in bonding molecular orbitals minus one-half of the number of electrons in anti-bonding molecular orbitals.

$$\text{Bond Order} = \frac{1}{2} \text{ (Bonding Electrons - Anti Bonding Electrons)} \qquad (7\text{-H-}2)$$

The bond order is significant for the following reason: An electron in an anti-bonding molecular orbital causes a decrease in the stability of the molecule, which thus, cancels the stabilization resulting from an electron in the corresponding bonding molecular orbital. The bond order indicates the number of pairs of electrons in bonding molecular orbitals for which there are no corresponding anti-bonding electrons, that is, the net pairs of bonding electrons constitutes the bond order. Hence, the bond orders of one, two, or three correspond to the VB concept of a single, double, or triple bond, respectively. The MO approach, unlike the VB approach, gives rise to the idea of partial or *fractional bonds*, indicated by bond orders such as one-half. On the other hand, the MO approach parallels the VB approach in that it also suggests that the bond stabilization is mainly due to the electrons from unfilled atomic orbitals. Two filled atomic orbitals, one from each atom, provide four electrons, enough to fill both a bonding and an anti-bonding molecular orbital.

Problem 7-H-1. Write the electronic configurations and determine the bond orders for the homonuclear diatomic molecules B_2 and C_2. Use the energy order given in equation (6-H-1). Answer: B_2: $(\sigma_{1s})^2(\sigma_{1s}^*)^2(\sigma_{2s})^2(\sigma_{2s}^*)^2(\sigma_{2p_x})^2$, bond order = 1;

C_2: $(\sigma_{1s})^2(\sigma_{1s}^*)^2(\sigma_{2s})^2(\sigma_{2s}^*)^2(\sigma_{2p_x})^2(\sigma_{2p_y})^1(\sigma_{2p_z})^1$, bond order = 2.

Problem 7-H-2. Obtain the bond dissociation energies (bond strengths) for F_2, O_2, and N_2 from Table 6, Section D, of the Appendix, and plot a graph of the bond dissociation energy versus bond order. Does a correlation exist? Answer: Yes, but it is not linear.

Table 7-H-1. Electronic Configurations of Homonuclear Diatomic Molecules and Ions

Number of electrons	Molecule or ion	Electron configuration	Number of bonding electrons	Number of anti-bonding electrons	Bond order
1	H_2^+	$(\sigma_{1s})^1$	1	0	1
2	H_2	$(\sigma_{1s})^2$	2	0	1
3	He_2^+	$(\sigma_{1s})^2(\sigma_{1s}^*)^1$	2	1	0.5
4	He_2	$(\sigma_{1s})^2(\sigma_{1s}^*)^2$	2	2	0
6	Li_2	$(\sigma_{1s})^2(\sigma_{1s}^*)^2(\sigma_{2\sigma})^2$	4	2	1
8	Be_2				
10	B_2				
12	C_2				
13	N_2^+	$(\sigma_{1s})^2(\sigma_{1s}^*)^2(\sigma_{2s})^2(\sigma_{2s}^*)^2(\sigma_{2px})^2(\pi_{2py})^2(\pi_{2pz})^1$	9	4	2.5
14	N_2	$(\sigma_{1s})^2(\sigma_{1s}^*)^2(\sigma_{2s})^2(\sigma_{2s}^*)^2(\sigma_{2px})^2(\pi_{2py})^2(\pi_{2pz})^2$	10	4	3
15	N_2^-	$(\sigma_{1s})2(\sigma_{1s}^*)^2(\sigma_{2s})^2(\sigma_{2s}^*)^2(\sigma_{2px})^2(\pi_{2py})^2(\pi_{2pz})^2$ $(\pi_{2py}^*)^1$	10	5	2.5
15	O_2^+	(same as N_2^-)			
16	O_2	$(\sigma_{1s})^2(\sigma_{1s}^*)^2(\sigma_{2s})^2(\sigma_{2s}^*)^2(\sigma_{2px})^2(\pi_{2py})^2(\pi_{2pz})^2$ $(\pi_{2py}^*)^1(\pi_{2pz}^*)^1$	10	6	2
17	O_2^-	$(\sigma_{1s})^2(\sigma_{1s}^*)^2(\sigma_{2s})^2(\sigma_{2s}^*)^2(\sigma_{2px})^2(\pi_{2py})^2(\pi_{2pz})^2$ $(\pi_{2py}^*)^2(\pi_{2pz}^*)^1$	10	7	1.5
17	F_2^+	(same as O_2^-)			
18	F_2	$(\sigma_{1s})^2(\sigma_{1s}^*)^2(\sigma_{2s})^2(\sigma_{2s}^*)^2(\sigma_{2px})^2(\pi_{2py})^2(\pi_{2pz})^2$ $(\pi_{2py}^*)^2(\pi_{2pz}^*)^2$	10	8	1
19	F_2^-	$(\sigma_{1s})^2(\sigma_{1s}^*)^2(\sigma_{2s})^2(\sigma_{2s}^*)^2(\sigma_{2px})^2(\pi_{2py})^2(\pi_{2pz})^2$ $(\pi_{2py}^*)^2(\pi_{2pz}^*)^2(\sigma_{2px}^*)^1$	10	9	0.5
19	Ne_2^+	(same as F_2^-)			
20	Ne_2	$(\sigma_{1s})^2(\sigma_{1s}^*)^2(\sigma_{2s})^2(\sigma_{2s}^*)^2(\sigma_{2px})^2(\pi_{2py})^2(\pi_{2pz})^2$ $(\pi_{2py}^*)^2(\pi_{2pz}^*)^2(\sigma_{2px}^*)^2$	10	10	0
22	Na_2	$(Ne_2)(\sigma_{3s})^2$	12	10	1
24	Mg_2	$(Ne_2)(\sigma_{3s})^2(\sigma_{3s}^*)^2$	12	12	0
26	Al_2				
28	Si_2				
30	P_2				
32	S_2	$(Ne_2)(\sigma_{3s})^2(\sigma_{3s}^*)^2(\sigma_{3p_x})^2(\pi_{3p_y})^2(\pi_{3p_z})^2$ $(\pi_{3p_y}^*)^1(\pi_{3p_z}^*)^1$	18	14	2

34	Cl_2	$(Ne_2)(\sigma_{3s})^2(\overset{*}{\sigma_{3s}})^2(\sigma_{3p_x})^2(\pi_{3p_y})^2(\pi_{3p_z})^2$ $(\overset{*}{\pi_{3p_y}})^2(\overset{*}{\pi_{3p_z}})^2$			
			18	16	1
35	Cl_2^-	$(Ne_2)(\sigma_{3s})^2(\overset{*}{\sigma_{3s}})^2(\sigma_{3p_x})^2(\pi_{3p_y})^2(\pi_{3p_z})^2$ $(\overset{*}{\pi_{3p_y}})^2(\overset{*}{\pi_{3p_z}})^2(\overset{*}{\sigma_{3p_x}})^1$			
			18	17	1
35	Ar_2^+	(same as Cl_2^-)			
36	Ar_2	$(Ne_2)(\sigma_{3s})^2(\overset{*}{\sigma_{3s}})^2(\sigma_{3p_x})^2(\pi_{3p_y})^2(\pi_{3p_z})^2$ $(\overset{*}{\pi_{3p_y}})^2(\overset{*}{\pi_{3p_z}})^2(\overset{*}{\sigma_{3p_x}})^2$			
			18	18	0

Properties of Homonuclear Diatomic Molecules

The most obvious predictions which result from the diatomic molecular electronic configurations derived in the previous section concern the stability or instability of the various homonuclear molecules and molecular ions. On the basis of the bond orders derived from the electron configurations given in Table 7-H-1, molecules such as H_2, N_2, O_2, and F_2 should exist as stable diatomic systems. Molecules such as He_2 and Ne_2 would not be expected to exist since their bond orders are zero. These conclusions, of course, are known experimentally to be true. Even more striking, however, is the prediction of the possible existence of the He_2^+ and Ne_2^+ molecular ions, based on bond orders of one-half. These molecular ions are well known by mass spectrometry and are quite stable in the gaseous phase.

In addition to the bond order, the number of unpaired electrons predicted by the electron configurations for a given homonuclear diatomic, is of great importance. The presence of unpaired electrons is generally an important factor in regard to the chemical properties of molecules and ions. In particular, the diradical nature of O_2 is most significant, since oxygen is in large abundace naturally and is of considerable importance both chemically and biologically. A glance at Table 7-H-1 will show that the electron configuration for O_2 predicts two unpaired electrons in the $\pi^*_{2p_y}$ and $\pi^*_{2p_z}$ molecular orbitals.

In addition to the prediction of chemical properties, the electronic configurations can predict certain physical properties such as paramagnetism. As mentioned earlier, unpaired electrons in a molecule lead to a paramagnetic character for the substance. According to the information in Table 7-H-1, oxygen is expected to be paramagnetic, while many other common gases, hydrogen and nitrogen, are not. This is in complete agreement with experimental observations.

Problem 7-H-3. Predict the stability or instability and the magnetic nature of B_2 and C_2 from the results of Problem 7-H-1. Answer: B_2: stable and diamagnetic; C_2: stable and paramagnetic.

Problem 7-H-4. Based on the electronic configuration, explain the inert nature of nitrogen. Answer: Nitrogen has a triple bond which is extremely stable and is therefore inert with respect to breaking the nitrogen-nitrogen bond.

Heteronuclear Diatomic Molecules

The subject of bonding in heteronuclear diatomic molecules can become quite complicated and, thus, will be discussed only briefly. The simplest MO approach to heteronuclear diatomic molecules is similar to that for homonuclear diatomic molecules except for two aspects: (1) the atomic orbitals used to form the molecular orbitals need not be of the same type, but they must be of nearly the same energy and the same symmetry, relative to the internuclear axis; (2) the contribution of each atomic orbital in a given molecular orbital is not equal, as in the case of homonuclear diatomic systems. The latter difference results in unequal "sharing" of the electrons in the molecular orbital and, consequently, leads to *polar bonds*.

One of the simplest examples is the diatomic molecule HF, hydrogen fluoride. The electronic configuration for the fluorine atom is:

$$F(1s^2, 2s^2, 2p^5) \text{ two filled 2p orbitals and one half-filled 2p orbital}$$

The hydrogen atom has several atomic orbitals available for combining with the **F** atomic orbitals. However, only the 1s orbital has an energy comparable to the lowest energy fluorine atomic orbital available for bonding, which is the third 2p orbital containing a single electron. For this reason the lowest energy molecular orbitals for **HF** are considered to be formed from non-equivalent linear combinations of the **H**(1s) and an **F**(2p) atomic orbital. The contributions of the atomic orbitals to the molecular orbitals are dependent on the relative electronegativities of the two atoms. The resulting bond in **HF** will be polar; hence, the **HF** molecule is expected to have a dipole moment.

The two molecular orbitals are shown in Figure 7-H-1.

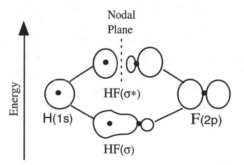

Figure 7-H-1. Hydrogen Fluoride Molecular Orbitals

Only the one valence electron in the half-filled 2p orbital of fluorine and the hydrogen electron are considered in the bonding. The other electrons of the fluorine atom are considered to remain essentially F atomic electrons in the **HF** molecule. The two bonding electrons occupy the lower energy molecular orbital with their spins paired. This approximation appears to be at least adequate in describing the ground state of the **HF** molecule.

I. Polyatomic Molecules

As we will see later, a complete MO treatment can become quite involved and complex for polyatomic molecules. If the molecule contains n atoms, each molecular orbital should be formed from a linear combination of atomic orbitals, including one from each of the *n* atoms. The resulting molecular orbitals are called polycentric or *group molecular orbitals*, since they encompass all of the atomic centers of the molecule. In some cases, this approach is necessary in order to adequately describe the molecule. However, in many instances, a further approximation known as *localized molecular orbitals* can provide a reasonably accurate description of the molecule.

Localized Molecular Orbital Approximation

In the localized MO approach, molecular orbitals are constructed for pairs of adjacent atoms in the molecule, employing, in each instance, the appropriate atomic orbitals on each of the two atoms. For each pair of adjacent atoms, the method is much the same as was described for the HF heteronuclear diatomic molecule. The result is a set of bicentric molecular orbitals for each pair of adjacent atoms in the molecule, which are said to be "bonded" to each other. Thus, the localized MO approach to polyatomic molecules yields results which are quite similar to the results of a VB treatment.

In some cases this appears satisfactory, since the resulting molecular structure is consistent with the properties of the molecule as observed experimentally. One of these properties which supports the localized MO concept is the constancy of the bond dissociation energy for a given type bond between two atoms in different molecules, for example, D_{C-H}, D_{C-C}, D_{C-O}, D_{O-H}. This was discussed in the thermochemistry section of Chapter 2, Section F.

Water. The water molecule, H_2O, provides a good example for illustrating the localized molecular orbital method. The electronic configuration for the oxygen atom is

$$1s^2, 2s^2, 2p^4, \text{ or more precisely, } 1s^2, 2s^2, 2p_z^2, 2p_x^1, 2p_y^1$$

One 2p atomic orbital is filled, while the other two 2p orbitals each contain a single electron. Hund's Rule applies equally in the case of atoms. Thus, the oxygen atom has two orbitals available for bonding, which are comparable in energy to the 1s atomic orbitals of the hydrogen atoms. If we designate the filled 2p orbital of oxygen as the $2p_z$ orbital, then the $2p_x$ and the $2p_y$ atomic orbitals will be the ones available for the bonding. One pair of localized molecular orbitals will be constructed from combinations of the oxygen $2p_x$ orbital and 1s orbital of the hydrogen atom, H_a, and the other pair will be constructed from combinations of the oxygen $2p_y$ orbital and the 1s orbital of the hydrogen atom H_b. The orientation of the atoms and bonding atomic orbitals is shown in Figure 7-I-1(a).

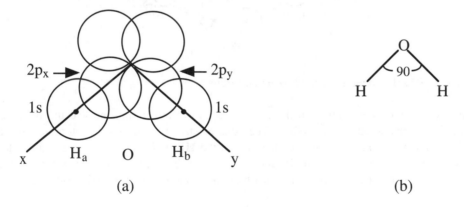

(a)   (b)

Figure 7-I-1. Predicted Molecular Structure of Water

The resulting sigma molecular orbitals for the oxygen atom and either hydrogen atom are quite similar to those for HF, shown in Figure 7-H-1. Since the oxygen atom contributes two electrons and each hydrogen atom contributes one, there are four electrons involved in the bonding. Hence, each of the bonding-localized molecular orbitals, one for each hydrogen atom and the oxygen atom, is occupied by a pair of electrons. A single covalent bond is said to exist between the oxygen atom and each hydrogen atom. These bonds are referred to as *sigma* bonds.

The resulting geometry of the bonded molecule is illustrated in Figure 7-I-1(b). The bonds are indicated by the lines joining each H to the O. The angle between the two internuclear axes is predicted to be 90°, and the reason for this is simple. The $2p_x$ and the $2p_y$ atomic orbitals of oxygen are symmetrical about the x and y axes which, of course, form an angle of 90°. As previously discussed, the greatest stability for a bond between two atoms corresponds to maximum possible overlap of the two atomic orbitals, all else being equal. Maximum overlap is attained between the H (1s) orbital and an O (2p) orbital, when the H (1s) orbital is centered on the axis of symmetry of the O(2p) orbital. Thus, the maximum overlap and, hence, maximum bond stability, is expected when one of the hydrogen nuclei is situated on the x axis and the other on the y axis.

Experimental results indicate that the angle between the H-O-H internuclear axes is approximately 104.5°, slightly greater than the predicted value. A more elaborate approach to the bonding, which considers such effects as the mutual repulsion between the two proton nuclei and possible hybridization (to be explained, shortly) of the oxygen atomic orbitals, can predict the angle more closely than can the naive approach presented here.

The bonds between the oxygen atom and each hydrogen atom are polar, as a consequence of the unequal electronegativities. The resulting dipole moment is directed toward the oxygen atom along a line in the plane of the molecule and bisecting the H-O-H angle. This qualitative prediction is in agreement with experiment.

Ammonia. As a second example of the application of the localized MO approach, we will consider the ammonia molecule, NH_3. The electronic configuration for the nitrogen atom, showing the three 2p atomic orbitals separately, is

$$1s^2, 2s^2, 2p_x^1, 2p_y^1, 2p_z^1$$

All three of the 2p orbitals, which are mutually perpendicular to each other, are available for bonding. The formation of sigma bonds can be visualized by imagining each of three hydrogen atoms approaching the nitrogen atom along a different one of the coordinate axes. Three sets of bicentric localized molecular orbitals, involving maximum overlap, are formed. Each set results from linear combinations of a different 2p orbital of the nitrogen atom and the 1s orbital of the hydrogen atoms. The details concerning the bicentric molecular orbitals are the same as explained in the case of water and hydrogen fluoride, and will not be further elaborated on.

The predicted geometry of the NH_3 molecule is illustrated in Figure 7-I-2. Again, as in the case of the water molecule, experimental measurements indicated that the H-N-H angles are slightly greater than 90°. The bonds formed between the nitrogen atom and each hydrogen atom are polar, as would be expected. The resulting dipole moment of the NH_3 molecule is directed toward the nitrogen atom, along a line which forms equal angles with each of the three N-H internuclear axes. This prediction is consistent with the fact that nitrogen is more electronegative than hydrogen and is in qualitative agreement with experiment.

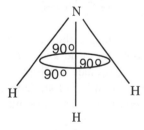

Figure 7-I-2. Predicted Molecular Structure of Ammonia

The examples discussed here illustrate the basis for an extremely important concept in molecular structure known as *directed valence*. As we have seen for both H_2O and NH_3 molecules, even the naive localized MO approximation predicts a specific geometry for a given polyatomic system. In the examples cited, this was a direct result of the directional character of p atomic orbitals, coupled with the intent of obtaining maximum stability in the bonds. In general, it can be said that the geometry of a polyatomic molecule is principally dependent on the directional character of the atomic orbitals which are involved in the bonding and the principle of maximum overlap.

Problem 7-I-1

(a) Predict the structure of PH_3, and indicate which atomic orbitals are involved in the bonds. Answer: 3p atomic orbitals of P overlap with the H_{1s} orbitals to form 3 bonds 90° apart.

(b) Repeat for PCl_3. (Hint: Recall from the discussion of homonuclear diatomics that sigma type molecular orbitals can be formed from two p atomic orbitals.)

Answer: (a) 3p atomic orbitals of P overlap with 1s orbitals of three H to form 3 bonds 90° apart (b) 3p orbitals of P overlap with 3p orbitals of three Cl to form 3 bonds 90° apart.

Hybridization and Carbon Compounds

sp³ Hybrid Atomic Orbitals. Let us turn our attention to the simplest hydrocarbon, methane, CH_4. Any explanation of the bonding in CH_4 must be capable of explaining the facts that the four C-H bond distances are equal and that the shape of the molecule is tetrahedral with the carbon nucleus at the center and each of the hydrogen nuclei at a corner. The angle between each pair of C-H internuclear axes is 109°28'. The geometry of CH_4 is shown in Figure 7-I-3, and should be familiar to you.

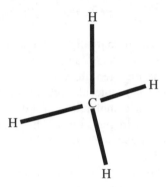

Figure 7-I-3. Tetrahedral Structure of Methane

In order to develop a bonding scheme which can explain the experimental facts, a concept known as *hybridization of atomic orbitals* must be postulated. The carbon electron configuration is

$$1s^2, 2s^2, 2p_x^1, 2p_y^1$$

Based on this electronic configuration, a carbon atom would be expected to form two bonds with an angle of 90° between the bond axes. However, if four equivalent atomic orbitals, known as sp^3 hybrid atomic orbitals, are formed by combining the 2s, $2p_x$, $2p_y$, $2p_z$ orbitals, carbon would then be able to form four equivalent bonds. There would be one electron available for each of the equivalent carbon orbitals. We might imagine a modified electronic configuration,

$$(1s)^2, (2sp^3)_1^1, (2sp^3)_2^1, (2sp^3)_3^1, (2sp^3)_4^1$$

A detailed discussion of the formation of the four hybrid wave functions can be found in various texts on valence and quantum chemistry. For our purposes, it will suffice to say that the four sp^3 hybrid atomic orbitals are identical in regard to the magnitude of their probability distributions for the electron, their axial symmetry, and are directed toward the corners of a tetrahedron centered about the carbon nucleus. Figure 7-I-4 shows a contour diagram for an sp^3 hybrid atomic orbital.

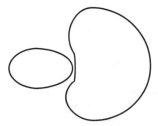

Figure 7-I-4. sp^3 Hybrid Atomic Orbital

Having formed the equivalent sp^3 hybrid orbitals for carbon, we can consider the formation of localized molecular orbitals between carbon and each of the hydrogens. Each localized molecular orbital is a linear combination of a carbon sp^3 orbital and a hydrogen 1s orbital, such that the hydrogen nucleus lies on the axis of the sp^3 orbital. Thus, four sets of sigma type localized molecular orbitals are formed, one set for each hydrogen atom and the carbon atom. One electron from the carbon atom and hydrogen electron are paired in the bonding molecular orbital of each set. The 1s electrons of carbon have been ignored, since they are of much lower energy and essentially atomic in nature—their principal contribution is shielding of the positive nucleus.

The justification for use of the sp^3 hybrid atomic orbitals in the bonding scheme is that it leads to results concerning the molecular structure of methane, as well as many other carbon-containing compounds, which are consistent with experimental facts.

sp^2 Hybrid Atomic Orbitals. The preceeding hybridization scheme is not appropriate for explaining the bonding in molecules such as ethylene. The geometric structure of the C_2H_4 molecule is shown in Figure 7-I-5.

Figure 7-I-5. Molecular Structure of Ethylene

C_2H_4 is *planar* in that both carbon nuclei and the four hydrogen nuclei all lie in the same plane. Moreover, each of the carbon atoms is bonded to only three other atoms, rather than to four, as in methane and other alkanes. In the terminology of the VB approach, the two carbon atoms are said to be joined to each other by a *covalent double bond.*

The localized molecular orbital approach describes the bonding in ethylene in the following manner. Designating the plane of the molecule as the xy plane, a set of three equivalent sp^2 *hybrid atomic orbitals* are constructed from the 2s, 2p$_x$, and 2p$_y$ atomic orbitals of each carbon atom. The shape of an sp^2 hybrid orbital is similiar, but not identical to that of an sp^3 hybrid orbital shown in Figure 7-I-3. Each sp^2 hybrid orbital has a large lobe and a small lobe (contour surfaces), which are symmetrical about an axis through the nucleus. As in the case of p and sp^3 orbitals, the sp^2 orbital has a nodal surface perpendicular to its axis of symmetry at the nucleus. The axes of the three sp^2 hybrid orbitals lie in the xy plane, forming angles of 120° with one another. It is important not to confuse the fact that the orbitals are three dimensional, as are all orbitals, and only their symmetry axes lie in a plane containing the nucleus. The sp^2 hybrid atomic orbital axes for a single carbon atom are illustrated in Figure 6-I-6 by solid arrows through the carbon nucleus. Each carbon atom has a 2p$_z$ atomic orbital, perpendicular to the xy plane, which is not involved in the hybridization.

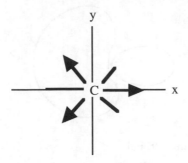

Figure 7-I-6. sp^2 Hybrid Orbital Axes

The electronic configuration for the "sp^2 carbon atom" might be imagined as

$$(1s)^2, (2sp^2)_1^1, (2sp^2)_2^1, (2sp^2)_3^1, (2p_z)^1$$

Thus, each carbon atom can provide four orbitals for bonding. The formation of localized molecular orbitals can be visualized with reference to the orbital orientation diagram in Figure 7-I-7.

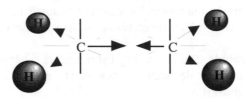

Figure 7-I-7. Formation of Ethylene Molecular Orbitals. Vertical lines represent axes of the 2p$_z$ atomic orbitals; the arrows represent axes of the sp^2 hybrid orbitals; and the spheres represent the hydrogen 1s atomic orbitals.

A set of sigma type localized molecular orbitals are formed from linear combinations of an sp^2 hybrid orbital from each carbon atom. Maximum overlap occurs when the axes of the two sp^2 hybrid orbitals are coincident. Four sets of sigma type localized molecular orbitals are formed from the four remaining sp^2 hybrid orbitals on the two carbon atoms and the 1s atomic orbitals of the four hydrogen atoms. Maximum overlap requires that the nucleus of each hydrogen atom lie on the axis of one of the sp^2 hybrid orbitals. The bonding molecular orbital of each of the five sets is occupied by a pair of electrons. This leaves two electrons, one in each of the carbon 2p$_z$ atomic orbitals. The axes of these two orbitals are parallel, thus, oriented properly for the formation of a π type molecular orbital. The bonding localized π molecular orbital is occupied by the remaining pair of electrons.

In summation, each hydrogen atom is bonded to a carbon atom by a single covalent bond resulting from a filled sigma type localized molecular orbital. The bonding between the two carbon atoms involves four electrons, a pair of electrons occupying a bonding σ type localized molecular orbital, and a pair of electrons occupying a bonding π type localized molecular orbital. This combination is the MO explanation of a *covalent double bond*.

sp Hybrid Atomic Orbitals. A final type of hybridization, which will be mentioned briefly, involves the construction of two equivalent *sp hybrid atomic orbitals* from the 2s and 2p$_x$ atomic orbitals. The two sp hybrid orbitals are both symmetrical about the x axis, but their larger lobes are on opposite sides of the nucleus, so that they are then available for the formation of sigma bonds with two other atoms, one on each side of the carbon atom. The 2p$_y$ and 2p$_z$ atomic orbitals of the carbon atom remain uninvolved, hence, available for the formation of π type localized molecular orbitals.

The bonding in the acetylene molecule, C$_2$H$_2$, is explained on the basis of sp hybridization. The bonding between the two carbon atoms involves a σ bond, resulting from the combination of an sp hybrid orbital from each carbon atom, and two π bonds, resulting from combination of the two $2p_y$ atomic orbitals and combination of the two $2p_z$ atomic orbitals. This type of bonding is referred to in the VB approach as a *covalent triple bond*. For orbital diagrams and further description of the binding in the acetylene molecule as well as other organic molecules, the reader is referred to any recent text in organic chemistry.

Problem 7-I-2. Describe the bonding in the propylene molecule, C_3H_6, indicating which atomic orbitals are involved in each bond and whether the localized molecular orbitals formed are σ or π. Answer: End carbon, which is bonded to three H's, has sp^3 hybridized orbitals which form three bonds with the H atoms by overlap with the H 1s atomic orbitals. Other two C atoms have sp^2 hybridized orbitals leaving one electron in a p-orbital perpendicular to the sp^2 orbitals. The bonding involving these two C atoms is similiar to that in ethylene. The central carbon is bonded to the C in the CH_3 group by overlap of a sp^2 with a sp^3 orbital of the C in the CH_3 group.

Problem 7-I-3. Describe the bonding in the linear CO_2 molecule to the same extent as in Problem 7-I-2. Answer: C has sp hybridized orbitals leaving one electron in $2p_y$ and one electron in a $2p_z$ orbital. sp orbitals form sigma bonds by overlap with the O $2p_x$ orbitals. The $2p_y$ orbital of C forms a π bond by overlap with a $2p_y$ of one O atom and the $2p_z$ forms a π bond by overlap with a $2p_z$ of the other O atom.

J. Delocalized Molecular Orbitals

Π Molecular Orbitals

There are certain molecules and ions in both organic and inorganic chemistry which have electrons associated with several or all of the atoms in the molecule or ion. Such electrons are referred to as delocalized electrons. Notable examples of such molecular systems are aromatic substances, of which benzene is a good example, and inorganic ions such as NO_2^-. Molecules and ions of this type have a common characteristic. If σ bonds are formed between the atoms, half-filled p atomic orbitals, oriented properly for the formation of π molecular orbitals, remain unused.

The benzene molecule, C_6H_6, is a particularly instructive example, and its geometric structure is well known to chemists. The six carbon nuclei and the six hydrogen nuclei lie in the same plane, as illustrated in Figure 7-J-1.

$$
\begin{array}{ccc}
 & H \qquad H & \\
 & C{-}C & \\
H{-}C & & C{-}H \\
 & C{-}C & \\
 & H \qquad H &
\end{array}
$$

Figure 7-J-1. Molecular Structure of Benzene

Furthermore, all six carbon-carbon bonds are known to be equivalent, as are the six carbon-hydrogen bonds. The angle formed by the intersection of any pair of internuclear axes, whether C-C-C or C-C-H , is equal to 120°. This fact, plus the coplanarity of the nuclei, suggest that carbon sp^2 hybridization is involved in the bonding.

The geometric structure of benzene is explained in either the localized MO approach, or the VB approach, by assuming that localized sigma bonds are formed between each adjacent pair of nuclei. The C-C sigma bonds result from the combination of sp^2 hybrid orbitals, one from each carbon atom, and the C-H sigma bonds each result from a combination of the remaining sp^2 hybrid atomic orbital of the carbon atom and the 1s atomic orbital of the hydrogen atom. This leaves each carbon atom with one electron available for bonding. The reader will recall that an sp^2 hybridized carbon atom has a half-filled $2p_z$ atomic orbital perpendicular to the plane containing the sp^2 hybrid orbital axes. The axes of the carbon $2p_z$ atomic orbitals are indicated by the solid vertical lines through each carbon nucleus in Figure 7-J-2. The hydrogen atoms are not shown in the figure.

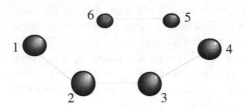

Figure 7-J-2. Carbon $2p_z$ Atomic Orbital Axes in Benzene

The VB approach, as the localized MO approach, would complete the description of the bonding in benzene by including the formation of additional bonds of the π type between alternate pairs of adjacent carbon atoms. Three localized π bonds can be formed between carbon atoms 1 and 2, 3 and 4, 5 and 6, or carbon atoms 2 and 3, 4 and 5, 6 and 1. This results in two possible localized bond schemes for the carbon portion of benzene, as illustrated in Figure 7-J-3 by the well known Kekule Structures. Each of the Kekule Structures involves three covalent double bonds between pairs of adjacent carbon atoms, and three covalent single bonds between the alternate pairs of adjacent carbon atoms. The VB approach assumes that a type of resonance exists between the two possible structures.

The MO approach to the bonding in benzene assumes a combination of both the σ localized molecular orbitals and π type delocalized molecular orbitals. This requires an important assumption, namely, that the sigma type molecular orbitals and the π type molecular orbitals can be separated. The delocalized pi molecular orbitals are of the highest energy and, therefore, govern many of the chemical and physical properties of the molecules. The approximation appears to be quite satisfactory in many instances and, moreover, it greatly simplifies the calculations. A rather extensive amount of theoretical work has been done on such compounds.

$$\bighexagon \quad \text{or} \quad \bighexagon$$

Figure 7-J-3. Kekule Structures for Benzene

The σ bonds between each adjacent pair of carbon atoms and each adjacent carbon atom and hydrogen atom, resulting from σ type localized molecular orbitals, were previously described. The remaining half-filled carbon $2p_z$ atomic orbitals are dealt with in an entirely different manner. The $2p_z$ atomic orbital of each carbon atom overlaps the $2p_z$ atomic orbitals of both its neighbors. Polycentric delocalized molecular orbitals are constructed from linear combinations, including all six of the carbon $2p_z$ atomic orbitals. Hence, the resulting polycentric molecular orbitals encompass the entire carbon skeleton. This is the reason for the terminology "delocalized molecular orbitals."

There are six possible linear combinations of the six $2p_z$ atomic orbitals. Thus, six delocalized molecular orbitals can be constructed---three bonding molecular orbitals, $\pi_1, \pi_2,$ and $\pi_3,$, and three anti-bonding molecular orbitals: $\pi_4^*, \pi_5^*,$ and π_6^*. These are illustrated in Figure 7-J-4. The shapes of the contour diagrams in this figure are not rigorously accurate, and only the top portion of each (looking down on the molecule) is shown. Each of the six delocalized molecular orbitals has a nodal plane coincident with the plane of the carbon nuclei, as would be expected from the nodal properties of the $2p_z$ atomic orbitals. In addition, the delocalized molecular orbitals have other nodal planes perpendicular to the plane of the molecule, indicated by the broken lines in Figure 7-J-4. The energy order of the delocalized molecular orbitals is also shown in this figure. The reader will note that π_2 and π_3 have the same energy, as do π_4^* and π_5^*.

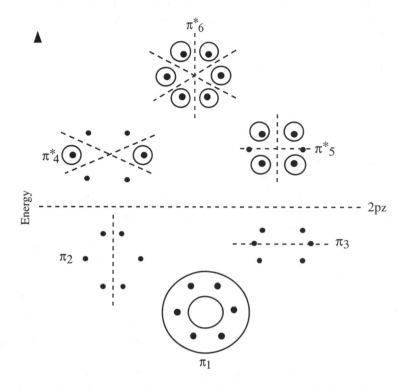

Figure 7-J-4. Delocalized Molecular Orbitals of Benzene. (The dots (•) denote the carbon nuclei.)

Since each delocalized molecular orbital can hold two electrons, the six $2p_z$ carbon electrons are paired in the three lowest energy orbitals, π_1, π_2, and π_3, resulting in the lowest possible energy for the system. The combination of the three molecular orbitals leads to a total wave function for the ground state. The total wave function is obtained by taking the product of the individual wave functions for each of the electrons. If the electron density described by the total wave function were represented by a three-dimensional contour surface, its shape would resemble two donuts, one above and one below the plane of the molecule. Moreover, the π electron density between each pair of adjacent carbon atoms is the same for the ground state. Thus, the delocalized MO approach results in equal contributions to the carbon-carbon bonds. The complete bonding scheme, including the σ type localized molecular orbitals, predict equivalent carbon-carbon bonds in agreement with experiment.

The electronic energy of the ground state of benzene will depend on the energies of the three occupied π delocalized molecular orbitals, as well as on the energies of the occupied σ localized molecular orbitals. It can be shown that the ground state energy predicted by the delocalized MO approach is considereably less than that predicted on the basis of either of the structures involving localized π molecular orbitals. The difference, which has been measured experimentally to be approximately 36 kcal/mole, is referred to as the *resonance energy* . (See Exercise 16, Chapter 3.) In other words, the bond stability of benzene is greater than would be expected if the carbon atoms were bonded by alternating double and single covalent bonds. This has been known by organic chemists for many years from experimental thermochemical studies on the hydrogenation of ethylene and benzene.

The VB approach explains the unusual stability of benzene, by assuming resonance between the two Kekule Structures. It can be shown mathematically that if the actual structure is represented by a simple combination of the two Kekule Structures, the resulting electronic energy is lower than it would be for either structure alone.

Problem 7-J-1. Describe the bonding in butadiene, $CH_2CHCHCH_2$. Answer: All C atoms have sp^2 hybridization and form C-C single bonds by overlap of the sp^2 orbitals. The remaining e^- on each C-atom is in a p-orbital perpendicular to the plane of the three sp^2 orbitals on each C-atom. π-bonds are formed by overlap of the p-orbitals between the C-atoms in the 1,2 and 3,4 positions. The H-atoms are bonded by overlap of the H-1s orbitals with the C-sp^2 orbitals.

Sigma/Pi Molecular Orbitals

In Section 7-I a Localized Molecular Orbital Approximation was used to account for various molecular structures. Hybridization was required in some cases in order to account for the structure of, in particular, carbon containing compounds. This approach is convenient since one can ascertain the gross structural features of molecules using somewhat simplistic principles. However, polyatomic molecules have molecular orbitals just as diatomic molecules, which were discussed in Section H. In the case of polyatomic molecules the molecular orbitals encompass all atoms of the molecule and it is not easy to visualize the shape of such orbitals. However, the molecular orbitals can be calculated using a computer to carry out the quantum mechanical solution. The procedure of calculation is too difficult to discuss quantitatively in this physical chemistry textbook, but the qualitative aspects of the calculation can be understood and appreciated.

Self Consistent Field (SCF) calculations can be performed using a commercially available computer program called HyperChem^{TM} (Hypercube, Inc., Waterloo, Canada). This program enables semi-empirical quantum mechanical calculations to be performed using any one of several sets of parameters. The Austin Model 1 or AM1 method, developed by the Dewar Group at the University of Texas at Austin, is included in this program and is generally thought to be the most accurate method. The basis for the AM1 method is the Neglect of Diatomic Differential Overlap (NDDO), and it uses many parameters such as the heats of formation and geometries of sample molecules to reproduce experimental quantities.

To calculate energy levels or eigenvalues for a molecule, a geometric optimization is first performed. This will calculate the set of Cartesian coordinates with a minimum potential energy for the molecule. Once the geometry is optimized, a single point calculation will give the static properties for the molecule including potential energy and its derivatives, electrostatic potential, molecular orbital energies and the coefficients of molecular orbitals for the ground or excited states. When used in conjunction with a modern personal computer, this program is capable of performing complex quantum mechanical calculations in a relatively short period of time.

The importance of using these calculations for molecules is best revealed by comparing the theoretical calculations with experimentally determined ionization potentials. You may recall from General Chemistry the ionization potential, or sometimes called the ionization energy, of an atom is the minimum energy required to remove an electron from the atom in the gas phase. Similarly, the ionization potential for a molecule is the lowest energy required to remove an electron from a molecule, where again the molecule is in the gas phase. The ionization potentials of the staight chain hydrocarbons, C_1 through C_7, along with 1-hexene and benzene are given in Table 7-J-1.

Table 7-J-1. Ionization Potentials for the Straight Chain Aliphatic Hydrocarbons from C_1 to C_7

Compound	I.P. (eV)	Vertical IP (calc.)
CH_4	12.6	13.3
C_2H_6	11.5	11.8
C_3H_8	11.1	11.3
C_4H_{10}	10.6	11.3
C_5H_{12}	10.35	11.1
C_6H_{14}	10.18	11.08
C_7H_{16}	9.90	11.07
$1\text{-}C_6H_{12}$	9.45	9.92
C_6H_6	9.24	9.6

Note in Table 7-J-1 that the ionization potentials of the straight chain hydrocarbons decrease as the carbon number increases. This suggests that the electron which is being removed comes from a delocalized molecular orbital which likely encompasses the entire molecule and is obviously of lower binding energy (higher orbital energy) as the carbon chain is increased. If the bonding in the normal aliphatic hydrocarbons was understood in terms of Localized Molecular Orbitals, there would be no rationale for the decrease in ionization potential as the chain length is increased. Consequently, it is obvious that a more exact molecular orbital calculation must be performed which encompasses all the atoms in the molecule.

In Table 7-J-2 we give the Hyperchem Molecular Orbital calculations for the normal hydrocarbons C_1 to C_7. In these calculations only the 2s and 2p electrons of carbon are considered in the bonding. The 1s electrons are held so tightly by the +6 nucleus that they are of much higher energy than the 2s and 2p electrons. The 1s electrons are located close to the nucleus and shield the +6 nucleus from the 2s and 2p electrons. For the H atoms the 1s electron is less strongly bound since the nuclear charge is only +1 and these electrons are considered in the molecular orbital calculations. The energy level diagrams resulting from the molecular orbital calculations are shown in Table 7-J-2.

The number of molecular orbitals calculated must equal the number of atomic orbitals used. For example, for CH_4 there are four carbon atomic orbitals since there are 4 electrons in the 2s and 2p orbitals and there are four 1s electrons from the 4 H-atoms. This gives a total of 8 orbitals and, consequently, there are 8 molecular orbitals generated. Since we add two electrons per orbital (Pauli Principle), only half of these molecular orbitals are occupied. In the ground state the electrons would occupy the most stable molecular orbitals. In Table 7-J-2 only the occupied orbitals between - 8 to -20 eV are shown. The highest occupied energy level in CH_4 is triply degenerate at -13.3 eV. A more stable molecular orbital for CH_4 (- 28.9 eV) is below -20 eV and is not shown in Table 7-J-2. The energy of the highest occupied molecular orbital is referred to as the vertical ionization potential, which is the energy required to remove an electron from the molecule with no change in configuration of the ion. Generally this is accomplished by the absorption of a photon which is a sufficiently fast process that the nuclei do not have a chance to move. The photons could also remove the electrons at the energy of - 28.9 eV, providing the photons have energies of this magnitude.

As defined earlier, the ionization potential of a molecule is defined as the lowest energy required to remove an electron from a molecule. In this process the positive ion is allowed to change its structure to give the lowest, most stable comfiguration. Obviously in order for the process to have its minimum energy, the electron being removed comes from the highest occupied energy orbital. The ionization potential, as defined in this way, is also referred to as the "adiabatic ionization potential" which differentiates it from the vertical ionization potential. Since the adiabatic ionization potential is the lowest energy required to remove the electron, it is lower in magnitude than the vertical ionization potential. The

Table 7-J-2. Occupied Upper Molecular Energy Levels for Alkanes C_1 to C_7

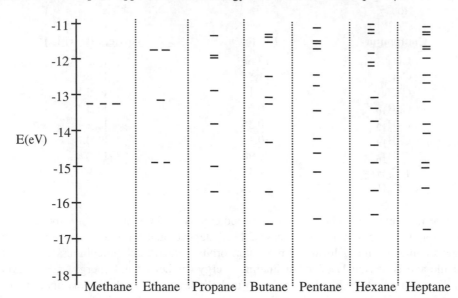

ionization potentials given in the first column of Table 7-J-1 are the adiabatic ionization potentials. Note that the ionization potential for CH_4 is 12.6 eV which is lower than the calculated vertical ionization potential of 13.3 eV shown in Table 7-J-2. Generally there is less of a difference between these ionization potentials for the remaining C_2 to C_7 hydrocarbons in this series. Commonly the vertical ionization potential is 0.2 - 0.3 eV greater than the adiabatic ionization potential.

In Table 7-J-1 the vertical ionization potentials for the C_1 to C_7 straight chain hydrocarbons are given as calculated by the Hyperchem program, column 3. Note that the vertical ionization potentials decrease as the chain length is increased, somewhat paralleling the decrease in experimental adiabatic ionization potentials. However, the decrease in the vertical ionization potential with increasing chain length is not as rapid as the decrease in the adiabatic ionization potential and the discrepancy between the two values for heptane becomes large, in excess of 1 eV.

In Table 7-J-3 we show the molecular orbital energy diagram for the unsaturated compounds 1-hexene, 2-hexene, and 3-hexene. In order to show the relative energies of the upper occupied orbitals more explicity, we have restricted the table to include energy levels from - 8.0 to - 20 eV. The saturated hydrocarbon hexane has been included for comparison. Note the obvious highest occupied energy level for the hexenes which is a result of the unsaturation. This is a π type orbital which is probably similiar to the formation by overlap of atomic orbitals in ethylene, which was discussed previously in Section 7-I. However, for the molecular orbitals for the hexenes, as given in Table 7-J-3, the orbitals extend over the entire molecule. As a consequence, the energy of this π orbital for 2-hexene and 3-hexene is of considerably higher energy than in 1-hexene. This difference of ~0.5 eV in the π orbital energy would be impossible to explain using the localized bond concept discussed in Section 7-I. Since the π orbital energies in the range - 9.5 to - 10.0 eV lie 1 to 1.5 eV above the highest occupied orbital in hexane, the ionization potentials for the hexenes are considerably less than for the corresponding saturated hydrocarbon. For example, the experimental adiabatic ionization potential for 1-hexene is 9.45 eV compared to 10.18 eV for hexane. According to the results of the molecular orbital calculations, the vertical ionization potential for 1-hexene would be 9.9 eV, higher than the experimental adiabatic ionization potential as expected.

To the right of the hexenes in Table 7-J-3 are the calculated molecular orbital energies for benzene. In this case there are two π orbitals at 9.6 eV and presumably they arise pri-

Table 7-J-3. Molecular Orbital Energy Diagrams of C_6 Unsaturated Hydrocarbons

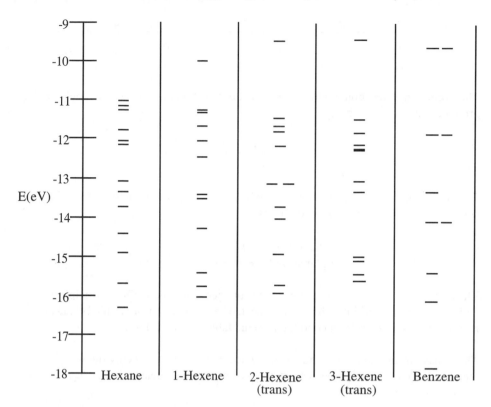

marily from the delocalized π electrons, as we discussed previously when we considered the π orbitals independent from the localized σ molecular bonds. Again it should be emphasized that the molecular orbital calculations shown in Table 7-J-3 consider all atoms in the molecule and higher energy π orbitals arise naturally from the calculations without any arbitrary separation of the π orbitals from the σ orbitals.

The second highest energy molecular orbital in benzene is also doubly degenerate at - 11.85 eV. According to our previous discussion, where we separated the π molecular orbitals from the localized σ bonded orbitals, we would expect the next lowest π orbital to be singly degenerate. Obviously this is not consistent with the doubly degenerate orbital at - 11.85 eV, if indeed the next lowest orbital is a π orbital. Possibly the next lowest energy orbital at - 13.4 eV , which is singly degenerate, could be associated with the next lowest π orbital and the higher energy orbital at - 11.85 eV could be a doubly degenerate σ orbital. In any event, the previous simplified concept of arbitrarily separating the localized σ orbitals from the π orbitals has its limitations and one should be exceedingly careful in using this simplistic model for quantitative evaluations.

K. Summary

The essential features of this chapter are:

1. The quantum mechanical solution to the hydrogen atom gives an eigenfunction which contains three quantum numbers, the allowed values of which are

    principal quantum number $= n = 1,2,3,....$
    azimuthal quantum number $= \ell = 0,1....(n - 1)$
    magnetic quantum number $= m = - \ell , - \ell +1, ...0,1...+ \ell$

Orbital designation is based upon the values of n and ℓ

n	1	2		3			4			
	0	0	1	0	1	2	0	1	2	3
	1s	2s	2p	3s	3p	3d	4s	4p	4d	4f

2. The orbital energies, eigenvalues, for the hydrogen atom orbitals depends only on the principal quantum number, n

$$E_n = -\frac{2\pi^2 me^4}{h^2}\frac{1}{n^2}$$

3. The shapes of the hydrogen atomic orbitals can be visualized by noting the nodal surfaces according to the following rules:

$$\text{\# radial nodes} = n - \ell - 1$$
$$\text{\# planar nodes} = \ell$$

4. Electronic configurations of atoms can be obtained using the Aufbau Principle where 2 electrons are placed in each atomic orbital. The filling order of atomic orbitals is readily evaluated by reference to the periodic table (Figure 7-D-3).

5. The emission spectra for He and Li can be accounted for by reducing the nuclear charge, Z, by a shielding constant, σ, to give an effective nuclear charge, Z_{eff}

$$Z_{eff} = Z - \sigma$$

For a given spectral series (common lower energy state) a single σ value can be assigned to all upper energy states and a σ value for the common lower energy state. The frequency of the transition in wave numbers is then given by

$$\bar{v} = -\left(R_H Z_U^2\right)\frac{1}{n_U^2} + \left(R_H Z_L^2 \frac{1}{4}\right)$$

where the lower state is $n_L = 2$. For the series $n^3D \to 2^3P$ the $\bar{v}$ values fit the above equation very precisely with a maximum deviation of only 0.002%. The poorest fit is for the transition $n^3S \to 2^3P$ where the maximum deviation is 0.1% for helium and 1% for lithium.

6. Molecular orbital theory of diatomic molecules is based upon the Aufbau Principle using the molecular orbitals of H_2^+. Bond order (BO) can be estimated by

$$BO = \frac{\text{\# bonding electrons - \# antibonding electrons}}{2}$$

The estimated BO is in reasonable agreement with the experimentally determined bond dissociation energy.

7. The geometry of molecules can be rationalized based upon overlap of orbitals of adjacent atoms. For atomic configurations with less than half-filled outer orbitals, hybridization is used to account for the resulting number of bonds and bond angles; e.g. sp^3 comes from $C \cdots s^2 p^2$ and all four sp^3 orbitals are identical except for orientation, which in this case gives an angle of ~109 degrees between sp^3 orbitals.

8. Delocalized π molecular orbitals can be used to account for the resonance stabilization in aromatic compounds such as benzene. The Aufbau Principle is used to add pairs of electrons to the π molecular orbitals starting from the lowest energy bonding orbitals.

9. Sigma bonds are also delocalized as evidenced by the decrease in ionization potential (IP) with increasing length of straight chain hydrocarbons. This change in IP with increasing chain length can be accounted for by semi-empirical calculations using Hyperchem^{TM}.

Exercises:

1. (a) Write the electronic configuration for silicon (Si).
   (b) Silicon reacts with fluorine to give SiF_4. Describe the nature of the bonding and the geometrical structure for this molecule.
   Answer: (a) $1s^2 2s^2 2p^6 3s^2 3p^2$ (b) 3s and 3p orbitals are hybridized to four sp^3 orbitals which overlap with F-2p orbitals to form four σ-bonds with a tetrahedral configuration.

2. (a) Sketch the probability density diagram for the hydrogen 4p and 4d orbitals. Show all model surfaces with dashed lines.
   (b) Compare the energies of the 4p and 4d orbitals.
   Answer: (a) 4p: 2 radial nodes and 1 planar node; 4d: 1 radial node and 2 planar nodes (b) The energies are equal.

3. Write the Schrödinger equation for Li^+.

   Answer: $\left[-\dfrac{\hbar^2}{2m} \left(\nabla_1^2 + \nabla_2^2 - \dfrac{3e^2}{r_1} - \dfrac{3e^2}{r_2} + \dfrac{e^2}{r_{12}} \right) \right] \psi = E\psi$.

4. (a) Write the electronic configuration for Gd based upon the filling order. (b) Write the electron configuration for Gd in the sequnce of orbital energies based on Figure 7-D-2. (c) Write the electron configuration for Gd^{3+} in the sequence of orbital energies. How many unpaired electrons are there in Gd^{3+}. Atomic Number of Gd is 64.
   Answer: (a) $1s^2 2s^2 2p^6 3s^2 3p^6 4s^2 3d^{10} 4p^6 5s^2 4d^{10} 5p^6 6s^2 6p^1 4f^7$
   (b) $1s^2 2s^2 2p^6 3s^3 3p^6 3d^{10} 4s^2 4p^6 4d^{10} 5s^2 5p^6 4f^7 6s^2 6p^1$
   (c) $1s^2 2s^2 2p^6 3s^3 3p^6 3d^{10} 4s^2 4p^6 4d^{10} 5s^2 5p^6 4f^7$; 7 unpaired electrons.

5. Write the Schrödinger Equation for He_2^+.
   Answer: $\left[-\dfrac{\hbar^2}{2m} \left(\nabla_1^2 + \nabla_2^2 + \nabla_3^2 \right) - \dfrac{2e^2}{r_{a1}} - \dfrac{2e^2}{r_{a2}} - \dfrac{2e^2}{r_{a3}} - \dfrac{2e^2}{r_{b1}} - \dfrac{2e^2}{r_{b2}} - \dfrac{2e^2}{r_{b3}} \right.$

   $\left. + \dfrac{e^2}{r_{ab}} + \dfrac{e^2}{r_{12}} + \dfrac{e^2}{r_{13}} + \dfrac{e^2}{r_{23}} \right] \psi = E\psi$.

6. Write the molecular orbital electronic configuration for Be_2^+ and predict the bond order. Answer: $(\sigma_{1s})^2 (\sigma 1s^*)^2 (\sigma_{2s})^2 (\sigma_{2s}^*)^1$; BO = 1/2.

7. Write the molecular orbital electronic configuration for C_2^- and predict the bond order. Answer: $(\sigma_{1s})^2 (\sigma 1s^*)^2 (\sigma_{2s})^2 (\sigma_{2s}^*)^2 (\sigma_{2px})^2 (\pi_{2py})^2 (\pi_{2pz})^1$; BO = 2.5.

8. The following reaction is known to occur in the gas phase:

    $$He^* + Ne \longrightarrow HeNe^+ + e^-$$

    where He* is an excited state for helium. Account for the fact that HeNe$^+$ can form, despite the fact that He and Ne are inert gases. Answer: HeNe$^+$ can form in a manner analogous to He$_2^+$ in that there is only one electron in the antibonding molecular orbital formed between He 1s and Ne 2p.

9. Describe the bonding and geometric structure of formaldehyde, H$_2$CO.
    Answer: Carbon has sp^2 hybridized orbitals which form 3 σ-bonds with the two H:1s orbitals and the O:2p orbital. The bonds lie in a plane with a bond angle of ~120^0. The remaining electrons in the 2p orbitals in C and O, which are perpendicular to the sp^2 orbital plane, form a π-bond between C and O.

10. Describe the bonding in acetonitrile, CH$_3$CN, using localized molecular orbitals.
    Answer: The CH$_3$ carbon has sp^3 hybridized orbitals whereas the CN carbon has sp hybridized orbitals. The C-C bond is formed by overlap of the sp^3 and sp orbitals. The C-H bonds are formed by overlap of the H:1s orbitals with three sp^3 orbitals. The two remaining p orbitals on C and N each contain an electron and these form two π-bonds, which in addition to the σ-bond formed between the C:sp orbital and the N:2p orbital, forms a triple bond.

11. Describe the bonding in the formate ion, HCOO$^-$, accounting for the fact that both C - O bonds are equivalent. Answer: Carbon has sp^2 hybridized orbitals which form a σ-bond by overlap with the H:1s orbitals and the other two form σ-bonds by overlap with the O:2p orbitals. The molecule would be planar with bond angles of 120^0. The remaing 2p orbitals of C and O contain 4 electrons which gives resonating structures in which the CO bonds alternate between single and double.

12. Benzene is planar, whereas cyclooctatetraene, C$_8$H$_8$, is nonplanar. Account for this fact on the basis of molecular bonding. (*Hint*: If C$_8$H$_8$ were planar, what would the C - C - C bond angle be?) Answer: The bond angle in C$_8$H$_8$, if planar, is 135^0, which is much greater than the 120^0 in benzene. If resonance were to occur in C$_8$H$_8$ as in C$_6$H$_6$ the structure should be planar or nearly planar. However, this would create too large a strain in the molecule since the sp^2 hybridized orbitals on each of the C-atoms prefer an angle of 120^0. As a result, the C$_8$H$_8$ is non-planar and has essentially alternating double and single bonds.

13. What is the electronic configuration of the oxygen atom? Would you expect the oxygen atom to be paramagnetic? Answer: O: 1s^2 2s^{2}2p^4. Yes, the four electrons in the p-orbitals have a configuration

    $$
    \begin{array}{ccc}
    p_x & p_y & p_z \\
    \uparrow\downarrow & \uparrow & \uparrow
    \end{array}
    $$

    Since there are two unpaired electrons, the O-atoms should be paramagnetic.

14. Write the electronic configuration for the fluorine atom. There should be one electron which is unpaired in a 2p orbital. Can you determine whether this is a 2p$_x$, 2p$_y$, or 2p$_z$ orbital ? Explain. Answer: F: 1s^2 2s^{2}2p^5. No. The 2p$_x$, 2p$_y$, 2p$_z$ orbitals are equivalent in energy and the electron can be placed in any of these orbitals.

15. Explain why electrons are added to the 5s orbital before the 4d in the Aufbau Principle. Answer: The 5s orbital is lower in energy. This results from the increased screening of the inner electrons on the 4d compared to the 5s orbital. In other words, the 5s has greater probability of having the electron near the nucleus than the 4d and is said to be more penetrating.

16. Compare the valence bond and molecular orbital descriptions of the bonding in O_2. How would they differ in their prediction? Answer: V.B. considers the formation of two bonds (1σ and 1π) from the 2p orbitals on O. All electrons are paired. M.O. constructs the molecular orbital configuration

$$(\sigma_{1s})^2(\overset{*}{\sigma_{1s}})^2(\sigma_{2s})^2(\overset{*}{\sigma_{2s}})^2(\sigma_{2px})^2(\pi_{2py})^2(\pi_{2pz})^2(\overset{*}{\pi_{2py}})^1(\overset{*}{\pi_{2pz}})^1$$

where there are two unpaired electrons in antibonding M.O.'s. M.O. theory would predict O_2 to be paramagnetic.

17. Describe the bonding in NO_2^- and account for the fact that the oxygens are equivalent. Is NO_2^- linear or nonlinear? Answer: Nitrogen $2p_x$ and $2p_y$ form sigma bonds with oxygen $2p_x$ orbitals. The remaining $2p_z$ orbital of nitrogen can form a π bond with the $2p_z$ of one oxygen atom. The additional e^- can be paired with the remaining e^- in the $2p_z$ orbital on the other oxygen. Resonance can then occur since either oxygen could be used to form the double bond. The O-N-O angle should be ~90°.

18. Using the criteria of maximum overlap, account for the fact that

$$D_{H_3C\text{-}CH_3} > D_{F\text{-}F} > D_{Li\text{-}Li}$$

Answer: The order of increasing directionality and hence order of increasing overlap is: $sp^3 > p > s$. Hence one would expect this order of the bond dissociation energies.

19. (a) Would you expect the heat of hydrogenation of ethylene to be less than, greater than, or equal to exactly half that of butadiene? Explain.
    (b) In Exercise 16 at the end of Chapter 2 it was shown that the heat of hydrogenation of the C = C bond in various compounds was about the same. Can you account for this on the basis of the localized Molecular Orbital Approximation.
    Answer: (a) Greater than half that of butadiene, since butadiene should have a small amount of resonance or delocalization energy. This increased resonance makes butadiene more stable than that expected from two isolated double bonds which in turn would lower its heat of hydrogenation. (b) According to the localized Molecular Orbital Approximation, all double bonds should be equivalent.

20. In Exercise 17 in Chapter 2 the C-H Bond Strength in CH_3-H was 438.4 kJ and only 367.6 kJ in $C_6H_5CH_2$-H. Can you account for this difference in bond strengths based upon molecular structure concepts learned in this chapter?
    Answer: The C-H bond strength in toluene is lower due to resonance stabilization of the electron in the p-orbital on the C in CH_2 with the benzene ring. There is little or no stabilization in the CH_3 after dissociation.

21. (a) Sketch the probability density diagrams for the hydrogen 2p and 3p orbitals (ψ^2).
    (b) Based on the diagrams in Part a, explain why carbon 2p orbitals on adjacent carbon atoms in a molecule form π bonds whereas silicon 3p orbitals do not.
    Answer: The 2p orbital has only one planar node whereas, the 3p orbital has a radial node in addition to the planar node (b) The 2p orbitals arranged side-by-side would normally have a large overlap. However, due to the spherical node in the 3p orbital

and the resulting curvature of the outer charge cloud, there is much less chance for 3p-3p orbital overlap.

22. Write the Schrodinger equation for Be^+.

Answer $\left[-\dfrac{\hbar^2}{2m}(\nabla_1^2 + \nabla_2^2 + \nabla_3^2) - \dfrac{4e^2}{r_1} - \dfrac{4e^2}{r_2} - \dfrac{4e^2}{r_3} + \dfrac{e^2}{r_{12}} + \dfrac{e^2}{r_{23}} + \dfrac{e^2}{r_{13}} \right]\psi = E\psi$

23. Using molecular orbial theory for homonuclear diatomics, predict the bond order and number of unpaired electrons in O_2^-. Answer: bond order = 1.5, one unpaired electron.

24. Since N and O are adjacent in the periodic table, the bonding in NO can be considered to be similar to that for a homonuclear diatomic. Predict the bond order and number of unpaired electrons in NO^+, NO, NO^-. Answer: NO^+: bond order = 3, no unpaired electrons; NO: bond order = 2.5, one unpaired electron; NO^-: bond order = 2, two unpaired electrons.

25. Assuming Cl bonded to Ar can be approximated by molecular orbital theory of homonuclear diatomics:
    (a) Predict the bond order and number of unpaired electrons in ArCl.
    (b) Account for the fact that an excited state for ArCl could have a higher bond order than the ground state.
    Answer: (a) bond order = 0.5, one unpaired electron (b) Excitation of an electron from an antibonding molecular orbital (such as $\sigma_{3p_x}^*$) to a higher energy bonding orbital (such as σ_{4s}) would give a higher bond order.

Appendix A
GRAPHICAL REPRESENTATION AND EVALUATION OF EXPERIMENTAL RESULTS

Most science students receive an adequate background in the theory of mathematics, but they are seldom exposed to its practical aspects. The purpose of this section is to acquaint the reader with the proper procedure for representing functions and data on graphs. In this first section it is assumed that the student will use linear or log graph paper to carry out the graphical solution. In the following section the use of the Microsoft Excel computer program for generating graphs will be presented for those students who have access and are familiar with operating a modern personal computer.

General Graphical Procedures

Before we can begin graphing a set of data or a function, we must decide on a suitable type of graph paper. The range of the experimental data, or the range over which we wish to graph a function, must be known. The choice of graph paper depends on two principal factors, the first of which is the general quality of the paper. Better graph paper has clear, more evenly-spaced rulings on the paper, and is of such a quality that resists stretching and wrinkling; thus, it is generally more permanent. The second factor to be considered is the size of the spacings and hence, the number of divisions along a given axis. If the axis is linear (equal spacings), then you divide the range of your data by the number of spacings or divisions to give you an estimate of the scale factor. The scale factor chosen is the *next highest convenient factor*. A scale factor that is *convenient* is one which allows you to graph a data point on the graph paper which you are using. Generally scale factors which are some factor of 10^n times 1, 2, 2.5, or 5 are convenient to use with linear graph paper with 10 subdivisions per major division. However, if there is too much difference between the estimate of the scale factor and the *next highest convenient factor*, then the data will not be distributed over the entire graph and accuracy is sacrificed. In this case, the size of the spacings would be inappropriate for this specific set of data.

We shall shortly consider graphs in which functions of the variables are plotted along each of the axes. For these cases graph paper is frequently available with spacings which reflect different functions. For example, a log function would have unequal spacings. If log y were to be plotted on this scale, y would correspond to the lower number scale shown in Figure A-1. For example, y = 2.85 is shown on the log scale (lower scale) by the arrow. If a linear scale were used, log y could be found from a log table or scientific calculator and this could be plotted on the linear scale, as shown by the arrow in the upper scale of Figure A-1.

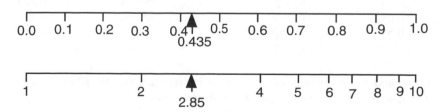

Figure A-1. Comparison of Linear and Log Scales

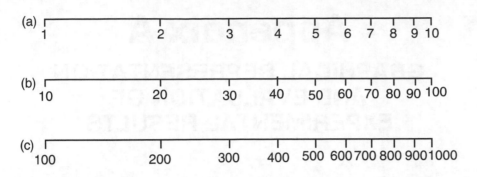

Figure A-2. Log Scale Divisions

The selection of a satisfactory scale factor still exists when using non-linear functions. For example, one log scale division could represent 1 to 10, as shown in Figure A-2(a), but it could also represent 10 to 100 or 100 to 1,000, and so forth, as shown in Figure A-2(b) and (c). If the data spans more than any of these ranges, then a graph paper with more than one range must be used. The number of ranges is referred to in terms of cycles. For example, a 3 cycle log scale is shown in Figure A-3. Ideally, it would be convenient for the student to have access to a wide selection of graph paper, but, unfortunately, the choice is frequently dictated by what is available.

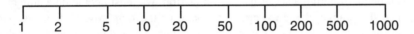

Figure A-3. 3 Cycle Log Scale

In order to illustrate the graphical procedure, let us suppose that we have a set of experimental data, in which there are two related variables. In this example, the variables are absorbance, A, as a function of time, for the reaction of an alkylperoxide anion (ROO) with tetranitromethane, $C(NO_2)_4$.

$$ROO^- + C(NO_2)_4 + H_2O \rightarrow C(NO_2)_3^- + NO_2^- + H^+ + ROH + O_2$$

The experimental values are taken from a paper by W. F. Sages and J. C. Hoffsommer [*J. Phys. Chem.*, **1969**, *73*, 4155]. (In the original article the term OD = optical density was used which is equivalent to the presently used A = absorbance.)

t(min)	A(L/mol-cm)	t(min)	A(L/mol-cm)
1.49	0.015	87.60	0.399
4.11	0.034	104.86	0.431
11.45	0.089	125.47	0.462
14.27	0.106	137.54	0.478
21.19	0.150	147.86	0.489
28.23	0.187	182.35	0.519
39.05	0.239	207.24	0.529
69.25	0.351	∞	0.559
81.57	0.384		

The units for t and A are given in parentheses. It is important that the units of every variable be shown, so that there is no chance of misinterpretation.

Let us graph A along the vertical axis (ordinate), and t along the horizontal axis (abscissa). The A values range essentially from zero to 0.559. The available graph paper has one axis with 25 major divisions and the other axis 18. Each of these divisions is further subdivided into 10 divisions. Since the first data point has small values of t and A, we will make the graph start at t = 0, A = 0. Considering the variable A, the scale factors would be

Range/# divisions	Convenient scale factor
$\frac{.559}{25} = .0224$	.025
$\frac{.559}{18} = .031$	.050

The first ratio is close to the scale factor of .025, whereas the second ratio differs considerably from the scale factor .050. In the latter case, only 60% of the scale would be used, so the axis with 25 divisions is selected to represent A. Since each of these divisions is further divided into 10, the scale factor per small division is

$$.0025 = 1 \text{ small division}$$

The choice of a proper scale factor for t is complicated by the fact that the range is infinite. However, the last point need not be plotted since it is at t = ∞, and we need only to show that A approaches the limit of 0.559 at large t. Taking the next to the last data point for our range,

Range/# divisions	Convenient scale factor
$\frac{207.24}{18} = 11.5$	20

The scale factor of 20 per large division will be satisfactory. Although only 60% of the scale will be used to represent the data, this will leave about 40% of the scale to show the approach to .559 at large t. The scale factor per small division is then

$$2.0 = 1 \text{ small division}$$

The graph should be constructed with a sharp lead pencil, with the axes properly identified and the scales clearly shown. The graph of A versus t is shown in Figure A-4. The absorbance scale (y-axis) has 25 (.025) divisions ranging from zero to 0.625 and the time scale (x-axis) has 18 (20) divisions ranging from zero to 360 as noted in Figure A-4. Note that the t -axis scale has been reduced by a factor of 100 in order to avoid the use of excess zeros in representing the scale. The factor of 100 is appropriately identified through the factor 10^{-2} multiplied by t. This should be $10^{-2}t$, rather than 10^2t, since the scale identification is represented on the linear scale. For example, the value of 1.00 on the abscissa scale is

$$10^{-2} t = 1.00 \text{ min}$$
$$t = 1.00 \times 10^2 \text{ min} = 100 \text{ min}$$

In other words, the value of 1.00 on the $10^{-2}t$ scale is actually 100 min. Frequently in the literature, the exponent with the opposite sign is used, so one should guard against this possible misinterpretation. The reader should examine the data to see if it is consistent with the author's identification of the scale.

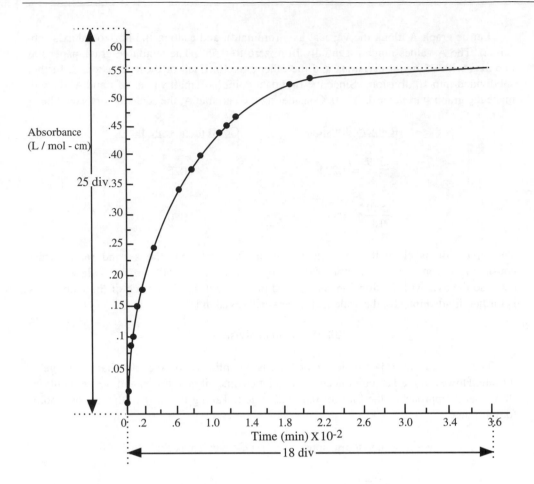

Figure A-4. Graph of Absorbance (A) Versus t (the curve approaches the dotted line asymptotically)

The broken line in Figure A-4 represents the upper limit of A = .559. The curve should approach this value asymptotically, and a smooth curve is generally drawn to best represent the trend of the data. A French curve or some other device is used for this purpose, but if necessary, the curve can be sketched in. The curve need not pass through every data point, since there are errors in each of the data points and they would not be expected to fall exactly on some smooth curve. The experimenter himself must make the decision where the smooth curve is drawn, and whether variations or trends in the data are real or the result of errors in the measurements.

In the previous example the graph started at the origin where both t = 0 and Absorbance = 0. However, in many situations the data does not extend down to zero and it is generally inappropriate to start the graph at zero. Such is the case with absolute temperature (T) measurements. At room temperature of say 25.0°C the absolute temperature is T = 25.0 + 273.1 = 298.1 K. If we were to raise the temperature to the boiling point of water (100.0°C), the absolute temperature would be T = 100.0 + 273.1 = 373.1 K. The temperature range would be 75 K and if we graphed data over this range we would want to start the graph in the vicinity of 298 K in order to utilize the entire graph paper.

Problem A-1. Suppose you had a table of absolute temperatures ranging from T = 300 K to T = 800 K. If you were to graph 1/T along the abscissa of your graph and the graph paper had 25 divisions, what magnitude of 1/T would you select per division and what would be the values of 1/T at the start and end of your graph? Answer: 1/T per division = 0.10 x 10^{-3} K^{-1}; start at 1/T = 1.00 x 10^{-3} K^{-1} and end at 1/T = 3.5 x 10^{-3} K^{-1}.

Problem A-2. Equilibrium constants (K) for the reaction

$$WO_{2(s)} + I_{2(g)} \rightleftarrows WO_2I_{2(g)}$$

have been determined at various temperatures [S. K. Gupta, *J. Phys. Chem.*, **1969**, *73*, 4086].

T(K)	K x 10^4
717	1.37
750	2.69
757	3.55
785	5.83
839	15.5
861	18.4
912	45.1
927	61.2

Graph K as the ordinate versus T as the abscissa.

In both thermodynamics and kinetics we will see that we will have occasion to graph the reciprocal of the absolute temperature (1/T). Previously we considered the absolute temperature range from 298.1 K to 373.1 K. The 1/T values would be

$$\frac{1}{298.1 \text{ K}} = 0.003355 \text{ K}^{-1} = 3.355 \times 10^{-3} \text{ K}^{-1}$$

$$\frac{1}{373.1 \text{ K}} = 0.002680 \text{ K}^{-1} = 2.680 \times 10^{-3} \text{ K}^{-1}$$

The range of 1/T values would then be $(3.355 \times 10^{-3} \text{K}^{-1} - 2.680 \times 10^{-3} \text{K}^{-1}) = 0.675 \times 10^{-3} \text{K}^{-1}$ and this is the quantity that we would use in deciding on an appropriate scale factor for graphing data over this range.

Problem A-3. Using the data from Problem A-2, graph ln K versus 1/T. Note that the data on this graph is best represented by a straight line. Answer: Graph should have a negative slope.

In the previous example we considered graphs of experimental data. We are frequently interested in examining the graph of a function of two variables. In this case, we must generate or calculate the data required for the graph. One assumes some value for one of the variables, and evaluates the other variable according to the equation. The number of data points which are required and the spacing between the data points is governed by the shape of the curve, the accuracy which is desired in the graph, and the range over which the function is to be examined.

For an example, let us consider the graph of the function

$$\left(P + \frac{n^2a}{V^2}\right)(V - nb) = nRT \qquad (A-1)$$

which is known as van der Waals' Equation of State. For simplicity, let us assume that n, the number of moles, equals 1, then

$$\left(P + \frac{a}{V^2}\right)(V - b) = RT \qquad\qquad (A\text{-}2)$$

where a and b are constants characteristic of the substance, R = gas constant = .08205 liter-atm/deg-mol, P = pressure, V = volume, and T = absolute temperature. Since we wish to examine how P varies with V at constant T (isotherm), we can readily solve for P.

$$P = \frac{RT}{V-b} - \frac{a}{V^2} \qquad\qquad (A\text{-}3)$$

For O_2, the constants are a = 1.362 liter^2-atm/mol^2 and b = .0319 liter/mol. We will graph the function over the range V = 0.05 liter to V = 0.2 liter, at T = 125 K. Values of V at equal intervals of .01 liter have been chosen, except in the range .05 liter - .06 liter, and the corresponding values of P evaluated according to equation (A-3). These values of V and P are shown in the table which follows.

V(liters)	P(atm)	V(liters)	P(atm)
0.0500	21.84	0.11	18.76
0.0550	-6.26	0.12	21.84
0.0575	-11.35	0.13	23.96
0.0600	-13.30	0.14	25.39
0.0625	-13.50	0.15	26.31
0.6500	-12.51	0.16	26.85
0.0675	-10.83	0.17	27.14
0.0700	-8.76	0.18	27.21
0.0800	0.42	0.19	26.96
0.0900	8.38	0.20	26.71
0.1000	14.40		

The calculation at V = .04 liter is shown in detail here, and in particular, the reader should note the dimensional analysis.

$$P = \frac{(1\ \text{mol})(.08205\ \text{L}\ \text{atm/K}\ \text{mol})(125\ \text{K})}{0.04\ \text{L} - (0.0319\ \text{L/mol})(1\ \text{mol})} - \frac{(1\ \text{mol})^2(1.362\ \text{L}^2\ \text{atm/mol}^2)}{(0.04\ \text{L})^2}$$

P = 1268 atm - 850 atm = 418 atm

Linear graph paper was used for this graph, with the P axis along the ordinate having 25 divisions, and the V axis along the abscissa with 18 divisions. The scale factors are then

V scale:

$$\frac{.20 - .05}{18} = \frac{.15}{18} = .00833 \text{ scale factor } .01/\text{div}$$

P scale:

$$\frac{27.2 - (-13.3)}{25} = \frac{41}{25} = 1.64 \text{ scale factor } 2.0/\text{div}$$

The graph of P versus V for van der Waals' Equation for O_2 is given in Figure A-5. Note the maximum at V = 0.18 liter, and the minimum at V = 0.06 liter. This will occur with this equation wherever

$$T < T_c = \frac{8a}{27Rb}$$

where T_c = critical temperature.

Problem A-4. A first order reaction has the following relationship between the concentration of reactant, [A], and the time, t.

$$[A] = [A]^o \, e^{-kt}$$

where $[A]^o$ = initial concentration of A (t = 0), k = rate constant. Graph $[A]/[A]^o$ versus time if k = 0.03465 min^{-1}. Let time range from zero to 60 minutes.

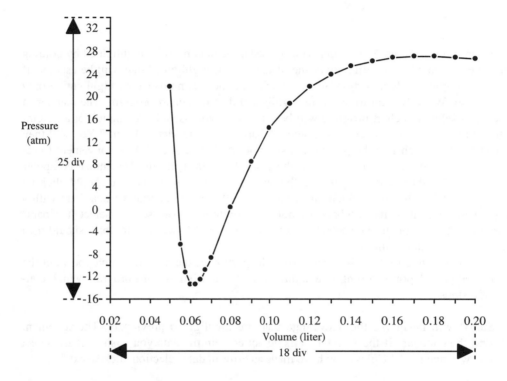

Figure A-5. Graph of P Versus V for van der Waals' Equation for O_2

Computer Generated Graphs Using Excel

Microsoft Excel is a very versatile program which can generate a variety of graphs. However, in this physical chemistry course we are interested primarily in two-dimensional graphs of X and Y and the discussion will be restricted to this pursuit. Let us first consider graphing of experimental data as was illustrated in Figure A-4. The data of t(min) and A(L/mol-cm) were given previously and these would be entered in the Excel Spreadsheet with the X-axis, in this case t (min), in the first column and the Y-axis, A(L/mol-cm), in the second column. In selecting this data for making an X-Y graph, Excel assumes the first column is X and subsequent columns to the right are Y data. Data can be selected in rows if so specified, but it is more convenient to enter data in columns. Since we have only one set of Y-values, there will be only one column to the right of the X-data. Your spreadsheet should look like this:

1.49	0.015
4.11	0.034
11.45	0.089
21.19	0.150
28.23	0.187
39.05	0.239
69.25	0.351
81.57	0.384
87.60	0.399
104.86	0.431
125.47	0.462
147.86	0.489
182.35	0.519
207.24	0.529

In order to make an X-Y graph with this data, the data must be highlighted by holding down the mouse while scrolling over the data. Only the highlighted data will be used, so if you wish to graph only a portion of the data this can be accomplished by highlighting only that portion. We will want to use your highlighted data in a chart program. The manner in which you select the chart program will be different if you use a PC computer or a Macintosh Computer. For the PC computer select "chart" under the Menu "Insert." You will then be asked if the graph is to be put on this sheet or "As a new sheet." It is convenient to select "on this sheet" and you can move the graph later. In using the Macintosh computer icons for the toolbars should appear at the bottom of your screen and you should click on the 5th icon from the right which has a bar graph appearance of four vertical bars with a magic wand over it. If the toolbars are not at the bottom of your screen, select "toolbars" under the "option" menu, then select "chart" and click on "show." The toolbars should then appear at the bottom of the screen.

The next step is to drag diagonally over the portion of the page in which you want the graph to appear. Upon so doing you should activate the ChartWizard program which consists of 5 steps:

1. Step 1 will designate the data range that you highlighted previously. The statement will then appear "If the selected cells do not contain the data you wish to chart, select a new range now." Unless you have made an error in data selection click "Next."

2. Step 2 of the ChartWizard allows you to select the type graph you desire and for this problem we want the "XY scatter" and you should click on this one and then click on "Next" to continue.

3. Step 3 of Chart Wizard allows you to select the type of XY scatter graph that you want. If we wish to graph only the data points, the graph to the left showing such a graph should be highlighted. If you want connecting straight lines between data points select the next one to the right. These are the only choices appropriate for our problem. Then click on "Next" to continue.

4. Step 4 will display a sample graph based upon our previous selections showing the choices you have made

Data Series in   ❏ Rows
○ Columns

Use first <u>1</u> Column(s) for X Data
Use first <u>0</u> Row(s) for legend Data

Since we will have a simple graph of one set of Y values versus X, we will not need a legend. In another problem you may be graphing more than one Y set of data and in this case a legend can be placed on the graph to designate the symbol used,(□, ○, etc) for each set of Y values. Alternatively, this designation can be made in the figure caption. Then click on "Next" to proceed to the final step.

5.   In Step 5 of ChartWizard you will be asked if you want a "legend," "chart title," and axis titles." You should type these in at this time or they can be added or changed later. Then click on the "finish" Key and the chart will appear with the appropriate labels. Do not be concerned about the size of the graph since it can be modified later. At this stage the screen should show the graph along side the data, as shown below:

1.49	0.015
4.11	0.034
11.45	0.089
21.19	0.15
28.23	0.187
39.05	0.239
69.25	0.351
81.57	0.384
87.6	0.399
104.86	0.431
125.47	0.462
147.86	0.489
182.35	0.519
207.24	0.529

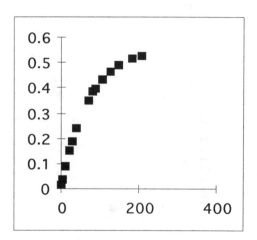

In order to refine the graph to our liking we can modify the scale of the axes and insert or change labels for the axes. To modify the scale simply point the arrow on the axis to be modifiesd and click the mouse. A new working graph will appear called Worksheet 1 Chart 1. Click on the X-axis again and a *Patterns* menu will appear that allows you to adjust the *Style* of your X-axis, which generally would be a solid line. Also the color and weight can be changed if so desired. The nature of the tick marks can be selected which I have selected as "cross" for the major, and "outside" for the minor, and "next to axis". The scale can then be adjusted by clicking on the *Scale* on the right of the screen. You now can select the minimum and maximum values, the same decision that one would make in using graph paper. Also you can select the *Major Unit* which will show the values of X along the axis at this spacing. The *Minor Unit* is the spacing of the minor tick marks but without numbers. I have selected 0 as the minimum, 350 as the maximum, Major Unit = 50, Minor Unit =10, and Y-axis crosses at 0.

If you wish to add a label to the X-axis, you can designate the *Font, Font Size*, and *Font Style* by selecting Font under the *Format* menu. Then go to the Chart menu and select *Attach Text*. Select X-axis and click OK. The position for the label will be designated by a sectioned box alongside the X-axis. You are to type the desired label for the X-axis and click on an open area of the spread sheet and a graph will appear with the X-axis label in place. Click on either axis and the Worksheet 1 Chart 1 graph will appear with the X-axis label also in place. Repeat this procedure for the Y-axis.

If you wish to place a title on the graph again go to the *Charts* menu and select *Attach Text*. Then select *Title* from this menu and a sectioned box will appear in the upper center portion of your graph. Type in your title, click outside the graph and the title will appear on your graph. For research publications a figure caption is placed beneath the graph and the title is not necessary. However, for a slide presentation it is convenient to have the title directly on the graph. If you use the same selections as given in the previous discussion, your graph should look like the following:

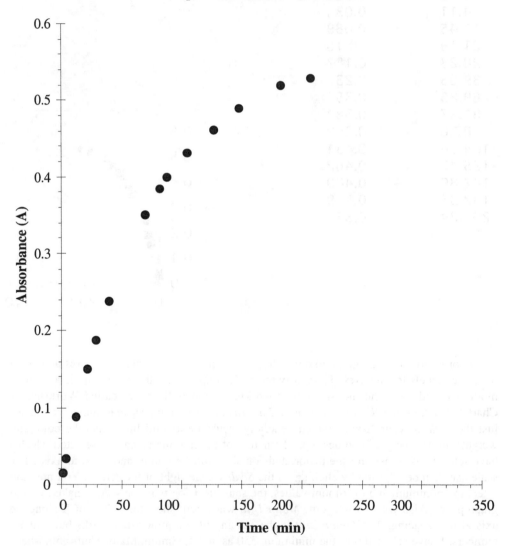

Previously we stated that it is sometimes helpful to graph a certain functional relationship between variables in order to have a better understanding of the equation. We used the van der Waals equation for one mole of gas, equation (A-3). Excel is also very useful in this regard in that a simple program can be written for the evaluation of a variable. For convenience we will rewrite the equation

$$P = \frac{RT}{V-b} - \frac{a}{V^2} \tag{A-3}$$

The constants "a" and "b" are constants which will vary from one compound to another. For this reason we may wish to change these to see how the graph of P will vary for different compounds. Also T is held constant and the resulting graph of P versus V is called an isotherm. We may also wish to vary T to see how it affects the equation. The gas constant R = 0.08205 L-atm/K-mol and remains constant for all gases. Since R will not be changed, we will put its value into the equation.

In order to program an equation there are certain symbols used to designate various mathematical operations. Many of these will be familiar to you from their name. We will limit these to the most commonly encountered in physical chemistry.

Operator	Description
+	addition
-	subtraction
*	multiplication
/	division
^	exponentionation
E( )	$10^{(\,)}$
Exp( )	$e^{(\,)}$
ln	logarithm to base e
log	logarithm to base 10
sin	sine of the angle
cos	cosine of the angle

In order to plot P versus V in equation (A-3) we need to designate values of V for which we want to calculate values of P. In Excel we would then make a column of V values as we did previously on page 302. However, before we do that we will designate the constants a, b, and T. The Excel spreadsheet would look like this:

a=	1.362
b=	0.0319
T=	125
V	P
0.05	
0.055	
0.0575	
0.06	
0.07	
0.08	
0.09	
0.1	
0.11	
0.12	
0.13	
0.14	
0.15	
0.16	
0.17	
0.18	
0.19	
0.2	

We then program the evaluation of P by clicking on the cell B6 which would contain the first calculated value of P. Above the spreadsheet we will program the equation as follows:

$$B6 = 0.08205*\$B\$3/(A6-\$B\$2)-\$B\$1/A6^2$$

An X and a check mark will appear between the B6 and the = sign designating that this is an equation. Note that we have placed a dollar sign before the cell column and the cell row for the constants a, b, and T in cells B1, B2, and B3. The $ will keep the location of these constants fixed. The variable V in A6 will vary as we go down the column of V values. In order to fill in the values of V we simply scroll down the right side of column B starting at B6. Point the arrow on the right lower corner of B6 and you should observe a cross form. Scroll down with the cross and column B will be filled in with the corresponding P values. Columns 1 and 2 containing V and P, respectively, are in the right positions to construct a graph of P versus V. As described previously, highlight columns A and B containing the V and P values, select the ChartWizard Program, as described previously, and continue to graph P versus V. Your initial graph should appear as shown on the following spreadsheet along with the data.

a=	1.362		
b=	0.0319		
T=	125		
V	P		
0.05	21.84364641		
0.055	-6.25442739		
0.0575	-11.3123043		
0.06	-13.3422301		
0.07	-8.76627029		
0.08	0.415150728		
0.09	8.379390578		
0.1	14.40572687		
0.11	18.76003958		
0.12	21.83267121		
0.13	23.95721369		
0.14	25.38763239		
0.15	26.31044313		
0.16	26.86127781		
0.17	27.13880794		
0.18	27.21515743		
0.19	27.14338465		
0.2	26.96279001		

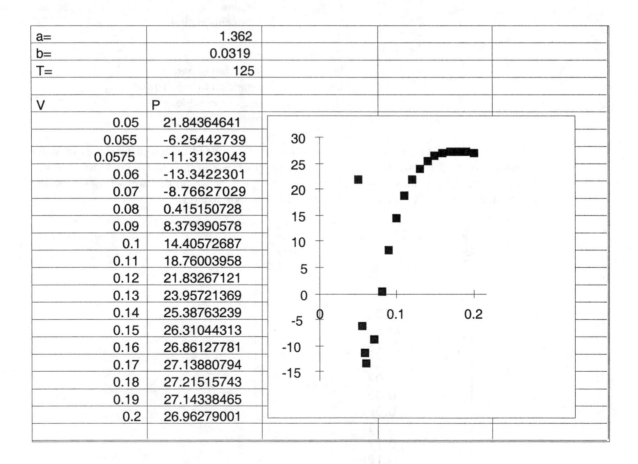

You can refine the graph as we discussed previously. If you wish to change the values of "a" and "b," simply change the values in cells B1 and B2 and a new graph will appear. Similarly if you wish to change the temperature, T, simply change the value of T in cell B3. However, the values for P and V and the graph will be <u>replaced</u> by the calculations using the new "a" and "b" or new T. If you want to retain a calculation/graph, you must save it as *save as*. In the following section mathematical operations will be performed which will convert an equation into a linear form. The mathematical calculation of the new X and Y variables in terms of the original variables can also be performed using the mathematical operation previously described. For example if we have K values in colmn B and wish to take the logarithm to the base "e" and place it in colmn D, we would perform the calculation: D1=ln(B1).

Linearization of Functions

In physical chemistry, we often examine the relationship between certain variables, but it would be advantageous to know the functional relationship between them. Frequently, the functional relationship is derived from some basic laws or theory of the subject and we wish to know whether the experimental data will fit or satisfy this function. In order to do this, we try to arrange the function into a linear relationship between the variables, and then graph these functions of the variables. If the resulting data appear to follow a linear function, we say that the data satisfy the proposed theoretical function.

As an example, we will reconsider the first example in Section A which consisted of the absorbance, A, as a function of time (t) for the reaction of an alkylperoxide anion with tetranitromethane. On the basis of classical chemical kinetics (Chapter 4), the following equation can be derived.

$$A = A^{\infty}(1 - e^{-kt}) \tag{A-4}$$

where A^{∞} represents the absorbance at $t = \infty$, and k = rate constant. This equation is derived on the hypothesis that the reaction is first order with respect to $C(NO_2)_4$. The reader can understand the significance of this statement after studying Chapter 4. The experimental conditions are adjusted so that other variables will be held constant, and agreement or disagreement of the experimental data with equation (A-4) will reveal whether the reaction is first order with respect to $C(NO_2)_4$.

In order to test whether the A and t data satisfy equation (A-4), we first rearrange equation (A-4) into a linear form. The reader will recall from analytic geometry that the general form for a linear function is

$$y = mx + b \tag{A-5}$$

where m = slope and b = intercept on a graph of y (ordinate) versus x (abscissa). Equation (A-4) can be set into a linear form by dividing by A^{∞}, rearranging to isolate the e^{-kt} term, and finally taking the logarithm.

$$\frac{A}{A^{\infty}} = (1 - e^{-kt})$$

$$\frac{A^{\infty} - A}{A^{\infty}} = e^{-kt}$$

$$\ln\left(\frac{A^{\infty} - A}{A^{\infty}}\right) = -kt$$

$$\ln\left(\frac{A^{\infty}}{A^{\infty} - A}\right) = kt \tag{A-6}$$

Comparing equation (A-6) with (A-5), we see that

$$y = \ln\left(\frac{A^\infty}{A^\infty - A}\right)$$
$$x = t$$
$$m = k$$
$$b = 0$$

and, thus, equation (A-6) is of the linear form, with zero intercept. The $\ln[A^\infty/(A^\infty - A)]$ values are tabulated as follows:

t(min)	A	$\dfrac{A^\infty}{A^\infty - A}$	$\ln\dfrac{A^\infty}{A^\infty - A}$
1.49	0.015	1.027	0.0266
4.11	0.034	1.065	0.0630
11.45	0.089	1.189	0.1731
14.27	0.106	1.234	0.2103
21.19	0.150	1.365	0.3112
28.23	0.187	1.503	0.4073
39.05	0.239	1.745	0.5568
69.25	0.351	2.685	0.9877
81.57	0.384	3.19	1.160
87.60	0.399	3.49	1.250
104.86	0.431	4.37	1.475
125.47	0.462	5.76	1.751
137.54	0.478	6.90	1.932
147.86	0.489	7.99	2.078
182.35	0.519	13.95	2.635
207.24	0.529	18.63	2.925
∞	0.559	--	--

The graph of $y = \ln[A^\infty/(A^\infty - A)]$ versus t is shown in Figure A-6.

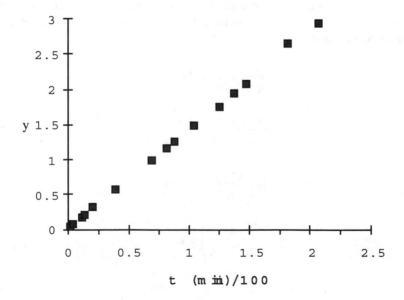

Figure A-6. Graph of $y = \ln[A^\infty/(A^\infty - A)]$ versus t

Note that the data represent a linear function very well, and the hypothesis that the reaction is first order with respect to $C(NO_2)_4$ is well supported. The only point which deviates significantly is the data point at $t = 182.35$ min. This is not unexpected, since the data points at large A will have the greatest error. The ratio $A^\infty/(A^\infty - A)$ has a large error, since $(A^\infty - A)$ becomes small at large A values. For example, at $t = 182.35$, the $A = 0.519$, and $(A^\infty - A) = 0.559 - 0.519 = .040$. Note that 0.040 has only two significant figures, and an error of ± 0.001 absorbance results in a $\pm 2.5\%$ error in the difference 0.040. This error is projected onto the ratio $A^\infty/(A^\infty - A)$ and then onto $\ln [A^\infty/(A^\infty - A)]$.

The graph of the linear function has another purpose, in addition to testing the validity or applicability of the proposed function. If the linear function is obeyed by the experimental data, the constants from the linear equation (slope and intercept) can be found and related to the constant terms in the linear function. In the case of equation (A-6), the intercept is zero, and the slope is related to k by

$$m = \text{slope} = k$$

The slope can be evaluated by reading off two points on the straight line graph. For greater accuracy, it is better to pick two points considerably far apart, for example, $t/100 = 0.2$ and $t/100 = 1.8$.

$$m = \frac{2.51 - 0.292}{(180 - 20) \text{ min}} = \frac{2.218}{160 \text{ min}} = 0.01386 \text{ min}^{-1}$$

The rate constant is then

$$k = m = 0.01386 \text{ min}^{-1}$$

$$k = 1.386 \times 10^{-2} \text{ min}^{-1}$$

The reader should always note the units of the slope and intercept, since they will be used in the units of any constants which are evaluated from m and b.

Frequently there are alternate linear forms that can be obtained from a given function. For example, equation A-6 can be re-written by taking the logarithm of the ratio and rearranging

$$\ln A^\infty - \ln (A^\infty - A) = kt$$

$$- \ln (A^\infty - A) = kt - \ln A^\infty$$

$$\ln (A^\infty - A) = - kt + \ln A^\infty \tag{A-7}$$

This is a linear equation where

$$y = \ln (A^\infty - A)$$

$$x = t$$

$$m = - k$$

$$b = \ln A^\infty$$

Another linear form could be obtained by dividing equation A-7 by t.

$$\frac{1}{t} \ln (A^\infty - A) = - k + \frac{1}{t} \ln A^\infty \tag{A-8}$$

This is also a linear equation where

$$y = \frac{1}{t} \ln (A^\infty - A) \qquad m = \ln A^\infty$$

$$x = \frac{1}{t} \qquad\qquad b = -k$$

Another example of linearization of a function is the Clausius-Clapeyron equation for vapor pressure, p.

$$p = C e^{-\Delta H_{vap}/RT}$$

In order to transform this equation into a linear form we take the logarithm

$$\ln p = \ln C + \ln (e^{-\Delta H_{vap}/RT})$$

Since the logarithm of an exponential is the function itself,

$$\ln p = \ln C - \frac{\Delta H_{vap}}{R} \frac{1}{T} \qquad\qquad\qquad\qquad (A-9)$$

This equation is of the linear form where

$$y = \ln p \qquad\qquad\qquad m = \frac{-\Delta H_{vap}}{R}$$

$$x = \frac{1}{T} \qquad\qquad\qquad b = \ln C$$

Problem A-5. By mathematical manipulation, transform the following functions into linear form and identify the slope (m) and intercept (b) with the constants for each of the equations. The variables are indicated in each case, and the remaining symbols can be assumed constant.
(a) $PV = nRT$ (P, V are variables)
(b) $k = A \exp[-E^*/RT]$ (k, T are variables)
(c) $P(V - nb) = nRT$ (P, V are variables)

Answer: (a) $P = nRT\frac{1}{V}$, m = nRT, b = 0 (b) $\ln k = -\frac{E^*}{R} \frac{1}{T} + \ln A$, $m = -\frac{E^*}{R}$, b = ln A

(c) $V = n RT\left(\frac{1}{P}\right) + nb$, m = nRT, b = nb.

Problem A-6. In Problem A-3, a linear graph of ln K versus 1/T was obtained. The equations which this data represent are

$$\Delta G^o = -RT \ln K \quad \text{and} \quad \Delta G^o = \Delta H^o - T\Delta S^o$$

Over relatively small temperature ranges, ΔH^o and ΔS^o can be considered constant.
(a) Eliminate ΔG^o between these equations and set the resulting equation into a linear function between ln K and 1/T.
(b) With this linear equation, use the graph in Problem A-3 to evaluate ΔH^o and ΔS^o. Let R = 8.314 joules/K-mole. The equilibrium constant, K, has no units.

Answer: (a) $\ln K = -\frac{\Delta H^o}{R}\left(\frac{1}{T}\right) + \frac{\Delta S^o}{R}$ (b) $\Delta H^o = -mR$, $\Delta S^o = bR$

Problem A-7. An important equation in biochemistry is the rate expression for enzyme catalysis of a substrate (S).

$$v = \frac{k_2[E]^o[S]}{[S] + K_S}$$

where v = velocity or rate of the reaction, $[E]^o$ = initial concentration of enzyme, $[S]$ = concentration of substrate, and k_2 and K_S are constants. The variables are v and $[S]$. Rearrange the equation into a linear function, and show how K_S can be evaluated.

Answer: $\dfrac{1}{v} = \dfrac{K_S}{k_2[E]^o}\dfrac{1}{[S]} + \dfrac{1}{k_2[E]^o}$; $K_S = \dfrac{m}{b}$

Appendix B
DIFFERENTIAL CALCULUS

Any treatment of physical chemistry which deals with the topics of thermodynamics and kinetics must be based upon at least an elementary knowledge of calculus. Although we intend to avoid complicated mathematical proofs in the text, it is essential that the presentations be based upon calculus and differential expressions. Sufficient examples and exercises will be presented so that the reader can acquire a sufficient understanding of the basic principles of calculus. However, a thorough study of the subject, for two to three semesters, is necessary for a comprehensive study of physical chemistry.

Definition of a Derivative

We can formally define a derivative, using the general functional expression

$$y = f(x)$$

where f is some function, such as x^3, sin x, ln x. A plot of y versus x gives the solid line in Figure B-1.

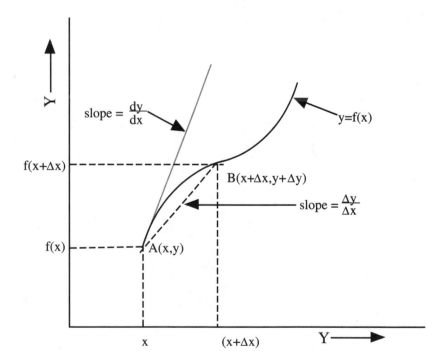

Figure B-1. Graphical Description of a Derivative

The derivative, dy/dx, is defined as

$$\frac{dy}{dx} = \lim_{\Delta x \to 0} \frac{\Delta y}{\Delta x} = \lim_{\Delta x \to 0} \frac{f(x + \Delta x) - f(x)}{\Delta x} \tag{B-1}$$

The term $\lim \Delta x \to 0$ is the limit of the quantity as Δx is made to approach zero. The value of y corresponding to $(x + \Delta x)$ can be evaluated from the function, $f(x + \Delta x)$, and is shown in Figure B-1, as the point, B, $[(x + \Delta x), f(x + \Delta x)]$. Just as in the example discussed previously, the broken line drawn through these points represents a chord, the slope of which is $\Delta y/\Delta x$.

$$\frac{\Delta y}{\Delta x} = \frac{f(x + \Delta x) - f(x)}{\Delta x} \tag{B-2}$$

Now, if we were to diminish the magnitude of Δx, the point B would approach point A. Simultaneously, the slope of the chord given by $\Delta y/\Delta x$ will increase and will eventually become the slope of the tangent line at A, when B reaches A. The slope of this tangent is the derivative, dy/dx, at the specific value of x designated in Figure B-1. As equation (B-1) states, it is the limit of the slope $\Delta y/\Delta x$, as Δx is made to approach zero. In general, the derivative is the slope of the tangent line at any point along the curve $f(x)$, and it represents the rate of change of y with respect to x at that point. The derivative is itself a function of x. In Figure B-2, we have plotted the derivative for the function drawn in Figure B-1. Note the minimum value in dy/dx, and the region in Figure B-1 to which it corresponds.

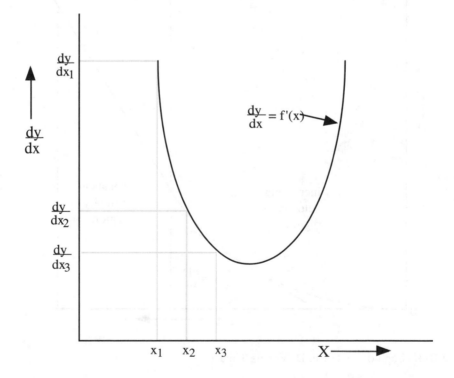

Figure B-2. Graph of Derivative dy/dx Versus x

In order to further illustrate the concept of differential calculus, let us consider the distance travelled, D, by a car. The velocity, v, at which the car travels is the first derivative of distance, D, with respect to time, t.

$$v = \frac{dD}{dt} \qquad\qquad (B-3)$$

The units for velocity are obviously distance divided by time such as miles/hr, kilometers/hr or meters/sec.

The acceleration, a, of the car is the first derivative of the velocity; i.e. the rate of change of velocity with time. The acceleration then is the <u>second</u> derivative of distance, D, with the time, t, since velocity is the first derivative of distance with time, t.

$$a = \frac{dv}{dt} = \frac{d^2D}{dt^2}$$

Let us assume that the car is at rest, $v = 0$, at time zero and for convenience we will assume that the distance is also zero at time zero. In order for the car to move we must begin accelerating and during this period the velocity will increase. This is shown in Figures B-3 through B-5. Note in Figure B-3 that the initial slope is zero in agreement with the assumption that $v = 0$ at $t = 0$. However, the slope continually increases with time as the car is accelerated. Finally when the desired velocity, v_f, is obtained the acceleration is reduced to zero. In Figure B-3 this occurs when the slope of the D versus t curve is constant as shown in the figure.

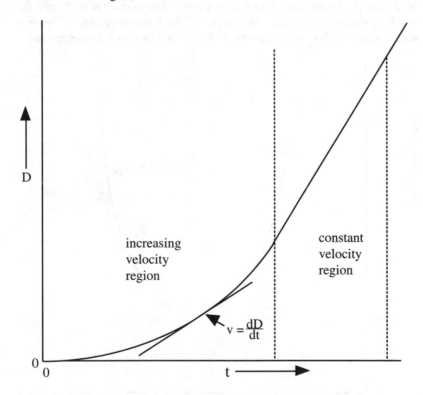

Figure B-3. Graph of Distance, D, Versus Time, t

In Figure B-4 we show the velocity increasing linearly under a constant acceleration until the desired v_f is obtained, after which the velocity remains constant at v_{max}. In Figure B-5 we show the acceleration period during which the acceleration is approximately constant until the v_{max} is attained, at which time the acceleration is reduced to zero.

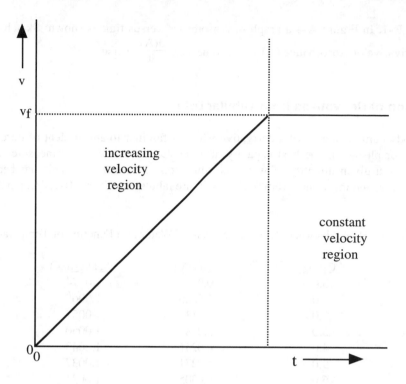

Figure B-4. Graph of Velocity, v, Versus Time, t

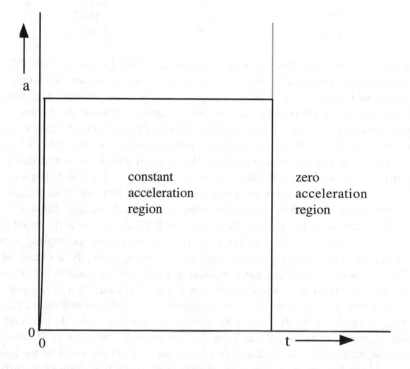

Figure B-5. Graph of Acceleration, a, Versus Time, t

Graphs similar to those shown in Figures B-3 to B-5 would be obtained if one examined the distance an object falls in a vacuum under a constant acceleration of gravity. However, in this case the acceleration would be continuous and the velocity would also be increasing continuously.

Problem B-1. In Figure A-4 a graph of Absorbance versus time is shown. Sketch a graph of the derivative of Absorbance (A) versus time, i.e., $\dfrac{d(A)}{dt}$ versus t.

Evaluation of Derivatives from Tabular Data

A thermodynamic example of a derivative which is familiar to any student of introductory chemistry or physics is the heat capacity of water. Experimentally, one measures the heat absorbed by a given quantity of water as a function of temperature. This has been done very precisely, and the values from 0°C to 10°C are tabulated in Table B-1, column 2.

Table B-1. Heat Absorbed, Q, by One Gram of Water as a Function of Temperature

$t°C$(deg)	Q(cal/g)	C(cal/g-deg)
0.0	0.0	1.0074
1.0	1.0069	1.0065
2.0	2.0130	1.0057
3.0	3.0184	1.0050
4.0	4.0231	1.0043
5.0	5.0271	1.0037
6.0	6.0305	1.0031
7.0	7.0334	1.0026
8.0	8.0358	1.0021
9.0	9.0377	1.0017
10.0	10.0392	1.0013

These values of Q are measurements of the heat absorbed by water from 0°C to the indicated temperature. In this data, we note that Q varies almost linearly with temperature; that is, a graph of Q versus t°C would be approximately a straight line. This necessarily results in the slope or derivative being approximately constant over this range of temperatures, which it is indeed, as shown in the third column containing heat capacity, C. In previous courses, the reader has probably assumed C to be constant for H_2O from 0°C to 100°C, and this is a good approximation. However, it should be noted that C is not actually constant, and with precise measurements, as shown in Table B-1, the variation of C can be observed well above any experimental error. The fact that C is not constant is significant, and points out the necessity for using differential calculus. Since C for H_2O does not vary greatly over the temperature range 0-10°C, the use of differential calculus may seem trivial. However, this is far from true in numerous examples, where the derivative changes quite rapidly with a change in t, or more generally, a change in x, the abscissa. Even for water differential calculus must be used, if precise results are desired.

As indicated in Figure B-2, the derivative at a specified value of x is the slope of the tangent to the curve at that point. If only the curve is known, and not the mathematical function, f(x), then the derivative must be evaluated from the slope. This would be the situation, for example, with the data in Table B-1, where only the curve of Q versus t°C is known from the data points. If the derivative is evaluated from the slope of the tangent to the curve, considerable error is generally encountered, since it is frequently difficult to accurately draw the tangent to the curve.

An alternate procedure is illustrated in Figure B-6, where the region of the specified value of x is expanded. The points A and B are the same as in Figure B-1, and point C,

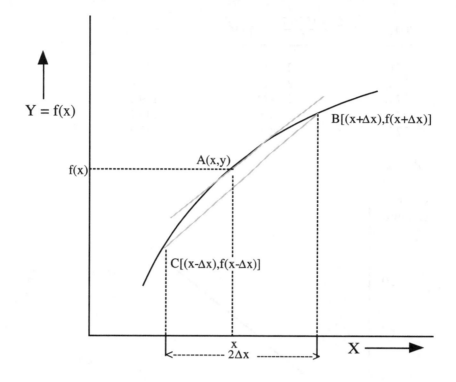

Figure B-6. Graphical Evaluation of the Derivative

is the point on the curve for the value of $(x - \Delta x)$. Now, if the chord is drawn through points B and C, the slope of this chord is seen to be nearly parallel to the tangent at A. If Δx is sufficiently small, the slope of the chord will equal the tangent at A, within the experimental error. The slope of the chord is given by

$$\frac{dy}{dx} = \frac{f(x + \Delta x) - f(x - \Delta x)}{2\Delta x} \tag{B-4}$$

Applying this equation to the data in Table B-1, the approximate derivative (or specific heat) at $2^\circ C$ is obtained, using $\Delta t = 1^\circ C$. The $Q_3{}^\circ C = f(x + \Delta x)$: $Q_1{}^\circ C = f(x - \Delta x)$; and

$$C_2{}^\circ C = \left(\frac{dQ}{dt}\right)2^\circ C = \frac{(3.0184 - 1.0069)\,cal/g}{2(1^\circ C)} = 1.0057\ cal/g\text{-}deg$$

Example B-1. As an example of a graphical evaluation of a derivative, let us consider the relationship between vapor pressure (p) and temperature (T). The specific relationship between p and T for hydrocyanic acid (HCN) in the range 265 K to 300 K is given by

$$p = 10^{\left[-\frac{1529.2}{T} + 7.7446\right]}$$

where p is expressed in units of torr. Suppose we wish to evaluate the derivative dp/dt at T = 273 K. We would first plot p versus T, and then find the tangent at 273 K. In order to graph the function in the region of 273 K, we need to evaluate some values of p, pertaining to specific values of T.

T(K)	$\dfrac{-1529.2}{T} + 7.7446$	p(torr)
271	2.1018	126.4
272	2.1226	132.6
273	2.1432	139.1
274	2.1636	145.7
275	2.1839	152.7

The slope is evaluated by reading off two sets of points on the tangent to the curve at 273 K.

$$\frac{dp}{dT} = slope = \frac{152.5 - 126.4}{275 - 271} = 6.5 \text{ torr/K}$$

Problem B-2. The equation of state for an ideal gas is

$$PV = nRT$$

where n = number of moles, R = .08205 liter-atm/K-mol.
(a) Plot V versus P for 40 grams of methane, from P = 1.0 atm to 10 atm, keeping the temperature constant at 200°C.
(b) Graphically evaluate the derivative dV/dP at P = 5 atm, when t is constant at 200°C.
Answer: (b) - 3.88 L/atm.

Quite often it is necessary to evaluate the derivative, when only values of the function itself can be obtained experimentally; in such cases it is necessary to carry out the differentiation graphically.

Formulas for Derivatives

If the function f(x) is known, then the derivative can be evaluated from formulas that have been developed using equation (B-1) as a starting point. These formulas can be found in any book on calculus, as well as in the Handbook of Chemistry and Physics. Of these numerous formulas, only six are of particular interest to us in this text. These are the constant, nth power, exponential, logarithm, and product, and sum formulas.

$$y = a = \text{constant} \qquad\qquad \frac{dy}{dx} = 0 \qquad\qquad (B\text{-}5)$$

$$y = x^n \qquad\qquad \frac{dy}{dx} = nx^{(n-1)} \qquad\qquad (B\text{-}6)$$

$$y = e^{ax} \qquad\qquad \frac{dy}{dx} = ae^{ax} \ (a = \text{constant}) \qquad\qquad (B\text{-}7)$$

$$y = \ln x \qquad\qquad \frac{dy}{dx} = \frac{1}{x} \qquad\qquad (B\text{-}8)$$

$$y = f(x)g(x) \qquad\qquad \frac{dy}{dx} = f(x)\frac{dg}{dx} + g(x)\frac{df}{dx} \qquad\qquad (B\text{-}9)$$

$$y = f(x) + g(x) \qquad\qquad \frac{dy}{dx} = \frac{df}{dx} + \frac{dg}{dx} \qquad\qquad (B\text{-}10)$$

Example B-2. A few examples of derivatives of functions are:

$$y = x^3 \qquad\qquad \frac{dy}{dx} = 3x^2$$

$$V = e^{bt} \ (b = \text{constant}) \qquad\qquad \frac{dV}{dt} = be^{bt}$$

$$z = (abc)e^c \ (a, b, c = \text{constants}) \qquad\qquad \frac{dz}{dx} = 0$$

$$y = a \ln x \ (a = \text{constant}) \qquad\qquad \frac{dy}{dx} = a\left(\frac{1}{x}\right) = \frac{a}{x}$$

$$y = x^3 \ln x \qquad\qquad \frac{dy}{dx} = x^3\left(\frac{1}{x}\right) + (\ln x)(3x^2)$$

Problem B-3. Evaluate the derivative of the following functions, where a = constant.

(a) $y = x^2 \quad \frac{dy}{dx} = ?$ 　　　　　　　　(b) $y = e^x \quad \frac{dy}{dx} = ?$

(c) $y = x^2 e^x \quad \frac{dy}{dx} = ?$ 　　　　　　(d) $z = ae^x \quad \frac{dz}{dx} = ?$

(e) $z = e^{ax} \quad \frac{dz}{dx} = ?$ 　　　　　　　(f) $z = ae^{ax} \quad \frac{dz}{dx} = ?$

(g) $y = x^{-4} \quad \frac{dy}{dx} = ?$ 　　　　　　　(h) $y = x^{-6}\ln x \quad \frac{dy}{dx} = ?$

(i) $\ln y = x \quad \frac{dy}{dx} = ?$ 　　　　　　　(j) $e^y = e^{ax} \quad \frac{dy}{dx} = ?$

(k) $y = ae^a \quad \frac{dy}{dx} = ?$ 　　　　　　　(l) $z = (a)^a \quad \frac{dz}{dx} = ?$

Answers: (a) $2x$ (b) e^x (c) $2xe^x + x^2e^x$ (d) ae^x (e) ae^{ax} (f) a^2e^{ax} (g) $-4x^{-5}$ (h) $-6x^{-7}\ln x + x^{-7}$ (i) e^x (j) a (k) 0 (l) 0.

The previous differentiation formulas, (B-5) through (B-10), can be applied directly to functions of x which are exactly of that form. However, if the function is more complicated, we must first introduce another concept from differential calculus with which we can apply formulas (B-5) through (B-10). If y is a function of v, and v, in turn, is a function of x,

$$y = g(v) \qquad \qquad \text{(B-11)}$$
$$v = h(x) \qquad \qquad \text{(B-12)}$$

then, from calculus, the derivative of y with respect to x is

$$\frac{dy}{dx} = \frac{dy}{dv}\frac{dv}{dx} \qquad \qquad \text{(B-13)}$$

The application of this differentiation formula to more complicated mathematical functions is quite straightforward, as will be illustrated in several examples. One must first break up the single function of y and x into the form of equations (B-11) and (B-12). The previous differentiation formulas can then be applied to the expressions of the form (B-11) and (B-12), and substituted into (B-13).

The previous differential formulas B-6 to B-8 would be

$$y = v^n \qquad\qquad \frac{dy}{dx} = nv^{n-1}\frac{dv}{dx} \qquad \qquad \text{(B-14)}$$

$$y = e^{av} \qquad\qquad \frac{dy}{dx} = ae^{av}\frac{dv}{dx} \qquad \qquad \text{(B-15)}$$

$$y = \ln v \qquad\qquad \frac{dy}{dx} = \frac{1}{v}\frac{dv}{dx} \qquad \qquad \text{(B-16)}$$

Example B-3. Suppose we have the function

$$y = (\ln x)^3$$

We note that this function is a log function raised to a power. We can differentiate both of these functions by equations (B-6) and (B-8); therefore, the function can be broken up accordingly:

$$y = v^3$$
$$v = \ln x$$

Using B-14,

$$\frac{dy}{dx} = 3v^2\frac{dv}{dx} = 3v^2\frac{1}{x} = 3(\ln x)^2\frac{1}{x}$$

After practice the student should be able to perform these operations easily, so that it will not be necessary to formally write down the function of v.

Some other examples will be given where the function v will not be formally defined, but will be obvious from the dy/dv.

$$y = e^{x^2} = e^v$$
$$v = x^2$$

$$\frac{dy}{dx} = e^v \frac{dv}{dx} = (e^v)(2x) = (e^{x^2})(2x)$$

$$y = e^{\ln x} = e^v$$
$$v = \ln x$$

$$\frac{dy}{dx} = e^v \frac{dv}{dx} = (e^v)\frac{1}{x} = (e^{\ln x})\left(\frac{1}{x}\right)$$

$$y = \ln x^2 = \ln v$$
$$v = x^2$$

$$\frac{dy}{dx} = \frac{1}{v}\frac{dv}{dx} = \left(\frac{1}{v}\right)(2x) = \left(\frac{1}{x^2}\right)(2x)$$

Problem B-4. Evaluate the derivatives of the following functions.

(a) $P = ae^{-b/T}$ (a, b = constants) $\frac{dP}{dT} = ?$

(b) $\psi = ae^{-(x-b)^2}$ (a, b = constants) $\frac{d\psi}{dx} = ?$

Answers: (a) $\frac{ab}{T^2} e^{-b/T}$ (b) $-2a(x-b)e^{-(x-b)^2}$

Differentials

Another calculus expression that we find useful in physical chemistry is the differential. The differential of y is defined as

$$dy = \frac{dy}{dx}\,dx \tag{B-17}$$

The quantity dx in (B-17) shows an infinitesimal change in x, which will reflect itself in dy, an infinitesimal change in y. The term dy/dx is the derivative which we have discussed previously. The term Δx, a small, but finite change in x, can be substituted for the differential quantity, dx, in (B-17), resulting in an approximate value of Δy (that is, $\Delta y \approx (dy/dx)\,\Delta x$).

Example B-4. Referring to the data in Table B-1, the change in heat, dQ, which corresponds to an increase in temperature from 4°C to 7°C, is given approximately by

$$\Delta Q \cong \frac{dQ}{dt}\Delta t = C\Delta t$$

where C has been defined as the heat capacity.

$$\Delta Q \cong \left(\frac{dQ}{dt}\right)_{4°C}(7°C - 4°C)$$

$$\Delta Q \cong (1.0043\,cal/g-deg)(3\,deg) = 3.0129\,cal/g$$

In this case, Δ is used to indicate some small but finite change in the variable, although it generally can be used to represent any finite change, not necessarily small.

Problem B-5. The volume of a sphere is given by

$$V = \frac{4}{3}\pi r^3$$

The differential of V is an expression of the change in volume, dV, corresponding to a differential change in radius, dr. The area of a sphere is $4\pi r^2$. What relationship does this have to the differential expression dV?

Answer: $4\pi r^2$; area of a sphere times dr gives the differential of V, dV.

Partial Derivatives

In our previous discussion of differential calculus, all of the expressions contained only two variables, with one exception. In Problem B-3, there were three variables—namely P, V, and T—since the volume of a gas is dependent on both the temperature and the pressure. However, one of the variables, T, was held constant, so the equation was reduced to only two variables. This situation occurs quite frequently in physical chemistry, where we have more than two variables. A derivative characterizes the variation of one variable with respect to only one variable, hence all others must be considered constant. For this reason, we define a **partial derivative**, which infers a partial rate of change, that is, a rate of change of the function with respect to only one independent variable, the others being held constant. We designate the constancy of the other variables by a subscript of that variable. For example, in the case of P, V, T variables,

$$\frac{dV}{dP} \text{ with T} = \text{constant} \qquad \text{(B-18)}$$

would be represented by the partial derivative designation,

$$\left(\frac{\partial V}{\partial P}\right)_T \qquad \text{(B-19)}$$

The partial derivative in (B-19) has the same meaning that was represented in another way in (B-18). The procedure for determining the partial derivatives of functions is identical to ordinary differentiation, with the single exception that the other variables must be held constant.

Example B-5.

(1) Evaluate $\left(\dfrac{\partial z}{\partial x}\right)_y$ and $\left(\dfrac{\partial z}{\partial y}\right)_x$ for the following functions.

$$z = x^2 y^3$$

$$\left(\frac{\partial z}{\partial x}\right)_y = y^3(2x)$$

$$\left(\frac{\partial z}{\partial y}\right)_x = x^2(3y^2)$$

(2) Evaluate $\left(\dfrac{\partial P}{\partial V}\right)_T$ and $\left(\dfrac{\partial P}{\partial T}\right)_V$ for the following function.

$$P = \frac{nRT}{V} = nRTV^{-1}$$

$$\left(\frac{\partial P}{\partial V}\right)_T = nRT(-1)V^{-2} = -\frac{nRT}{V^2}$$

$$\left(\frac{\partial P}{\partial T}\right)_V = \frac{nR}{V}$$

Problem B-6. From the Ideal Gas Equation of State given in Problem B-2, obtain the analytical expression for $(\partial V/\partial P)_T$, and evaluate this derivative for 40 g CH_4 at 200°C and P = 5 atm. Compare with the graphical result obtained in Problem B-3. Answer: $-\dfrac{nRT}{P^2}$ = - 3.88 L/atm.

Problem B-7. The volume of a cylinder is given by

$$V = \pi r^2 \ell$$

where r = radius and ℓ = length of the cylinder.

   (a) Obtain expressions for $(\partial V/\partial r)_\ell$ and $(\partial V/\partial \ell)_r$. How are these related to the surface of the cylinder?
   (b) If ℓ = 5 cm and r = 2 cm, calculate the change in volume if (i) ℓ were increased by .01 cm, r = constant; (ii) r were increased by .01 cm, ℓ = constant.

Answers: (a) $\left(\dfrac{\partial V}{\partial r}\right)_\ell = 2\pi r \ell$, $\left(\dfrac{\partial V}{\partial \ell}\right)_r = \pi r^2$ (b) (i) 0.1256 cm^3 (ii) 0.628 cm^3.

Maxima, Minima, and Points of Inflection

If a function, f(x), displays a maximum, a minimum, or a point of inflection, the first derivative of the function, f'(x), is zero. In Figure B-7 we illustrate f(x) curves which display a maximum, a minimum, and a point of inflection.

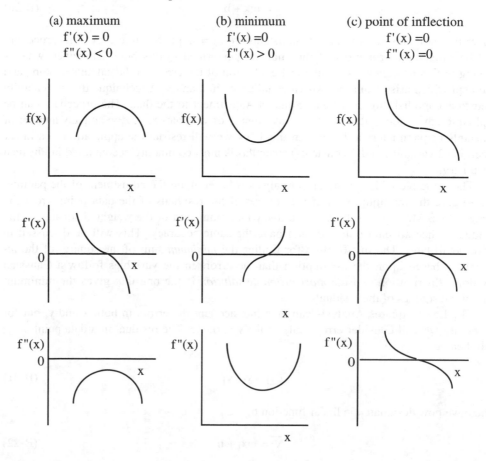

(a) maximum	(b) minimum	(c) point of inflection
f'(x) = 0	f'(x) = 0	f'(x) = 0
f''(x) < 0	f''(x) > 0	f''(x) = 0

Figure B-7. (a) f(x) Has a Maximum, (b) f(x) Has a Minimum (c) f(x) Has a Point of Inflection

As shown in Figure B-7(a), the derivative (slope) of f(x) goes from positive to negative as we go through the maximum; however, the second derivative [slope of f'(x)] is <u>negative</u> throughout. Conversely, as shown in Figure B-7(b), the derivative (slope) of f(x) goes from negative to positive as we go through the minimum. The second derivative [slope of f'(x)] is <u>positive</u> throughout. Finally, as shown in Figure B-7(c), the slope of f(x) is negative before the point of inflection, zero at the point of inflection, and again negative at large x values. Since the first derivative, f'(x) in Figure B-7(c) shows a maximum, the derivative of the first derivative, the second derivative f''(x) goes through zero at the point of inflection.

Problem B-8. Given the functions

(a) $y = f(x) = 10 - 10x + x^2$
(b) $y = f(x) = 10 + 10x - x^2$

determine if these functions have a maximum, minimum, or point of inflection. If there is one of these, evaluate the value of x at which it occurs. Answer: (a) minimum at x = 5 (b) maximum at x = 5.

Least Squares Analysis

In Appendix A we described a graphical procedure for the evaluation of the slope and intercept of a linear function

$$y = mx + b \tag{B-20}$$

where the variables are x,y and the constants are m = slope, b = intercept. This procedure will be satisfactory for most of the analyses performed in this book. However, what is lacking in this procedure is a statistical evaluation of the errors in the calculated slope and intercept which arise from the experimental errors in x and y. A technique that is generally used to accomplish this is a Least Squares Adjustment to the data. The procedure can be applied to any function thay contains any number of variables x, y, z,⋯ and any number of constants or parameters a, b, c⋯. In this book we will restrict the application to a linear function (2 variables and 2 constants) since this is most commonly encountered in Physical Chemistry.

The principle of Least Squares Analysis is based upon the adjustment of the parameters to give the <u>minimum</u> sum of the squares of the residuals of the data points from the straight line. More precisely we should say minimum sum of the <u>weighted</u> squares of the residuals since all data points do not have the same accuracy. This will be discussed in more detail later. The justification for finding the *minimum* sum of the squares of the residuals is based upon the assumption that the errors in the variables follow a Gaussian (random) distribution and the *most probable solution* is the one that gives the minimum sum of the squares of the residuals.

The Least Squares Analysis can take into account the errors in both x and y, but for simplicity we will consider errors only in the y variable. The residual in a data point x_i, y_i will then be

$$\delta y_i = y_i - \overline{y_i} \tag{B-21}$$

where we now designate the linear function by

$$\overline{y_i} = mx_i + b \tag{B-22}$$

If the data points have different precision, and this has been established experimentally, then we must consider this in the adjustment of the straight line. In other words we wish to make the straight line closer (smaller residuals) to the most precise data points and vice versa for the least precise data points. We accomplish this by using a <u>weighting factor</u>, W_i, for each data point

$$W_i = \frac{\sigma_o^2}{\sigma_i^2} \tag{B-22}$$

where σ_i^2 is the variance (square of the standard deviation) for the i-th data point. σ_o^2 is simply a scaling factor to make the W_i values a convenient range. It is called the variance of unit weight since the weight is unity when the variance, σ_i^2 becomes equal to σ_o^2. σ_o^2 is essentially factored out and can be made unity if so desired. However, if the y_i measurements are extremely large or small, the variance $\sigma_{y_i}^2$ will also be extremely large or small and it may be necessary to select σ_o^2 so that the weighting factors range about 1.00. If we use weighting factors, then we wish to minimize the sum of the squares of the <u>weighted residuals</u>

$$S = \Sigma W_i \delta_{y_i}^2 = \Sigma W_i (y_i - mx_i - b)^2 \tag{B-23}$$

The minimization procedure is then carried out by taking the derivatives of $S = \Sigma W_i \delta_{y_i}^2$ with respect to the parameters (constants) m and b and setting these derivatives to zero to obtain the minimum value. As you will recall, when the derivative is zero the function is going through a maximum, minimum, or point of inflection. In this case it will be a minimum which can be shown by examining the second derivative which is positive. Taking the first derivatives of (B-23) and setting them equal to zero to find the minimum, we obtain

$$\frac{\partial S}{\partial m} = 0 = \Sigma W_i(2)(y_i - mx_i - b)(-x_i) \tag{B-24}$$

$$\frac{\partial S}{\partial b} = 0 = \Sigma W_i(2)(y_i - mx_i - b)(-1) \tag{B-25}$$

We now need to solve these two equations (B-24) and (B-25) simultaneously to find the values of m and b. The 2's can be factored out and the summation of the separate terms can be performed giving

$$\begin{cases} -\Sigma W_i x_i y_i + m\Sigma W_i x_i^2 + b\Sigma W_i x_i = 0 \\ -\Sigma W_i y_i + m\Sigma W_i x_i + b\Sigma W_i = 0 \end{cases} \tag{B-26}$$

or

$$\begin{cases} m\Sigma W_i x_i^2 + b\Sigma W_i x_i = \Sigma W_i x_i y_i \\ m\Sigma W_i x_i + b\Sigma W_i = \Sigma W_i y_i \end{cases} \tag{B-27}$$

Any of several techniques can be used to solve these two equations for m and b. However, for the statistical analysis it is convient to solve the equations using matrices. If you are not familiar with matrices, you can go directly to equations (B-29) through (B-32) for the final results. In matrix form equations (B-27) are

$$\begin{bmatrix} \Sigma W_i x_i^2 & \Sigma W_i x_i \\ \Sigma W_i x_i & \Sigma W_i \end{bmatrix} \begin{bmatrix} m \\ b \end{bmatrix} = \begin{bmatrix} \Sigma W_i x_i y_i \\ \Sigma W_i y_i \end{bmatrix} \tag{B-28}$$

If we multiply each side of (B-28) by the inverse of the 2x2 matrix, we obtain

$$\begin{bmatrix} m \\ b \end{bmatrix} = \begin{bmatrix} \Sigma \dfrac{W_i}{D} & -\Sigma \dfrac{W_i x_i}{D} \\ -\Sigma \dfrac{W_i x_i}{D} & \Sigma \dfrac{W_i x_i^2}{D} \end{bmatrix} \begin{bmatrix} \Sigma W_i x_i y_i \\ \Sigma W_i y_i \end{bmatrix}$$

where D is the value of the determinant $|D|$.

$$|D| = \begin{vmatrix} \Sigma W_i x_i^2 & \Sigma W_i x_i \\ \Sigma W_i x_i & \Sigma W_i \end{vmatrix} = (\Sigma W_i)(\Sigma W_i x_i^2) - (\Sigma W_i x_i)^2$$

The expressions for m and b can be obtained by multiplying out the matrices

$$m = \frac{\Sigma W_i \Sigma W_i x_i y_i - \Sigma W_i x_i \Sigma W_i y_i}{D} \tag{B-29}$$

$$b = \frac{-\Sigma W_i x_i \Sigma W_i x_i y_i + \Sigma W_i x_i^2 \Sigma W_i y_i}{D} \tag{B-30}$$

The standard deviations of m and b are given by

$$\sigma_m = \left(\sigma_0^2 \frac{\Sigma W_i}{D} \right)^{1/2} \tag{B-31}$$

$$\sigma_b = \left(\sigma_0^2 \frac{\Sigma W_i x_i^2}{D} \right)^{1/2} \tag{B-32}$$

If the variances are known for each of the data, equations (B-29) and (B-30) can be used to calculate m and b and the standard deviations in m and b can be evaluated from equations (B-31) and (B-32). In order to evaluate the goodness of fit of this data to the straight line, we calculate the "external variance", σ_{ext}^2, from S (sum of weighted squares of residuals) and the degrees of freedom

$$\sigma_{ext}^2 = \frac{S}{n-2}$$

where n = number of data points. We then compare σ_{ext}^2 with $\sigma_{int}^2 = \sigma_0^2$ = variance of unit weight. σ_{ext}^2 can be formally compared to σ_{int}^2 using the F-test. Generally if σ_{ext}^2 is on the order of σ_{int}^2, we conclude that the data fit the linear function within the precision of the data.

If on the other hand, the variances of the data points are not known, we commonly assume the data points have the same precision (unit weights) and estimate the error in the m and b by

$$\sigma_m = \left(\sigma_{ext}^2 \frac{\Sigma W_i}{D} \right)^{1/2}$$

$$\sigma_b = \left(\sigma_{ext}^2 \frac{\Sigma W_i x_i^2}{D} \right)^{1/2}$$

The previous procedure is preferred, but in the absence of the variances in the individual data points, the latter is a practical, alternative procedure.

One final word of caution should be expressed regarding the Least Squares Analysis of a linear function. As we discussed in Appendix A, equations can be put into a linear form where x and y are some function of the variables. For example, the Arrhenius Equation

$$k = A\, e^{-E^*/RT}$$

was put into a linear form by taking the logarthims of both sides of the equation

$$\ln k = \ln A - \frac{E^*}{R}\left(\frac{1}{T}\right)$$

where $y = \ln k$ and $x = 1/T$. The variance in lnk is related to the variance in k by

$$\sigma_{\ln k}^2 = \left(\frac{d\ln k}{dk}\right)^2 \sigma_k^2$$

$$\sigma_{\ln k}^2 = \left(\frac{1}{k}\right)^2 \sigma_k^2$$

$$\sigma_{\ln k} = \left(\frac{\sigma_k}{k}\right)$$

In other words the standard deviation in lnk is equal to the <u>relative</u> standard deviation in k; i.e. σ_k/k. If you have reason to believe that the error in k, σ_k^2, is constant, then you should use a weighting factor to give the proper error in lnk:

$$W_i = \frac{\sigma_o^2}{\sigma_{\ln k}^2} = \frac{\sigma_o^2}{\left(\dfrac{\sigma_k}{k_i}\right)^2}$$

If we set $\sigma_o^2 = \sigma_k^2$, we simplify the weighting factors to be

$$W_i = k_i^2$$

However, generally the values of k_i vary greatly with temperature and the experimental conditions are such that the <u>relative error</u> in k remains approximately constant and this conveniently reflects in a <u>constant</u> error in lnk and constant weighting factors for lnk.

We will consider the application of least squares adjustment to a linear function to familiarize you with the use of equations (B-29) through (B-32) and the evaluation of weighting factors.

Example B-6. Use the data of absorbance, A, versus time, t, on page 298 in a least squares adjustment to equation (A-7). Assume the variance in A, σ_A^2, is constant. Since the A values are given to the third place past the decimal point, we will assume $\sigma_A = \pm\, 0.001$.

Solution: Equation (A-7) is in a linear form

$$\ln(A^\infty - A_i) = -kt_i + \ln A^\infty \tag{A-7}$$
$$y_i = mx_i + b$$

The propagation of the variance of A (a constant) upon y_i is obtained from

$$\sigma_{y_i}^2 = \left(\frac{dy_i}{dA_i}\right)^2 \sigma_A^2$$

Since

$$y_i = \ln(A^\infty - A_i)$$

$$\frac{dy_i}{dA_i} = \frac{-1}{(A^\infty - A_i)}$$

and

$$\sigma_{y_i}^2 = \left[\frac{-1}{(A^\infty - A_i)}\right]^2 \sigma_A^2$$

The weighting factors are then

$$W_i = \frac{\sigma_o^2}{\sigma_{y_i}^2} = \frac{\sigma_o^2}{\frac{1}{(A^\infty - A_i)^2}\sigma_A^2}$$

Since σ_A^2 is a constant, it is convenient to set

$$\sigma_o^2 = \sigma_A^2 = (0.001)^2$$

and

$$W_i = (A^\infty - A_i)^2$$

We now can evaluate the summations in equations (B-29) through (B-32) where

$$x_i = t_i, \quad y_i = \ln(A^\infty - A_i)$$

$x_i =$ $\qquad$ $W_i =$ $\qquad$ $y_i =$

t_i	A_i	$(A^\infty-A_i)^2$	$\ln(A^\infty-A_i)$	$W_i x_i$	$W_i x_i^2$	$W_i x_i y_i$	$W_i y_i$
1.49	0.015	0.29594	- 0.60881	0.44095	0.65702	- 0.26846	- 0.18017
4.11	0.034	0.27563	- 0.64436	1.13284	4.65597	- 0.72996	- 0.17760
11.45	0.089	0.22090	- 0.75502	2.5293	28.96054	- 1.90968	- 0.16678
14.27	0.106	0.20521	- 0.79186	2.92835	41.78751	- 2.31884	- 0.16250
21.19	0.15	0.16728	- 0.89404	3.90647	75.1113	- 3.16907	- 0.14956
28.23	0.187	0.13838	- 0.98886	3.90647	110.2796	- 3.86295	- 0.13684
39.05	0.239	0.10240	- 1.13943	3.99872	156.15	- 4.55626	- 0.11668
69.25	0.351	0.04326	- 1.57022	2.99576	207.456	- 4.70399	- 0.06793
81.57	0.384	0.03063	- 1.7297	2.49849	203.8018	- 4.35479	- 0.05339
87.60	0.399	0.0256	- 1.83258	2.24256	196.4483	- 4.10967	- 0.0469
104.86	0.431	0.01638	- 2.05573	1.71761	180.1082	- 3.53096	- 0.03367
125.47	0.462	0.00941	- 2.33304	1.18067	148.139	- 2.75456	- 0.02195
137.54	0.478	0.00656	- 2.51331	0.90226	124.0972	- 2.26767	- 0.01649
147.86	0.489	0.0049	- 2.65926	0.72451	107.1266	- 1.92667	- 0.01303
182.24	0.519	0.0016	- 3.21888	0.29178	53.2024	- 0.93914	- 0.00515
207.24	0.529	0.0009	- 3.50656	0.18652	38.6536	- 0.65403	- 0.00316
∞	0.559						
Sums:		1.54328		31.2214	1676.635	- 42.0567	- 1.3518

With the summations shown under the above columns of data, the slope, m, can be calculated from equation (B-29).

$$|D| = (\Sigma W_i)(\Sigma W_i x_i^2) - (\Sigma W_i x_i)^2 = (1.54328)(1676.635) - (31.2214)^2 = 1612.74$$

$$m = \frac{\Sigma W_i \Sigma W_i x_i y_i - \Sigma W_i x_i \Sigma W_i y_i}{D} = \frac{(1.54328)(-42.0567) - (31.2214)(-1.3518)}{1612.74}$$

$$m = -0.01407 \text{ min}^{-1}$$

The error (standard deviation) in the slope can be calculated from equation (B-31)

$$\sigma_m = \left(\sigma_o^2 \frac{\Sigma W_i}{D} \right)^{1/2} = \left((0.001)^2 \frac{1.54328}{1612.74} \right)^{1/2} = 0.00003 \text{ min}^{-1}$$

The rate constant is then

$$k = -m \pm \sigma_m = 0.01407 \pm 0.00003 \text{ min}^{-1}$$

Problem B-9. Calculate the intercept and the error (standard deviation) in the intercept for the analysis performed in Example B-6 using equations (B-30) and (B-32). Answer: $b = \ln A^\infty = -0.59117 \pm 0.0102$; $A^\infty = 0.5537$.

Example B-6 is an excellent analysis for showing the importance of weighting factors. Note the large change in the weighting factor from 0.29594 for the first data point to 0.00090 for the last data point. This reflects the greater precision in the $y_i = \ln(A^\infty - A_i)$ values for low absorbance values. Note the influence that the weighting factors have on the $W_i x_i$, $W_i x_i^2$, $W_i x_i y_i$, $W_i y_i$ columns of data. Generally the values of these quantities increase with increasing absorbance, go through a maximum and then decrease.

Problem B-10. In Example B-6 we assumed that $\sigma_A^2 = $ constant. However, very likely with most spectrophotometers the error in A increases as the absorbance increases. The absorbance, A, is actually determined by measuring the transmittance, T, for a sample and A is evaluated electronically or by software with the following equation

$$A = -\log T = \frac{-\ln T}{2.303}$$

From the very nature of the measurement, one would expect that the error in T (σ_T^2) is constant, independent of the value of T. Evaluate the propagation of error from T to A by

$$\sigma_A^2 = \left(\frac{dA}{dT} \right)^2 \sigma_T^2$$

Substitute this expression for σ_A^2 into the equation for $\sigma_{y_i}^2$ given in Example B-6 and derive the expression for the weighting factor. Assume the variance of unit weight = $\sigma_o^2 = \frac{\sigma_T^2}{(2.303)^2}$. Answer: $W_i = (A^\infty - A_i)^2 \, 10^{-2A}$.

Appendix C
INTEGRAL CALCULUS

The process of integration is the reverse of differentiation, that is, we start with the derivative and seek the original function. This may not, however, appear to be the case from the formal definition of an integral.

Definition of an Integral

The formal definition of an integral of y(x) over x, is given by

$$\int_a^b y \, dx = \lim_{\Delta x \to 0} \sum_{x=a}^b y_i \Delta x = \lim_{\Delta x \to 0} \sum_{x=a}^b f(x_i) \Delta x \qquad \text{(C-1)}$$

We can illustrate the process of integration graphically and, as we will see, it can be interpreted as an area under a curve. If we were to graph y versus x using the formula C-1 we would consider the summation in steps progressing to larger x values. Let us consider the sum in C-1 as a sum of small, but finite Δx values. We will consider taking the lim $\Delta x \longrightarrow 0$ later. The sum would then be

$$\sum_{x=a}^b y_i \Delta x = y_a \Delta x + y_{(a + \Delta x)} \Delta x + y_{(a + 2\Delta x)} \Delta x + \dots y_{(b - \Delta x)} \Delta x \qquad \text{(C-2)}$$

Each of the successive terms in the summation is the area of a rectangle since the base of each rectangular section is Δx and it is multiplied by the height (y_i). In the first term we have the height of the rectangle as y_a and base $\Delta x = (a + \Delta x) - (a)$. The second term has the height of the rectangle $y_{a + \Delta x}$ and base $\Delta x = (a + 2\Delta x) - (a + \Delta x)$, etc. The process occurs until we reach x = b.

The sum shown in equation C-2 can be interpreted in terms of a sum of the areas of rectangles as depicted in Figure C-1. If you wish you can also take the sum in equation C-2 as the <u>area</u> under the step-curve in Figure C-1.

Now if the size of Δx were decreased one can visualize that the rectangles would become more narrow and the size of the step would be smaller. Of course the number of rectangles required in going from x = a to x = b would be increased. In the limit as $\Delta x \to 0$ the stepsize would be infintesimally small and the step-curve would eventually become a continuous function of y versus x and the area under this smooth curve would be the integral from a to b as defined in equation C-1. Such a smooth curve is shown along side the step-function in Figure C-1. The area under this smooth curve is the integral of y dx from x = a to x = b

$$\int_a^b y \, dx = \lim_{\Delta x \to 0} \sum_{x=a}^b y_i \Delta x = \text{area under curve of y = f(x) versus x from x = a to x = b} \qquad \text{(C-3)}$$

The integral, as depicted in equation (C-1), has definite limits from x = a to x = b, thus the name <u>definite integral</u>. The integral can also be taken without specifying the limits and this is called an indefinite integral. With an indefinite integral it is necessary to add a constant of integration, C. The reason for which will be apparent after a subsequent discussion.

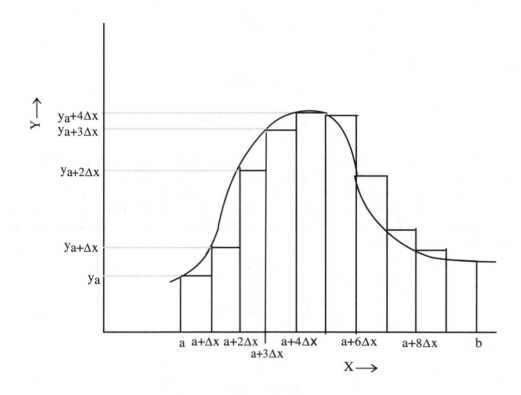

Figure C-1. Graph of y Versus x Using Incremental Δx's

Integration Formulas

As in the case of differentiation there are numerous formulas which have been derived for the integration of specific functions. Again, these can be found in any comprehensive text on calculus, or in Lange's and Chemical Rubber Handbooks. Of these formulas, we will have occasion to use only three.

$$\int x^n dx = \frac{x^{n+1}}{n+1} + C \quad (n \neq -1) \tag{C-4}$$

$$\int \frac{dx}{x} = \ln x + C \tag{C-5}$$

$$\int e^x dx = e^x + C \tag{C-6}$$

A unique application of the integral of a polynomial, equation (C-4), is commonly encountered. This is the case when n = 0 and

$$\int dx = \int x^0 dx = \frac{x^{0+1}}{0+1} + C = x + C$$

In other words,

$$\int dx = x + C$$

Example C-1.

$$\int T^2 dT = \frac{T^3}{3} + C$$

$$\int e^t dt = e^t + C$$

$$\int \frac{d(PV)}{PV} = \ln(PV) + C$$

The formulas (C-4), (C-5), and (C-6), are more general than they may appear, since x is any variable. In fact, the variable represented by x in these equations may be a function of another variable. These integral formulas can be applied in such cases, as long as dx represents the differential of the function, x. In order to clarify this point further, we can rewrite equations (C-4), (C-5), (C-6) in terms of another variable, v, which in turn is a function of x.

$$v = g(x)$$

$$\int v^n dv = \frac{v^{n+1}}{n+1} + C$$

$$\int \frac{dv}{v} = \ln v + C$$

$$\int e^v dv = e^v + C$$

These integral formulas are equivalent to (C-4), (C-5), (C-6); however, the use of these formulas may be simpler, since the functional dependence of v on x is shown explicitly.

Example C-2.

$$\int (x+b)^2 dx = \frac{(x+b)^3}{3} + C$$

where

$$v = (x+b)$$
$$dv = dx$$

$$\int e^{bx} b \, dx = e^{bx} + C$$

where

$$v = bx$$
$$dv = b \, dx$$

In some cases, the equation must be multiplied by a constant in order to make the differential, dv. For example,

$$\int e^{bx} dx$$

is similar to the example above, however, dx must be multiplied by b in order that bdx = dv. Therefore, we would multiply the differential by b and, in order to preserve the equality, multiply the integral by 1/b.

$$\frac{1}{b} \int e^{bx} b \, dx = \frac{1}{b} e^{bx} + C$$

where

$$v = bx$$
$$dv = b \, dx$$

Similarly,

$$\int e^{x^2} x \, dx = \frac{1}{2} \int e^{x^2} 2x \, dx = \frac{1}{2} e^{x^2} + C$$

where

$$v = x^2$$
$$dv = 2x \, dx$$

Problem C-1. Integrate the following expressions:

(a) $\int e^{2x} 2 dx$ (b) $\int e^{2x} dx$

(c) $\int x^5 dx$ (d) $\int T^{-2} dT$

(e) $\int\limits_{P_1}^{P_2} \frac{dP}{P}$ (f) $\int x \, dx$

(g) $\int nRT \frac{dV}{V}$ (h) $\int bT \, dT$
    (n, R, T = constants)   (b = constant)

Answers: (a) $e^{2x} + C$ (b) $\frac{e^{2x}}{2} + C$ (c) $\frac{x^6}{6} + C$ (d) $-\frac{1}{T} + C$ (e) $\ln \frac{P_2}{P_1}$ (f) $\frac{x^2}{2} + C$

(g) $n RT \ln V + C$ (h) $\frac{bT^2}{2} + C$.

Integration of Differential Equations

In thermodynamics and kinetics the theory defines a derivative or partial differential derivative. If we wish to obtain the equation relating the variables, we must integrate the differential expression involving this derivative. Recall in Appendix B the differential expression, Equation (B-17)

$$dy = \frac{dy}{dx} dx \qquad (B-17)$$

After the derivative is substituted into Equation (B-17) we obtain a *differential equation* which relates the differentials dx and dy. The solution to differential equations is a subject in itself and involves various techniques designed to solve various forms of differential equations. The simplest technique in the solution of differential equations is one called *separation of variables* whereby one algebraically transforms the y variables to the side of the equation containing dy, and the x variables to the other side containing dx. The solution to the differential equation then involves integration of both sides of the equation to obtain the relationship between y and x. If an indefinite integration is performed, only one constant of integration needs to be specified. If a definite integral is performed, the initial value of x must correspond to the inital value of y and the final value of x must correspond with the final value of y.

Example C-3. Suppose we knew that the rate of change of the molar concentration of A, [A], with time, t, was given by

$$-\frac{d[A]}{dt} = k[A]^2 \qquad (C-7)$$

In order to find the relationship between [A] and t, we first use the differential expression which is similiar to equation (B-17)

$$d[A] = \frac{d[A]}{dt} \, dt \tag{C-8}$$

we then substitute equation (C-7) into equation (C-8)

$$d[A] = -k[A]^2 \, dt \tag{C-9}$$

We then separate the variables by dividing by $[A]^2$

$$\frac{d[A]}{[A]^2} = -k \, dt \tag{C-10}$$

Then both sides of the differential equation are integrated

$$\int \frac{d[A]}{[A]^2} = \int [A]^{-2} \, d[A] = \int -k \, dt$$

which gives

$$\frac{[A]^{-1}}{-1} = -kt + C$$

$$\frac{-1}{[A]} = -kt + C \tag{C-11}$$

As an alternative, we could integrate the expression between definite limits (definite integral), say $(t_1, [A]_1)$ to $(t_2, [A]_2)$. Then the integral would be

$$\int_{[A]_1}^{[A]_2} \frac{d[A]}{[A]^2} = \int_{t_1}^{t_2} -k \, dt \tag{C-12}$$

which gives

$$-\frac{1}{[A]_2} + \frac{1}{[A]_1} = -k \, (t_2 - t_1) \tag{C-12}$$

Problem C-2. Obtain the relationship between P and T, if

$$\frac{d(\ln P)}{d(1/T)} = \text{constant} = a$$

What kind of graph would you make involving functions of P and T which could be used to easily evaluate the constant, a? This type of graph will be used extensively in the chapters on thermodynamics and kinetics to evaluate important parameters.

Answer: Graph ln P versus 1/T, slope should equal "a."

From the basic definitions of a derivative, equation (B-1), and an integral, equation (C-1), the operations of differentiation and integration do not seem to be inverse operations; i.e., if we differentiate f(x) to give $\frac{df}{dx}$, integration of the derivative by substitution into equation (B-17) gives the function f(x) back again but with a constant of integration. Application of the differentiation and integration formulas will illustrate this inverse relationship.

Problem C-3. In Problem B-2, we differentiated some equations. Integrate some of these results (a, b, d, e, f, g, i, j, k, and l) to regenerate the original functions. This should convince the reader that integration is the reverse of differentiation. Answer: You should obtain the original functions plus an arbitrary constant of integration.

Graphical/Numerical Integration

As in the case of differentiation, we frequently need a numerical evaluation of an integral. This would arise, for example, when the y's in equation (C-1) were known only as a series of numerical values. For this purpose, the summation in equation (C-1) is depicted as the area under the curve, as shown in Figure C-1. There are several ways to obtain the area under the curve in order to numerically evaluate the integral. One of these is called Simpson's Rule which assumes a quadratic function between three successive data points. However, the formula for Simpson's Rule can only be used when the data points are spaced in equal intervals of x.

A simple but frequently quite adequate approximation is to draw straight lines between successive data points, thus making each segment into a trapezoid. The area then is the sum of these trapezoidal segments. This is shown graphically in Figure C-2. For simplicity of nomenclature, let $y_i = f(x_i)$. The area of each trapezoidal area is the average of the two unequal, parallel sides,

$$\frac{(y_i + y_{i+1})}{2}$$

multiplied by the base, Δx_i. Letting ΔA_i represent the elemental trapezoidal areas,

$$\Delta A_0 = \frac{(y_a + y_1)}{2} \, \Delta x_0 \qquad\qquad (C\text{-}13)$$

$$\Delta A_1 = \frac{(y_1 + y_2)}{2} \, \Delta x_1$$

The first trapezoidal area, ΔA_0, is represented in Figure C-2.

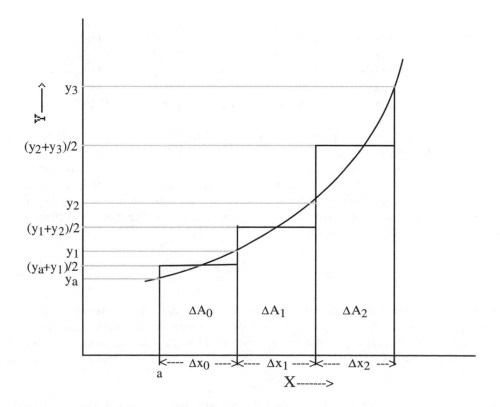

Figure C-2. Trapezoidal Rule for Numerical Integration

The average value,

$$\frac{y_a + y_1}{2}$$

drawn from a to $(a + \Delta x_0)$, generates two triangles which are equal and hence, have equal areas. The trapezoidal area, A_0, is then also represented by the area of the rectangle of height,

$$\frac{y_a + y_1}{2}$$

and base, Δx_0. Use of the trapezoidal rule for integration can be shown in tabular form, as given in Table C-1. Let us consider the data of Table B-1, except that we will start with the heat capacity data at various temperatures and will generate the heat absorbed, Q, as a function of temperature. For purpose of illustration only, the data at certain temperatures will be used, so that the Δx, or in this case, Δt, interval will be variable, the most general case. In Table C-1, columns three and four are generated in order to calculate the elemental areas ΔQ_i found in column five. Finally, Q_i is the summation of the ΔQ_i, up to the ith point, and of course, represents the heat absorbed by H_2O in order to raise the temperature from 0 °C to the specified temperature, t_i. If Q were an absolute quantity rather than a change, then a constant of integration would have to be added to the values in column six.

Table C-1. Heat Capacity of Water as a Function of Temperature

t°C	C_i(cal/g-K)	$\dfrac{C_i + C_{i-1}}{2}$	Δt°C	ΔQ_i	Q_i
0.0	1.0074				
1.0	1.0065	1.00695	1.0	1.00695	1.00695
4.0	1.0043	1.00540	3.0	3.01620	4.02315
6.0	1.0031	1.00370	2.0	2.00740	6.03055
7.0	1.0026	1.00285	1.0	1.00285	7.03340
9.0	1.0017	1.00215	2.0	2.00430	9.03770
10.0	1.0013	1.00150	1.0	1.00150	10.03920

A simple numerical integration procedure involves drawing the best smooth curve through the data points and directly measuring the area under this curve. This can be accomplished by simply counting the squares under the curve, when the curve is plotted on finely ruled graph paper. The area can also be determined by simply cutting out the area under the curve and weighing it. The weight/unit area is then determined by weighing out a piece of the graph paper of known area, for example, a rectangular piece, and comparing the two. The integral is then given by

$$\text{integral} = \text{area} = \frac{(\text{weight of integral})}{(\text{weight/unit area})} \qquad (C\text{-}14)$$

The units of the integral will then be the units of the unit area. This is seemingly a crude technique; however, it is surprisingly precise if homogeneous graph paper is used. Mechanical devices called planimeters are also available. In addition, more sophisticated numerical techniques can be used, and the reader is referred to a treatise on Numerical Analysis.

Problem C-4. The following heat capacities for aluminum were obtained at constant pressure.

t^oC	C(cal/g-K)	t^oC	C(cal/g-K)
-250	0.0039	-50	0.1914
-233	0.0165	0	0.2079
-200	0.076	20	0.214
-150	0.1367	100	0.225
-100	0.1676		

(a) Evaluate the heat (Q) required to raise the temperature of 10 grams of aluminum from -250°C to 100°C.

(b) Plot Q absorbed by the aluminum versus temperature. Analogous to situations mentioned in Problem B-1, numerical integration must often be carried out to evaluate thermodynamic parameters.

Answer: (a) ≈555.3 cal (b) Numerically integrate from - 250°C to various temperatures to find Q as a function of T.

Before leaving the subject of integration it should be pointed out that quite frequently, we want the integral equation of a partial derivative, for example, $(\partial z/\partial x)y$. In other words, we want to integrate the differentials associated with this partial derivative. This can be accomplished using the techniques described previously, if the variable to be held constant is indeed constant. For example, with the derivative $(\partial z/\partial x)y$, if we state that y is to be held constant, z will depend only on x.

$$dz = \left(\frac{\partial z}{\partial x}\right)_y dx = \frac{dz}{dx} \, dx \tag{C-15}$$

The integration is then performed as before, except that the previous constant of integration becomes some general function of y, C(y); that is,

$$z = z(x) + C(y) \tag{C-16}$$

where z(x) is some function of x. The reason for this is that y is held constant in the process, and therefore, C(y) is also a constant. The justification of C(y) is readily seen by taking the partial of (C-16) with respect to x, at constant y. Equation (C-15) is obtained again, thus showing that (C-16) is a solution of (C-15).

Example C-4. In Example B-5, we obtained an expression for a partial derivative from the Ideal Gas Equation of State, namely,

$$\left(\frac{\partial P}{\partial T}\right)_V = \frac{nR}{V}$$

If we state that V = constant,

$$dP = \frac{dP}{dT} \, dT$$

$$\int dP = \int \frac{nR}{V} dT = \frac{nR}{V} \int dT$$

$$P = \frac{nR}{V} \, T + C(V)$$

This gives us the Ideal Gas Equation of State back again, except for the added term C(V). Since P = 0 at T = 0 for an ideal gas, we can evaluate C(V) from the equation

$$0 = \left(\frac{nR}{V}\right) 0 + C(V)$$

$$C(V) = 0$$

In this specific equation, C(V) is zero, and the equation becomes our Ideal Gas Equation State.

$$P = \left(\frac{nR}{V}\right) T$$

Problem C-5. The following equation can be derived from thermodynamics:

$$\left(\frac{\partial \ln K}{\partial T}\right)_P = \frac{\Delta H^o}{RT^2}$$

where K is the equilibrium constant. Assume ΔH^o is constant and derive the expression between K and T, when the pressure P is held constant. R is the gas constant, as previously defined.

Answer: $\ln K = -\dfrac{\Delta H^o}{RT} + C$.

Problem C-6. The chemical reaction,

$$A \rightarrow Products$$

was studied by measuring the concentration of A as a function of time. The following results were obtained.

t(min)	[A](moles/L)
0.0	1.00
1.0	0.667
2.0	0.500
3.0	0.400
4.0	0.333
5.0	0.286
6.0	0.250
7.0	0.222
8.0	0.200
9.0	0.182
10.0	0.166

(a) Plot [A] versus t, and calculate the slope, d[A]/dt, graphically, at t = 1, 3, 5, 7, 9 min.
(b) Make two graphs with the results of (a): (1) d[A]/dt versus [A], (2) d[A]/dt versus $[A]^2$.
(c) With the graphs of (b), what can you conclude about n in

$$-\frac{d[A]}{dt} = k[A]^n$$

(d) With your value of n in part (c), you now have an expression for d[A]/dt. With this expression for d[A]/dt, substitute into the differential expression

$$d[A] = \frac{d[A]}{dt} dt$$

Show that the definite integral of this equation from the lower limit

$$t = 0$$
$$[A] = [A]^o$$

to the upper limit

$$t = t$$
$$[A] = [A]$$

is

$$\frac{1}{[A]} = kt + \frac{1}{[A]^o}$$

(e) Show that a graph between the variables 1/[A] and t would be linear. (Hint: y = mx + b is a form for a linear function.) Evaluate k from such a graph.
Answer: (c) n = 2 (e) m = k.

Integration Using Derivatives—Graphical Representation

In Section B we described the derivative, $\frac{dy}{dx}$, of a function, y = f(x), as the *slope* at some specified point on a graph of y versus x. Previously in this section we viewed the integral, $\int y dx$, as the *area* under a curve of y versus x. We know that differentiation/integration are inverse operations; i.e., if we have a function y = f(x), differentiation gives

$$\frac{dy}{dx} = \frac{d\,f(x)}{dx} = f'(x)$$

where the prime denotes differentiation, and integration gives the function f(x) back

$$\int f'(x)dx = f(x) + C$$

From the graphical interpretation of a derivative as a slope and an integral as an area under a curve, it is not obvious that differention/integration are inverse operations. At this time we wish to view the graphical representation in a manner that will clearly show that integration is the inverse function of differentiation.

Since it is easy to visualize velocity as the derivative of distance with time

$$v(t) = \frac{d\,D(t)}{dt} = D'(t)$$

we will use these quantities to show the integration of a derivative, v(t) = D'(t). In Section B, Figures B-3 through B-5, we showed graphs of distance (D), velocity (v), and acceleration (a) as a function of time. We assumed that acceleration increased rapidly (almost instantaneously) to a constant value until the desired velocity, v_f, was reached. For the present discussion let us assume

$$a = acceleration = constant$$

during the period t = 0 to the time when v = v_f. The integration of acceleration over time should give us velocity.

$$v(t) = \int v'(t)dt = at + C$$

Since a(t) = constant, it can be taken out of the integral and we find that v(t) is a linear function of time, t,

$$v(t) = a \int dt = at + C$$

where C = constant of integration. Previously in Section B we assumed the velocity was zero at time zero, thus

$$v(0) = a\,(0) + C$$

and the constant of integration is

$$C = 0$$

and

$$v(t) = at \qquad\qquad\qquad (C\text{-}17)$$

That is, the velocity increases linearly with time during this interval from t = 0 to v = v_f.

The graphical interpretation of the integration process is quite simple in this case since the derivative is a constant. Recall our basic definition of an integral, equation C-1

$$\int y\, dx = \lim_{\Delta x \to 0} \sum y_i \Delta x \qquad\qquad (C\text{-}1)$$

where the limits of x = a to x = b have been left off since we want an indefinite integral. In this case we are replacing y with the derivative of y

$$\frac{dy}{dx} = y'(x)$$

to show that integration of the derivative will give the function y = f(x) back. Equation C-1 can then be written as

$$\int y'\, dx = \lim_{\Delta x \to 0} \sum y_i{}' \Delta x \qquad\qquad (C\text{-}18)$$

Returning to the integration of acceleration to give velocity, we replace y with v and x with t

$$\int v'\, dt = \lim_{\Delta t \to 0} \sum v_i{}' \Delta t \qquad\qquad (C\text{-}19)$$

Since $v_i' = a_i$

$$\int y'\, dx = \lim_{\Delta t \to 0} \sum a_i \Delta t \qquad\qquad (C\text{-}20)$$

In this graphical representation of integration we will view a_i as a slope since it is equal to v' = $\frac{dv}{dt}$. We will start the integration at v = 0 and draw a_i = slope for the first segment as shown in Figure C-3. As we did before, we will first consider Δt as small but finite segments and then consider the limit as Δt = 0. According to the summation in equation (C-19) the slope a_i is multiplied by Δt. This gives a value of $\Delta v_1 = a_1 \Delta t$, as shown in Figure C-3. In the second segment we draw the slope as a_2 and again multiply by Δt to give $\Delta v_2 = a_2 \Delta t$, as shown in Figure C-3. This procedure would be continued in which successive segments of $a_i \Delta t$ would be added. In this integration the acceleration is assumed to be constant in the region v < v_f, so the slopes are all equal $a_1 = a_2 = a_3$ ... and the integral is a straight line with slope a. This is consistent with the integration which we performed before, equation (C-17). In the graphical representation shown in Figure C-3 we should be considering the limit as Δx becomes zero. One can visualize that if Δt is decreased to 1/2 the value used in Figure C - 3 the integration would still be linear with the same slope. Obviously as Δt approaches zero we will obtain the same linear relationship. As we will see shortly, the integration of velocity to give distance will change as Δt becomes smaller.

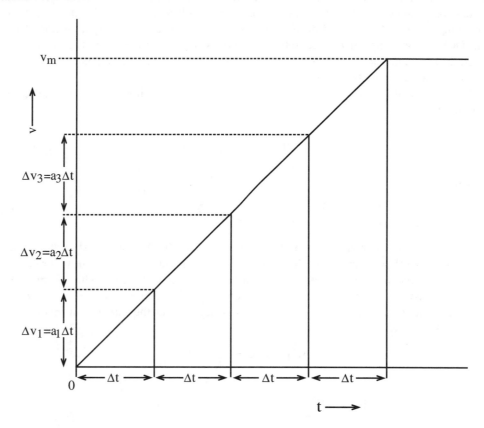

Figure C-3. Integration of Acceleration (a) to Give Velocity (v) as a Function of Time

The integration of velocity to give distance would be given by

$$D(t) = \int D'\, dt = \int v\, dt.$$

Since v is changing linearly with time when $v < v_f$

$$v = at \qquad\qquad\qquad (C\text{-}17)$$

$$D(t) = \int D'\, dt = \int at\, dt = \frac{at^2}{2} + C \qquad\qquad (C\text{-}21)$$

When the velocity reaches the v_f = constant, then D changes linearly with time

$$D(t) = \int D'\, dt = \int v_f dt = v_f t + C \qquad\qquad (C\text{-}22)$$

A graph of D(t) using equation (C-21) when $v < v_f$ and equation (C-22) when $v > v_f$ will look like the graph in Figure B-3. Equation (C-21) is a quadratic equation where the distance increased with the square of the time, t. However, note in equation (C-21) the constant of integration C. C is the distance, D, at t = 0 since

$$D(0) = \frac{a(0)^2}{2} + C,$$

$$C = D(0).$$

In the graph of Figure B-3 we assumed that the distance was zero at $t = 0$, but this was an arbitrary assumption. The distance could be referenced to some other point and $D(0)$ could be positive or negative. The $D(0)$ would be a simple displacement along the D-axis at $t = 0$. The same curve of $D(t)$ would be obtained but it would be displaced by a quantity $D(0)$. This $D(0)$ is the constant of integration that arises in equation (C-21). The constant of integration in equation (C-22) would have to be defined in another manner, such as the distance when the velocity $v = v_f$.

The graphical integration to give distance as a function of time would be specified by the basic definition

$$\int D'\, dt = \lim_{\Delta t \to 0} \sum D_i' \Delta t \qquad (C\text{-}23)$$

Since $D_i' = v_i$

$$D(t) = \int D'\, dt = \lim_{\Delta t \to 0} \sum v_i \Delta t \qquad (C\text{-}24)$$

Again we view the derivative as a slope starting at $t = 0$. We must first assume some starting point $D(0) = C$ and draw the first slope from this point as shown in Figure (C-4). As before, at first we will consider the summation over small, but finite changes in x, Δx, and later consider the limit as Δx goes to zero. The first slope is $v(0) = v_1$, and it is assumed to be zero at $t = 0$. Therefore the first segment in the integration involves no change in D.

$$\Delta D_1 = v_1\, \Delta t = 0\, \Delta t = 0$$

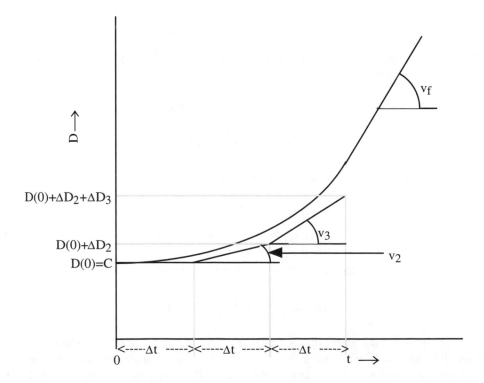

Figure C-4. Integration of Velocity (v) to Give Distance (D) as a Function of Time

However, at the end of the first Δt, $t = \Delta t$ and the velocity will have some value $v(\Delta t) = v_2$. Consequently, when we multiply the slope $= v_2$ by the increment Δt there will be a change in D for the second segment.

$$\Delta D_2 = v_2 \, \Delta t$$

In the third segment the velocity will be $v(2\Delta t) = v_3$ and it will increase over v_2 since $v = $ at. Again multiplying the slope v_3 by the increment Δt we obtain the next increase in D, ΔD_3

$$\Delta D_3 = v_3 \, \Delta t.$$

Obviously this procedure continues and we generate a function D(t) which is the sum of successive increments of ΔD_i.

$$D = \sum v_i \, \Delta t = \sum \Delta D_i$$

Now as we take the limit of very small Δt values, as designated in equation (C-24), we obtain a *smooth* curve of D(t). This is shown in Figure C-4 as the smooth, continuous curve above the semented curve where the Δt's were small but finite. The D(t) will continue to curve upward until the final velocity, v_f, is reached. When v_f has been reached the distance changes linearly with time as shown in equation (C-22).

Appendix D
TABLES OF DATA

Table 1. Conversion Factors and Fundamental Constants

Fundamental Constants

Gas Constant:
   $R = 0.08205$ L-atm/K-mol
   $= 1.987$ cal/K-mol
   $= 8.3142 \times 10^7$ erg/K-mole $= 8.3142$ J/K-mol
Avogadro's Number: $N = 6.022 \times 10^{23}$ molecules/mol
Boltzmann's Constant: $k = R/N = 1.3804 \times 10^{-16}$ erg/ K-molecule
   $= 1.3804 \times 10^{-23}$ J/K-molecule
Planck's Constant: $h = 6.626 \times 10^{-27}$ erg-sec $= 6.626 \times 10^{-34}$ J-sec
Speed of Light $= c = 2.997925 \times 10^{10}$ cm/sec $= 2.997925 \times 10^8$ m/sec
Mass of the Electron $= 9.10 \times 10^{-28}$ g $= 9.10 \times 10^{-31}$ kg
Unit of Charge $= 4.8 \times 10^{-10}$ esu $= 1.602 \times 10^{-19}$ coulomb
   1 esu $= $ (erg-cm)$^{1/2}$

Energy Units

1 erg $= 1$ g-cm^2/sec^2
1 Joule(J) $= 1$ kg-m^2/sec^2

1 Joule(J) $= 10^7$ erg
1 cal $= 4.184$ J

The following table gives the conversion factors between different energy units which are commonly used in physical chemistry. The vertical; column to the left of the table gives the energy unit, and the appropriate conversion factor to another energy unit can be found in the table under the column identifying the other energy unit. For example,

1 L-atm/mol $= 0.024218$ kcal/mol
1 kcal/mol $= 4.338 \times 10^{-2}$ eV/molecule

Note that some energy units are given on a per mole basis where as others are given on a per molecule or particle basis, such as atom, ion, or photon. The reason for this is that the different energy units are commonly associated with either the mole or molecule basis. If the coversion factor between the energy units is desired independent of the amount of material, Avogadro's Number must be used to account for the per mole versus per molecule basis. For example the conversion factor between kcal and eV is

$$1 \text{ kcal/mol} = \frac{4.338 \times 10^{-2}\,\text{eV}}{\text{molecule}} \left(6.022 \times 10^{23} \frac{\text{molecules}}{\text{mol}} \right) = 2.612 \times 10^{22} \text{ eV/mol}$$

Obviously the kcal unit is much larger than an eV unit.

	$\dfrac{\text{liter-atm}}{\text{mol}}$	$\dfrac{\text{kcal}}{\text{mol}}$	$\dfrac{\text{erg}}{\text{mol}}$	$\dfrac{\text{eV}}{\text{molecule}}$	$\dfrac{\text{cm}^{-1}}{\text{molecule}}$
$1\ \dfrac{\text{liter-atm}}{\text{mol}}$	1	0.024218	1.013×10^9	1.050×10^{-3}	8.474×10^{-3}
$1\ \dfrac{\text{kcal}}{\text{mol}}$	4.129×10^1	1	4.184×10^{10}	4.338×10^{-2}	349.96
$1\ \dfrac{\text{erg}}{\text{mol}}$	9.869×10^{-10}	2.390×10^{-11}	1	1.036×10^{-12}	8.361×10^{-9}
$1\ \dfrac{\text{eV}}{\text{molecule}}$	952.2	23.05	9.648×10^{11}	1	8067.
$1\ \dfrac{\text{cm}^{-1}}{\text{molecule}}$	118.0	2.58×10^{-3}	1.196×10^8	1.2395×10^{-4}	1

Commonly Used Conversion Factors

1 atm = 1.01325 bar
1 bar = 10^5 Pascals (Pa)
1 Pa = 1 Newton (N)/m^2
1 atm = 101,325 Pa = 0.101325 MPa
1 bar = 750 mmHg
1 atm = 760 torr
1 torr = 1mm Hg
1 micron(μ) = 1 x 10^{-3} torr = 1 x 10^{-4} cm Hg
1 atm = 14.7 lbs/in^2
1 atm = 7.6 x 10^5 μ
1 in = 2.54 cm
1 m = 39.37 in
1 L = 1.057 qt
1 lb = 453.6 g
1 ft^3 = 2832 L
1 A^o = 1 x 10^{-8} cm
1 nm = 1 x 10^{-9} m
1 A^o = 10 nm

Table 2. Standard Thermodynamic Properties of Non-carbon Containing Substances at 25°C

Molecular Formula	Name	State	ΔH_f^o(kJ/mol)	ΔG_f^o(kJ/mol)	S^o(J/K-mol)	C_P(J/K-mol)
Ag	Silver	cry	0.0	0.0	42.6	25.4
		gas	284.9	246.0	173.0	20.8
Ag$^+$	(m = 1)	aq	105.58	77.124	72.68	21.8
AgBr	Silver Bromide	cry	-100.4	-96.9	107.1	52.4
AgCl	Silver Chloride	cry	-127.0	-109.8	96.3	50.8
AgF	Silver Fluoride	cry	-204.6			
AgI	Silver Iodide	cry	-61.8	-66.2	115.5	56.8
AgNO$_3$	Silver Nitrate	cry	-124.4	-33.4	140.9	93.1
Al	Aluminum	cry	0.0	0.0	28.3	24.4
AlBr$_3$	Aluminum Tribromide	cry	-527.2			101.7
AlCl$_3$	Aluminum Trichloride	cry	-704.2	-628.8	110.7	91.8
AlF$_3$	Aluminum Trifluoride	cry	-1510.4	-1431.1	66.5	75.1
AlI$_3$	Aluminum Triiodide	cry	-313.8	-300.8	159.0	98.7
As	Arsenic (gray)	cry	0.0	0.0	35.1	24.6
AsBr$_3$	Arsenic Tribromide	cry	-197.5			
AsCl$_3$	Arsenic Trichloride	liq	-305.0	-259.4	216.3	
AsF$_3$	Arsenic Trifluoride	liq	-821.3	-774.2	181.2	126.6
As$_2$	Arsenic (As$_2$)	gas	222.2	171.9	239.4	35.0
Au	Gold	cry	0.0	0.0	47.4	25.4
Au$_2$	Gold (Au$_2$)	gas	515.1			36.9
B	Boron (Rhombic)	cry	0.0	0.0	5.9	11.1
BCl$_3$	Boron Trichloride	liq	-427.2	-387.4	206.3	106.7
BF$_3$	Boron Trifluoride	gas	-1136.0	-1119.4	254.4	
BH$_3$	Borane (BH$_3$)	gas	100.0			
B$_2$H$_6$	Diborane	gas	35.6	86.7	232.1	56.9
Ba	Barium	cry	0.0	0.0	62.8	28.1
		gas	180.0	146.0	170.2	20.8
BaCl$_2$	Barium Chloride	cry	-858.6	-810.4	123.7	75.1
BaI$_2$	Barium Iodide	cry	-602.1			
Br	Bromine	gas	111.9	82.4	175.0	20.8
BrCl	Bromine Chloride	gas	14.6	-1.0	240.1	35.0
BrF	Bromine Fluoride	gas	-93.8	-109.2	229.0	33.0
BrF$_3$	Bromine Trifluoride	liq	-300.8	-240.5	178.2	124.6
		gas	-255.6	-229.4	292.5	66.6
BrF$_5$	Bromine Pentafluoride	liq	-458.6	-351.8	225.1	
		gas	-428.9	-350.6	320.2	99.6
BrH	Hydrogen Bromide	gas	-36.3	-53.4	198.7	29.1

Molecular

Formula	Name	State	ΔH_f^o(kJ/mol)	ΔG_f^o(kJ/mol)	S^o(J/K-mol)	C_P(J/K-mol)
BrI	Iodine Bromide	gas	40.8	3.7	258.8	36.4
BrK	Potassium Bromide	cry	-393.8	-380.7	95.9	52.3
BrNa	Sodium Bromide	cry	-361.1	-349.0	86.8	41.4
Br_2	Bromine (Br_2)	liq	0.0	0.0	152.2	75.7
		gas	30.9	3.1	245.5	36.0
Br_2Ca	Calcium Bromide	cry	-682.8	-663.6	130.0	
Br_2Hg_2	Mercury Bromide (Hg_2Br_2)	cry	-206.9	-181.1	218.0	
Br_2Pb	Lead Bromide ($PtBr_2$)	cry	-278.7	-261.9	161.5	80.1
Ca	Calcium	cry	0.0	0.0	41.6	25.9
		gas	177.8	144.0	154.9	20.8
$CaCl_2$	Calcium Chloride	cry	-795.4	-748.8	108.4	72.9
CaO	Calcium Oxide	cry	-634.9	-603.3	38.1	42.0
Cl^-	(m = 1)	aq	-167.159	-131.26	56.6	136.4
ClF	Chlorine Fluoride	gas	-50.3	-51.8	217.9	32.1
ClF_3	Chlorine Trifluoride	liq	-189.5			
		gas	-163.2	-123.0	281.6	63.9
ClH	Hydrogen Chloride	gas	-92.3	-95.3	186.9	29.1
ClH_4N	Ammonium Chloride	cry	- 314.4	- 202.9	94.6	84.1
ClI	Iodine Chloride	liq	-23.9	-13.6	135.1	
		gas	17.8	-5.5	247.6	35.6
ClK	Potassium Chloride	cry	-436.5	-408.5	82.6	51.3
Cl2	Chlorine (Cl2)	gas	0.0	0.0	223.1	33.9
Cl_2Hg	Mercury Chloride ($HgCl_2$)	cry	-224.3	-178.6	146.0	
Cl_2Hg_2	Mercury Chloride (Hg_2Cl_2)	cry	-265.4	-210.7	191.6	--
Cl_2Mg	Magnesium Chloride	cry	-641.32	-591.8	89.62	71.38
Cl_3P	Phosphorus Trichloride	liq	-319.7	-272.3	217.1	
		gas	-287.0	-267.8	311.8	71.8
Cl_5P	Phosphorus Pentachloride	cry	-443.5			
		gas	-374.9	-305.0	364.6	112.8
Cs	Cesium	cry	0.0		85.2	32.2
		gas	76.5	49.6	175.6	20.8
CsF	Cesium Fluoride	cry	-553.5	-525.5	92.8	51.1
CsI	Cesium Iodide	cry	-346.6	-340.6	123.1	52.8
F	Fuorine	gas	79.4	62.3	158.8	22.7
F^-	(m = 1)	aq	-332.6	-278.8	-13.8	-106.7
FH	Hydrogen Fluoride	liq	-299.8			
		gas	-271.1	-273.2	173.7	29.13
FI	Iodine Fluoride	gas	-95.7	-118.5	236.2	33.4
F_2	Fluorine (F_2)	gas	0.0	0.0	202.8	31.3

Molecular Formula	Name	State	ΔH_f^o(kJ/mol)	ΔG_f^o(kJ/mol)	S^o(J/K-mol)	C_P(J/K-mol)
F_3N	Nitrogen Trifluoride	gas	-132.1	-90.6	260.8	53.4
F_3P	Phosphorus Trifluoride	gas	-958.4	936.9	273.1	68.8
F_5P	Phosphorus Pentafluoride	gas	-1594.4	-1520.7	300.8	84.8
H	Hydrogen	gas	218.0	203.3	114.7	20.8
H^+	Hydrogen ion (m=1)	aq	0.0	0.0	0.0	0.0
HI	Hydrogen Iodide	gas	26.5	1.7	206.6	29.2
HNO_2	Nitrous Acid	gas	-79.5	-46.0	254.1	45.6
H_2	Hydrogen (H_2)	gas	0.0	0.0	130.7	28.8
H_2N	Amidogen (NH_2)	gas	184.9	194.6	195.0	33.9
HO^-	Hydroxyl (m = 1)	aq	-229.99	-157.29	-10.75	-148.5
H_2O	Water	liq	-285.8	-237.1	70.0	75.3
		gas	-241.8	-228.6	188.8	33.6
H_2O_2	Hydrogen Peroxide	liq	-187.8	-120.4	109.6	89.1
H_2S	Hydrogen Sulfide	gas	-20.6	-33.4	205.8	34.2
H_2Se	Hydrogen Selenide	gas	29.7	15.9	219.0	34.7
H_2Te	Hydrogen Telluride	gas	99.6			
H_3N	Ammonia	gas	-45.9	-16.4	192.8	35.1
	(undissociated) (m = 1)	aq	-80.29	-26.457	111.3	---
H_4N^+	(m = 1)	aq	-132.5	-79.16	113.4	79.9
H_4IN	Ammonium Iodide	cry	-201.4	-112.5	117.0	
$H_4N_2O_2$	Ammonium Nitrite	cry	-256.5			
$H_4N_2O_3$	Ammonium Nitrate	cry	-365.6	-183.9	151.1	139.3
H_5NO	Ammonium Hydroxide	liq	-361.2	-254.0	165.6	154.9
$H_8N_2O_4S$	Ammonium Sulfate	cry	-1180.9	-901.7	220.1	187.5
Hg	Mercury	liq	0.0	0.0	75.9	28.0
		gas	61.4	31.8	175.0	20.8
Hg_2	Mercury (Hg_2)	gas	108.8	68.2	288.1	37.4
I	Iodine	gas	106.8	70.2	180.8	20.8
I^-		gas	-196.6			
I^-	(m = 1)	aq	-55.19	-51.59	111.3	-142.3
IK	Potassium Iodide	cry	-327.9	-324.9	106.3	52.9
ILi	Lithium Iodide	cry	-270.4	-270.3	86.8	51.0
INa	Sodium Iodide	cry	-287.8	-286.1	98.5	52.1
I_2	Iodine (Rhombic)	cry	0.0	0.0	116.1	54.4
		gas	62.4	19.3	260.7	36.9
K	Potassium	cry	0.0	0.0	64.7	29.6
		gas	89.0	60.5	160.3	20.8
KNO_2	Potassium Nitrite	cry	-369.8	-306.6	152.1	107.4
KNO_3	Potassium Nitrate	cry	-494.6	-394.9	133.1	96.4
K_2	Potassium (K_2)	gas	123.7	87.5	249.7	37.9

Molecular Formula	Name	State	ΔH_f^o(kJ/mol)	ΔG_f^o(kJ/mol)	S^o(J/K-mol)	C_P(J/K-mol)
Li	Lithium	cry	0.0	0.0	29.1	24.8
		gas	159.3	126.6	138.8	20.8
$LiNO_2$	Lithium Nitrite	cry	-372.4	-302.0	96.0	
$LiNO_3$	Lithium Nitrate	cry	-483.1	-381.1	90.0	
Li_2	Lithium (Li_2)	gas	215.9	174.4	197.0	36.1
Mg	Magnesium	cry	0.0	0.0	32.7	24.9
		gas	147.1	112.5	148.6	20.8
Mg^{2+}	(m = 1)	aq	-466. 85	-454.8	-138.1	--
MgO	Magnesium Oxide	cry	-601.6	-569.3	27.0	37.2
$Mg(OH)_2$	Magnesium Hydroxide (m = 1)	aq	-926.84	-769.44	-159.8	--
MgS	Magnesium Sulfide	cry	-346.0	-341.8	50.3	45.6
N	Nitrogen	gas	472.7	455.5	153.3	20.8
$NNaO_3$	Sodium Nitrate	cry	-467.9	-367.0	116.5	92.9
NO_2	Nitrogen Dioxide	gas	33.2	51.3	240.1	37.2
N_2	Nitrogen (N_2)	gas	0.0	0.0	191.6	29.1
N_2O	Nitrogen Oxide	gas	82.1	104.2	219.9	38.5
N_2O_3	Nitrogen Trioxide	liq	50.3			
		gas	83.7	139.5	312.3	65.6
N_2O_4	Nitrogen Tetroxide	liq	-19.5	97.5	209.2	142.7
		gas	9.2	97.9	304.3	77.3
N_2O_5	Nitrogen Pentoxide	cry	-43.1	113.9	178.2	143.1
		gas	11.3	115.1	355.7	84.5
Na	Sodium	cry	0.0	0.0	51.3	28.2
		gas	107.5	77.0	153.7	20.8
NaO_2	Sodium Superoxide (NaO_2)	cry	-260.2	218.4	115.9	72.1
Na_2	Sodium (Na_2)	gas	142.1	103.9	230.2	37.6
Na_2O	Sodium Oxide	cry	-414.2	-375.5	75.1	69.1
Na_2S	Sodium Sulfide	cry	-364.8	-349.8	83.7	
O	Oxygen	gas	249.2	231.7	161.1	21.9
OS	Sulfur Oxyde (SO)	gas	6.3	-19.9	222.0	30.2
OZn	Zinc Oxide	cry	-350.5	-320.5	43.7	40.3
O_2	Oxygen (O_2)	gas	0.0	0.0	205.2	29.4
O_2S	Sulfur Dioxide	liq	-320.5			
		gas	-296.8	-300.1	248.2	39.9
O_2Se	Selenium Dioxide	cry	-225.4			
O_2Si	Silicon Dioxide	cry	-910.7	-856.3	41.5	44.4
	(α-quartz)	gas		-322.0		
O_3	Ozone	gas	142.7	163.2	238.9	39.2
O_3S	Sulfur Trioxide	cry	-454.5	-374.2	70.7	
		liq	-441.0	-373.8	113.8	
		gas	-395.7	371.1	256.8	50.7
O_3W	Tungsten Oxide (WO_3)	cry	-842.9	-764.0	75.9	73.8

Molecular Formula	Name	State	ΔH_f^o(kJ/mol)	ΔG_f^o(kJ/mol)	S^o(J/K-mol)	C_P(J/K-mol)
P	Phosphorus (white)	cry	0.0	0.0	4.1	23.8
	Phosphorus (red)	cry	-17.6		22.8	21.2
	Phosphorus (black)	cry	-39.3			
		gas	316.5	280.1	163.2	20.8
P_2	Phosphorus (P_2)	gas	144.0	103.5	218.1	32.1
S	Sulfur (Rhombic)	cry	0.0	0.0	32.1	22.6
SZn	Zinc Sulfide (wurtzite)	cry	-192.6			
S_2	Sulfur (S_2)	gas	128.6	79.7	228.2	32.5
Se	Selenium	cry	0.0	0.0	42.4	25.4
		gas	227.1	187.0	176.7	20.8
Se_2	Selenium (Se_2)	gas	146.0	96.2	252.0	35.4
Si	Silicon	cry	0.0		18.8	20.0
		gas	450.0	405.5	168.0	22.3
Si_2	Silicon (Si_2)	gas	594.0	536.0	229.9	34.4
Te	Tellurium	cry	0.0	0.0	49.7	25.7
Te_2	Tellurium (Te_2)	gas	168.2	118.0	268.1	36.7
Ti	Titanium	cry	0.0	0.0	30.7	25.0
W	Tungsten	cry	0.0	0.0	32.6	24.3
Zn	Zinc	cry	0.0	0.0	41.6	25.4
		gas	130.4	94.8	161.0	20.8

Table 3. Standard Thermodynamic Properties of Carbon Containing Substances at 25°C

Molecular Formula	Name	State	ΔH_f^o(kJ/mol)	ΔG_f^o(kJ/mol)	S^o(J/K-mol)	C_P(J/K-mol)
Ag	Silver	cry	0.0	0.0	42.6	25.4
C	Carbon (graphite)	cry	0.0	0.0	5.7	8.5
	Carbon (diamond)	cry	1.9	2.9	2.4	6.1
CBr_2ClF	Dibromochloro-fluoromethane	gas			342.8	82.4
CBr_2Cl_2	Dibromodichloro-methane	gas			347.8	87.1
CBr_2F_2	Dibromodifluoro-methane	gas			325.3	77.0
CBr_3Cl	Tribromochloro-methane	gas			357.8	89.4
CBr_3F	Tribromofluoro-methane	gas			345.9	84.4
CBr_4	Tetrabromomethane	cry	18.8	47.7	212.5	144.3
		gas	79.0	67.0	358.1	91.2
$CCaO_3$	Calcium Carbonate (calcite)	cry	-1207.6	-1129.1	91.7	83.5
	Calcium Carbonate (aragonite)	cry	-1207.8	-1128.2	88.0	82.3
$CClF_3$	Chlorotrifluoro-methane	gas	-706.3			66.9
$CClN$	Cyanogen Chloride	liq	112.1			
		gas	138.0	131.0	236.2	45.0
CCl_2F_2	Dichlorodifluoro-methane	gas	-477.4	-439.4	300.8	72.3
CCl_2O	Carbonyl Chloride	gas	-219.1	-204.9	283.5	57.7
CCl_3F	Trichlorofluoro-methane	liq	-301.3	-236.8	225.4	121.6
		gas	-268.3			78.1
CCl_4	Tetrachloromethane	liq	-128.2			130.7
CFN	Cyanogen Fluoride	gas			224.7	41.8
CF_3	Trifluoromethyl	gas	-477.0	-464.0	264.5	49.6
CIN	Cyanogen Iodide	cry	166.2	185.0	96.2	
CN	Cyanide (CN)	gas	437.6	407.5	202.6	29.2
CO	Carbon Monoxide	gas	-110.5	-137.2	197.2	29.1
COS	Carbon Oxysulfide	gas	-142.0	-169.2	231.6	41.5
CO_2	Carbon Dioxide	gas	-393.5	-394.4	213.8	37.1
CS	Carbon Sulfide	gas	234.0	184.0	210.6	29.8
CS_2	Carbon Disulfide	liq	89.0	64.6	151.3	76.4
$CHBr_3$	Tribromomethane	liq	-28.5	-5.0	220.9	130.7
		gas	17.0	8.0	330.9	71.2
$CHClF_2$	Chlorodifluoro-methane	gas	-482.6		280.9	55.9
$CHCl_2F$	Dichlorofluoro-methane	gas			293.1	60.9
$CHCl_3$	Trichloromethane	liq	-134.5	-73.7	201.7	114.2
		gas	-103.1	6.0	295.7	65.7
CHN	Hydrogen Cyanide	liq	108.9	125.0	112.8	70.6
		gas	135.1	124.7	201.8	35.9
CH_2	Methylene	gas	390.4	372.9	194.9	33.8

Molecular Formula	Name	State	ΔH_f^o(kJ/mol)	ΔG_f^o(kJ/mol)	S^o(J/K-mol)	C_P(J/K-mol)
CH_2Cl_2	Dichloromethane	liq	-124.1		177.8	101.2
		gas	-95.6		270.2	51.0
CH_2F_2	Difluoromethane	gas	-452.2		246.7	42.9
CH_2O	Formaldehyde	gas	-108.6	-102.5	218.8	35.4
CH_2O_2	Formic Acid	liq	-424.7	-361.4	129.0	99.0
		gas	-378.6			
CH_3	Methyl	gas	145.7	147.9	194.2	38.7
CH_3Br	Bromomethane	liq	-59.4			
		gas	-35.5	-26.3	246.4	42.4
CH_3Cl	Chloromethane	gas	-81.9		234.6	40.8
CH_3F	Fluoromethane	gas			222.9	37.5
CH_3I	Iodomethane	liq	-12.3		163.2	126.0
		gas	14.7		254.1	44.1
CH_3NO	Formamide	liq	-254.0			
CH_3NO_2	Nitromethane	liq	-113.1	-14.4	171.8	106.6
		gas	-74.7	-6.8	275.0	57.3
CH_3NO_3	Methyl Nitrate	liq	-159.0	-43.4	217.1	157.3
CH_4	Methane	gas	-74.4	-50.3	186.3	35.3
CH_4O	Methanol	liq	-239.1	-166.6	126.8	81.1
		gas	-201.5	-162.6	239.8	43.9
CH_5N	Methylamine	liq	-47.3	35.7	150.2	102.1
		gas	-22.5	32.7	242.9	50.1
C_2Ca	Calcium Carbide	cry	-59.8	-64.9	70.0	62.7
C_2F_4	Tetrafluoroethylene	cry	-820.5			
		gas	-658.9		300.1	80.5
C_2HCl_3	Trichloroethylene	liq	-43.6		228.4	124.4
		gas	-8.1		324.8	80.3
C_2HCl_3O	Trichloroacetal-	liq	-236.2			151.0
	dehyde	gas	-196.6			
C_2H_2	Acetylene	gas	228.2	210.7	200.9	43.9
$C_2H_2Cl_2$	cis-1,2-Dichloro-	liq	-26.4		198.4	116.4
	ethylene	gas	4.6		289.6	65.1
$C_2H_2Cl_2$	trans-1,2-Dichloro-	liq	-23.1	27.3	195.9	116.8
	ethylene	gas	6.2	28.6	290.0	66.7
C_2H_2O	Ketene	liq	-67.9			
		gas	-47.5	-48.3	247.6	51.8
$C_2H_2O_4$	Oxalic Acid	cry	-821.7		109.8	91.0
$C_2H_3Cl_3$	1,1,1-Trichloro-	liq	-177.4		227.4	144.3
	ethane	gas	-144.6		323.1	93.3
$C_2H_3Cl_3$	1,1,2-Trichloro-	liq	-191.5		232.6	150.9
	ethane	gas	-151.2		337.2	89.0
C_2H_3N	Acetronitrile	liq	31.4	77.2	149.6	91.4
		gas	64.3	81.7	245.1	52.2
C_2H_4	Ethylene	gas	52.5	68.4	219.6	43.6
C_2H_4O	Acetaldehyde	liq	-191.8	-127.6	160.2	89.0
		gas	-166.2	-138.8	263.7	55.3
$C_2H_4O_2^-$	Acetate Ion (m = 1)	aq	-486.01	-369.4	86.6	-6.3
$C_2H_4O_2$	Acetic Acid	liq	-484.5	-389.9	159.8	123.3
		gas	-432.8	-374.5	282.5	66.5
	(m = 1)	aq	-485.76	-396.64	178.7	--

Molecular

Formula	Name	State	ΔH_f^o(kJ/mol)	ΔG_f^o(kJ/mol)	S^o(J/K-mol)	C_P(J/K-mol)
$C_2H_4O_2$	Methyl Formate	liq	-386.1			119.1
		gas	-355.5		285.3	64.4
C_2H_5NO	Acetamide	cry	-317.0		115.0	91.3
C_2H_6	Ethane	gas	-83.8	-31.9	229.6	52.6
C_2H_6O	Dimethyl Ether	gas	-184.1	-112.6	266.4	64.4
C_2H_6O	Ethanol	liq	-277.7	-174.8	160.7	112.3
		gas	-235.1	-168.5	282.7	65.4
C_2H_6OS	Dimethyl Sulfoxide	liq	-204.2	-99.9	188.3	153.0
C_2H_7N	Dimethylamine	liq	-43.9	70.0	182.3	137.7
		gas	-18.5	68.5	273.1	70.7
C_2H_7N	Ethylamine	liq	-74.1			130.0
		gas	-47.5	36.3	283.8	71.5
C_3H_6	Propene	liq	1.7			
		gas	20.0			
C_3H_6	Cyclopropane	gas	53.3			
C_3H_6O	Acetone	liq	-248.1		199.8	126.3
		gas	-217.3		297.6	75.0
C_3H_8	Propane	gas	-104.7			
C_3H_8O	1-Propanol	liq	-302.6		193.6	143.9
		gas	-255.1		322.6	85.6
C_3H_8O	2-Propanol	liq	-318.1		181.1	156.5
		gas	-272.8		309.2	89.3
C_3H_9N	Trimethylamine	liq	-45.7		208.5	137.9
C_4H_8	1-Butene	liq	-20.5		227.0	118.0
		gas	0.1			
C_4H_8	cis-2-Butene	liq	-29.7		219.9	127.0
		gas	-7.1			
C_4H_8	trans-2-Butene	liq	-33.0			
		gas	-11.4			
C_4H_8	Isobutene	liq	-37.5			
		gas	-16.9			
C_4H_8	Cyclobutene	liq	3.7			
		gas	28.4			
C_4H_8	Methylcyclopropane	liq	1.7			
C_4H_8O	Butanal	liq	-239.2		246.6	163.7
C_4H_{10}	Butane	liq	-146.6			140.9
		gas	-125.6			
C_4H_{10}	Isobutane	liq	-153.5			
		gas	-134.2			
$C_4H_{10}O$	2-methyl-2-propanol	liq	-359.2		193.3	218.6
		gas	-312.5		326.7	113.6
$C_4H_{10}O$	Methyl Propyl Ether	liq	-266.0		262.9	165.4
		gas	-238.2			
$C_4H_{10}O$	Isopropyl Methyl Ether	liq	-278.7		253.8	161.9
		gas	-252.0			
$C_4H_{11}N$	Isobutylamine	liq	-132.6			183.2
		gas	-98.7			
$C_4H_{11}N$	sec-Butylamine	liq	-137.5			
		gas	-104.9			
$C_4H_{11}N$	tert-Butylamine	liq	-150.6			192.1
		gas	-120.9			

Molecular Formula	Name	State	ΔH_f^o(kJ/mol)	ΔG_f^o(kJ/mol)	S^o(J/K-mol)	C_P(J/K-mol)
$C_4H_{11}N$	Diethylamine	liq	-103.7			169.2
		gas	-72.5			
C_5H_{10}	cis-2-Pentene	liq	-53.7		258.6	151.7
		gas	-27.6			
C_5H_{10}	trans-2-Pentene	liq	-58.2		256.5	157.0
		gas	-31.9			
C_5H_{10}	Cyclopentane	liq	-105.1		204.5	128.8
		gas	-76.4			
C_5H_{12}	Pentane	liq	-173.5			167.2
		gas	-146.9			
C_5H_{12}	Isopentane	liq	-178.5		260.4	164.8
		gas	-153.7			
C_5H_{12}	Neopentane	liq	-190.2			
		gas	-168.1			
C_6F_6	Hexafluorobenzene	liq	-991.3		280.8	221.6
		gas	-955.4			
C_6H_5Br	Bromobenzene	liq	60.9		219.2	154.3
C_6H_5Cl	Chlorobenzene	liq	11.0			150.1
C_6H_5F	Fluorobenzene	liq	-150.6		205.9	146.4
C_6H_5I	Iodobenzene	liq	117.2		205.4	158.7
$C_6H_5NO_2$	Nitrobenzene	liq	12.5			185.8
		gas	67.5			
C_6H_6	Benzene	liq	49.0			136.3
		gas	82.6	-9.5		
C_6H_7N	Aniline	liq	31.3			191.9
		gas	87.5	-7.0	317.9	107.9
C_6H_{12}	Cyclohexane	liq	-156.4			154.9
		gas	-123.4			
C_6H_{12}	1-Hexene	liq	-74.2		295.2	183.3
		gas	-43.5			
C_6H_{12}	cis-2-Hexene	liq	-83.9			
		gas	-52.3			
C_6H_{12}	trans-2-Hexene	liq	-85.5			
		gas	-53.9			
$C_7H_6O_2$	Benzoic Acid	cry	-385.2		167.6	146.8
		gas	-294.1			
C_7H_8	Toluene	liq	12.4			157.3
		gas	50.4			
C_7H_9N	o-Methyl Aniline	gas	56.4	167.6	351.0	130.2
		liq	-6.3			
C_7H_9N	m-Methyl Aniline	gas	54.6	165.4	352.5	125.5
		liq	-8.1			
C_7H_9N	p-Methyl Aniline	cry	-23.5			
		gas	55.3	167.7	347.0	126.2
C_7H_{14}	Cycloheptane	liq	-156.6			
		gas	-118.1			
C_7H_{14}	Methylcyclohexane	liq	-190.1			184.8
		gas	-154.7			
C_8H_{10}	o-Xylene	liq	-24.4			186.1
		gas	19.1			
C_8H_{10}	m-Xylene	liq	-25.4			183.0
		gas	17.3			

Molecular Formula	Name	State	ΔH_f^o(kJ/mol)	ΔG_f^o(kJ/mol)	S^o(J/K-mol)	C_P(J/K-mol)
C_8H_{10}	p-Xylene	liq	-24.4			181.5
		gas	18.0			
C_9H_{12}	o-Ethyltoluene	liq	-46.4			
		gas	1.3			
C_9H_{12}	m-Ethyltoluene	liq	-48.7			
		gas	-1.8			
C_9H_{12}	p-Ethyltoluene	liq	-49.8			
		gas	-3.2			
C_9H_{12}	1,2,3-Trimethyl-benzene	liq	-58.5		267.9	216.4

Table 4. Thermodynamic Properties as a Function of Temperature

T(K)	C_P^0(J/K-mol)	S^0(J/K-mol)	ΔH_f^0(kJ/mol)	ΔG_f^0(kJ/mol)

1. Argon Ar(g)

T(K)	C_P^0(J/K-mol)	S^0(J/K-mol)	ΔH_f^0(kJ/mol)	ΔG_f^0(kJ/mol)
298.15	20.786	154.845	0.000	0.000
300	20.786	154.973	0.000	0.000
400	20.786	160.953	0.000	0.000
500	20.786	165.591	0.000	0.000
600	20.786	169.381	0.000	0.000
700	20.786	172.585	0.000	0.000
800	20.786	175.361	0.000	0.000
900	20.786	177.809	0.000	0.000
1000	20.786	179.999	0.000	0.000
1100	20.786	181.980	0.000	0.000
1200	20.786	183.789	0.000	0.000
1300	20.786	185.453	0.000	0.000
1400	20.786	186.993	0.000	0.000
1500	20.786	188.427	0.000	0.000

2. Dibromine Br_2

T(K)	C_P^0(J/K-mol)	S^0(J/K-mol)	ΔH_f^0(kJ/mol)	ΔG_f^0(kJ/mol)
298.15	36.057	245.467	30.910	3.105
300	36.074	245.690	30.836	2.933
332.5	36.340	249.387	pressure = 1 bar	
400	36.729	256.169	0.000	0.000
500	37.082	264.406	0.000	0.000
600	37.305	271.188	0.000	0.000
700	37.464	276.951	0.000	0.000
800	37.590	281.962	0.000	0.000
900	37.697	286.396	0.000	0.000
1000	37.793	290.373	0.000	0.000
1100	37.883	293.979	0.000	0.000
1200	37.970	297.279	0.000	0.000
1300	38.060	300.322	0.000	0.000
1400	38.158	303.146	0.000	0.000
1500	38.264	305.782	0.000	0.000

3. Hydrogen Bromide HBr(g)

T(K)	C_P^0(J/K-mol)	S^0(J/K-mol)	ΔH_f^0(kJ/mol)	ΔG_f^0(kJ/mol)
298.15	29.141	198.697	-36.290	-53.360
300	29.141	198.878	-36.333	-53.466
400	29.220	207.269	-52.109	-55.940
500	29.454	213.811	-52.484	-56.854
600	29.872	219.216	-52.844	-57.694
700	30.431	223.861	-53.168	-58.476
800	31.063	227.965	-53.446	-59.214
900	31.709	231.661	-53.677	-59.921
1000	32.335	235.035	-53.864	-60.604
1100	32.919	238.145	-54.012	-61.271
1200	33.454	241.032	-54.129	-61.925
1300	33.938	243.729	-54.220	-62.571
1400	34.374	246.261	-54.291	-63.211
1500	34.766	248.646	-54.348	-63.846

T(K)	C_P^o(J/K-mol)	S^o(J/K-mol)	ΔH_f^o(kJ/mol)	ΔG_f^o(kJ/mol)

4. Carbon Oxide CO(g)

T(K)	C_P^o(J/K-mol)	S^o(J/K-mol)	ΔH_f^o(kJ/mol)	ΔG_f^o(kJ/mol)
298.15	29.141	197.658	-110.530	-137.168
300	29.142	197.838	-110.519	-137.333
400	29.340	206.243	-110.121	-146.341
500	29.792	212.834	-110.027	-155.412
600	30.440	218.321	-110.157	-164.480
700	31.170	223.067	-110.453	-173.513
800	31.898	227.277	-110.870	-182.494
900	32.573	231.074	-111.378	-191.417
1000	33.178	234.538	-111.952	-200.281
1100	33.709	237.726	-112.573	-209.084
1200	34.169	240.679	-113.228	-217.829
1300	34.568	243.430	-113.904	-226.518
1400	34.914	246.005	-114.594	-235.155
1500	35.213	248.424	-115.291	-243.742

5. Carbon Dioxide CO_2 (g)

T(K)	C_P^o(J/K-mol)	S^o(J/K-mol)	ΔH_f^o(kJ/mol)	ΔG_f^o(kJ/mol)
298.15	37.135	213.783	-393.510	-394.373
300	37.220	213.784	-393.511	-394.379
400	41.328	225.305	-393.586	-394.656
500	44.627	234.895	-393.672	-394.914
600	47.327	243.278	-393.791	-395.152
700	49.569	250.747	-393.946	-395.367
800	51.442	257.492	-394.133	-395.558
900	53.008	263.644	-394.343	-395.724
1000	54.320	269.299	-394.568	-395.865
1100	55.423	274.529	-394.801	-395.984
1200	56.354	279.393	-395.035	-396.081
1300	57.144	283.936	-395.265	-396.159
1400	57.818	288.196	-395.488	-396.219
1500	58.397	292.205	-395.702	-396.264

6. Methane CH_4 (g)

T(K)	C_P^o(J/K-mol)	S^o(J/K-mol)	ΔH_f^o(kJ/mol)	ΔG_f^o(kJ/mol)
298.15	35.695	186.369	-74.600	-50.530
300	35.765	186.590	-74.656	-50.381
400	40.631	197.501	-77.703	-41.827
500	46.627	207.202	-80.520	-32.525
600	52.742	216.246	-82.969	-22.690
700	58.603	224.821	-85.023	-12.476
800	64.084	233.008	-86.693	-1.993
900	69.137	240.852	-88.006	8.677
1000	73.746	248.379	-88.996	19.475
1100	77.919	255.607	-89.698	30.358
1200	81.682	262.551	-90.145	41.294
1300	85.067	269.225	-90.367	52.258
1400	88.112	275.643	-90.390	63.231
1500	90.856	281.817	90.237	74.200

T(K)	C_P^o(J/K-mol)	S^o(J/K-mol)	ΔH_f^o(kJ/mol)	ΔG_f^o(kJ/mol)

7. Acetylene C_2H_2 (g)

T(K)	C_P^o	S^o	ΔH_f^o	ΔG_f^o
298.15	44.036	200.927	227.400	209.879
300	44.174	200.927	227.397	209.770
400	50.388	214.814	227.161	203.928
500	54.751	226.552	226.846	198.154
600	58.121	236.842	226.445	192.452
700	60.970	246.021	225.968	186.823
800	63.511	254.331	225.436	181.267
900	65.831	261.947	224.873	175.779
1000	67.960	268.995	224.300	170.355
1100	69.909	275.565	223.734	164.988
1200	71.686	281.725	223.189	159.672
1300	73.299	287.528	222.676	154.400
1400	74.758	293.014	222.203	149.166
1500	76.077	298.218	221.774	143.964

8. Ethylene C_2H_4 (g)

T(K)	C_P^o	S^o	ΔH_f^o	ΔG_f^o
298.15	42.883	219.316	52.400	68.358
300	43.059	219.582	52.341	68.457
400	53.045	233.327	49.254	74.302
500	62.479	246.198	46.533	80.887
600	70.673	258.332	44.221	87.982
700	77.733	269.770	42.278	95.434
800	83.868	280.559	40.655	103.142
900	89.234	290.754	39.310	111.036
1000	93.939	300.405	38.205	119.067
1100	98.061	309.556	37.310	127.198
1200	101.670	318.247	36.596	135.402
1300	104.829	326.512	36.041	143.660
1400	107.594	334.384	35.623	151.955
1500	110.018	341.892	35.327	160.275

9. Ethane C_2H_6 (g)

T(K)	C_P^o	S^o	ΔH_f^o	ΔG_f^o
298.15	52.487	229.161	-84.000	-32.015
300	52.711	229.487	-84.094	-31.692
400	65.459	246.378	-88.988	-13.473
500	77.941	262.344	-93.238	5.912
600	89.188	277.568	-96.779	26.086
700	99.136	292.080	-99.663	46.800
800	107.936	305.904	-101.963	67.887
900	115.709	319.075	-103.754	89.231
1000	122.552	331.628	-105.105	110.750
1100	128.553	343.597	-106.082	132.385
1200	133.804	355.012	-106.741	154.096
1300	138.391	365.908	-107.131	175.850
1400	142.399	376.314	-107.292	197.625
1500	145.905	386.260	-107.260	219.404

T(K)	C_P^o(J/K-mol)	S^o(J/K-mol)	ΔH_f^o(kJ/mol)	ΔG_f^o(kJ/mol)

10. Benzene C_6H_6 (g)

T(K)	C_P^o	S^o	ΔH_f^o	ΔG_f^o
298.15	82.430	269.190	82.880	129.750
300	83.020	269.700	82.780	130.040
400	113.510	297.840	77.780	146.570
500	139.340	326.050	73.740	164.260
600	160.090	353.360	70.490	182.680
700	176.790	379.330	67.910	201.590
800	190.460	403.860	65.910	220.820
900	201.840	426.970	64.410	240.280
1000	211.430	448.740	63.340	259.890
1100	219.580	469.280	62.620	277.640
1200	226.540	488.690	62.200	299.320
1300	232.520	507.070	62.000	319.090
1400	237.680	524.490	61.990	338.870
1500	242.140	541.040	62.110	358.640

11. Naphthalene $C_{10}H_8$ (g)

T(K)	C_P^o	S^o	ΔH_f^o	ΔG_f^o
298.15	131.920	333.150	150.580	224.100
300	132.840	333.970	150.450	224.560
400	180.070	378.800	144.190	250.270
500	219.740	423.400	139.220	277.340
600	251.530	466.380	135.350	305.330
700	277.010	507.140	132.330	333.950
800	297.730	545.520	130.050	362.920
900	314.850	581.610	128.430	392.150
1000	329.170	615.550	127.510	421.700
1100	341.240	647.500	127.100	450.630
1200	351.500	677.650	126.960	480.450
1300	360.260	706.130	127.060	509.770
1400	367.780	733.110	127.390	539.740
1500	374.270	758.720	127.920	568.940

12. Formaldehyde H_2CO (g)

T(K)	C_P^o	S^o	ΔH_f^o	ΔG_f^o
298.15	35.387	218.760	-108.700	-102.667
300	35.443	218.979	-108.731	-102.630
400	39.240	229.665	-110.438	-100.340
500	43.736	238.900	-112.073	-97.623
600	48.181	247.270	-113.545	-94.592
700	52.280	255.011	-114.833	-91.328
800	55.941	262.236	-115.942	-87.893
900	59.156	269.014	-116.889	-84.328
1000	61.951	275.395	-117.696	-80.666
1100	64.368	281.416	-118.382	-76.929
1200	66.453	287.108	-118.966	-73.134
1300	68.251	292.500	-119.463	-69.294
1400	69.803	297.616	-119.887	-65.418
1500	71.146	302.479	-120.249	-61.514

T(K)	C_P^o(J/K-mol)	S^o(J/K-mol)	ΔH_f^o(kJ/mol)	ΔG_f^o(kJ/mol)

13. Acetone C_3H_6O (g)

T(K)	C_P^o(J/K-mol)	S^o(J/K-mol)	ΔH_f^o(kJ/mol)	ΔG_f^o(kJ/mol)
298.15	74.517	295.349	-217.150	-152.716
300	74.810	295.809	-217.233	-152.339
400	91.755	319.658	-222.212	-129.913
500	107.864	341.916	-226.522	-106.315
600	122.047	362.836	-230.120	-81.923
700	134.306	382.627	-233.049	-56.986
800	144.934	401.246	-235.350	-31.673
900	154.097	418.860	-237.149	-6.109
1000	162.046	435.513	-238.404	19.707
1100	168.908	451.286	-239.283	45.396
1200	174.891	466.265	-239.827	71.463
1300	180.079	480.491	-240.120	97.362
1400	184.556	493.963	-240.203	123.470
1500	188.447	506.850	-240.120	149.369

14. Chlorotrifluoromethane $CClF_3$ (g)

T(K)	C_P^o(J/K-mol)	S^o(J/K-mol)	ΔH_f^o(kJ/mol)	ΔG_f^o(kJ/mol)
298.15	66.886	285.419	-707.800	-667.238
300	67.111	285.834	-707.810	-666.986
400	77.528	306.646	-708.153	-653.316
500	85.013	324.797	-708.170	-639.599
600	90.329	340.794	-707.975	-625.901
700	94.132	355.020	-707.654	-612.246
800	96.899	367.780	-707.264	-598.642
900	98.951	379.317	-706.837	-585.090
1000	100.507	389.827	-706.396	-571.586
1100	101.708	399.465	-705.950	-558.126
1200	102.651	408.357	-705.505	-544.707
1300	103.404	416.604	-705.064	-531.326
1400	104.012	424.290	-704.628	-517.977
1500	104.512	431.484	-704.196	-504.660

15. Dichlorodifluoromethane CCl_2F_2 (g)

T(K)	C_P^o(J/K-mol)	S^o(J/K-mol)	ΔH_f^o(kJ/mol)	ΔG_f^o(kJ/mol)
298.15	72.476	300.903	-486.000	-447.030
300	72.691	301.352	-486.002	-446.788
400	82.408	323.682	-485.945	-433.716
500	89.063	342.833	-485.618	-420.692
600	93.635	359.500	-485.136	-407.751
700	96.832	374.189	-484.576	-394.897
800	99.121	387.276	-483.984	-382.126
900	100.801	399.053	-483.388	-369.429
1000	102.062	409.742	-482.800	-356.799
1100	103.030	419.517	-482.226	-344.227
1200	103.786	428.515	-481.667	-331.706
1300	104.388	436.847	-481.121	-319.232
1400	104.874	444.602	-480.588	-306.799
1500	105.270	451.851	-480.065	-294.404

T(K)	C_P^o(J/K-mol)	S^o(J/K-mol)	ΔH_f^o(kJ/mol)	ΔG_f^o(kJ/mol)

16. Dichlorine Cl_2 (g)

T(K)	C_P^o(J/K-mol)	S^o(J/K-mol)	ΔH_f^o(kJ/mol)	ΔG_f^o(kJ/mol)
298.15	33.949	223.079	0.000	0.000
300	33.981	223.290	0.000	0.000
400	35.296	233.263	0.000	0.000
500	36.064	241.229	0.000	0.000
600	36.547	247.850	0.000	0.000
700	36.874	253.510	0.000	0.000
800	37.111	258.450	0.000	0.000
900	37.294	262.832	0.000	0.000
1000	37.442	266.769	0.000	0.000
1100	37.567	270.343	0.000	0.000
1200	37.678	273.617	0.000	0.000
1300	37.778	276.637	0.000	0.000
1400	37.872	279.440	0.000	0.000
1500	37.961	282.056	0.000	0.000

17. Hydrogen Chloride HCl (g)

T(K)	C_P^o(J/K-mol)	S^o(J/K-mol)	ΔH_f^o(kJ/mol)	ΔG_f^o(kJ/mol)
298.15	29.136	186.902	-92.310	-95.298
300	29.137	187.082	-92.314	-95.317
400	29.175	195.468	-92.587	-96.278
500	29.304	201.990	-92.911	-97.164
600	29.576	207.354	-93.249	-97.983
700	29.988	211.943	-93.577	-98.746
800	30.500	215.980	-93.879	-99.464
900	31.063	219.604	-94.149	-100.145
1000	31.639	222.907	-94.384	-100.798
1100	32.201	225.949	-94.587	-101.430
1200	32.734	228.774	-94.760	-102.044
1300	33.229	231.414	-94.908	-102.645
1400	33.684	233.893	-95.035	-103.235
1500	34.100	236.232	-95.146	-103.817

18. Fluorine F (g)

T(K)	C_P^o(J/K-mol)	S^o(J/K-mol)	ΔH_f^o(kJ/mol)	ΔG_f^o(kJ/mol)
298.15	22.746	158.750	79.380	62.280
300	22.742	158.891	79.393	62.173
400	22.432	165.394	80.043	56.332
500	22.100	170.363	80.587	50.340
600	21.832	174.368	81.046	44.246
700	21.629	177.717	81.442	38.081
800	21.475	180.595	81.792	31.862
900	21.357	183.117	82.106	25.601
1000	21.266	185.362	82.391	19.308
1100	21.194	187.386	82.654	12.986
1200	21.137	189.227	82.897	6.642
1300	21.091	190.917	83.123	0.278
1400	21.054	192.479	83.335	-6.103
1500	21.022	193.930	83.533	-12.498

T(K)	C_P^o(J/K-mol)	S^o(J/K-mol)	ΔH_f^o(kJ/mol)	ΔG_f^o(kJ/mol)

19. Difluorine F_2 (g)

T(K)	C_P^o(J/K-mol)	S^o(J/K-mol)	ΔH_f^o(kJ/mol)	ΔG_f^o(kJ/mol)
298.15	31.304	202.790	0.000	0.000
300	31.337	202.984	0.000	0.000
400	32.995	212.233	0.000	0.000
500	34.258	219.739	0.000	0.000
600	35.171	226.070	0.000	0.000
700	35.839	231.545	0.000	0.000
800	36.343	236.365	0.000	0.000
900	36.740	240.669	0.000	0.000
1000	37.065	244.557	0.000	0.000
1100	37.342	248.103	0.000	0.000
1200	37.588	251.363	0.000	0.000
1300	37.811	254.381	0.000	0.000
1400	38.019	257.191	0.000	0.000
1500	38.214	259.820	0.000	0.000

20. Hydrogen Fluoride HF (g)

T(K)	C_P^o(J/K-mol)	S^o(J/K-mol)	ΔH_f^o(kJ/mol)	ΔG_f^o(kJ/mol)
298.15	29.137	173.776	-273.300	-275.399
300	29.137	173.956	-273.302	-275.412
400	29.149	182.340	-273.450	-276.096
500	29.172	188.846	-273.679	-276.733
600	29.230	194.169	-273.961	-277.318
700	29.350	198.683	-274.277	-277.852
800	29.549	202.614	-274.614	-278.340
900	29.827	206.110	-274.961	-278.785
1000	30.169	209.270	-275.309	-279.191
1100	30.558	212.163	-275.652	-279.563
1200	30.974	214.840	-275.988	-279.904
1300	31.403	217.336	-276.315	-280.217
1400	31.831	219.679	-276.631	-280.505
1500	32.250	221.889	-276.937	280.771

21. Hydrogen H (g)

T(K)	C_P^o(J/K-mol)	S^o(J/K-mol)	ΔH_f^o(kJ/mol)	ΔG_f^o(kJ/mol)
298.15	20.786	114.716	217.998	203.276
300	20.786	114.845	218.010	203.185
400	20.786	120.824	218.635	198.149
500	20.786	125.463	219.253	192.956
600	20.786	129.252	219.867	187.639
700	20.786	132.457	220.476	182.219
800	20.786	135.232	221.079	176.712
900	20.786	137.680	221.670	171.131
1000	20.786	139.870	222.247	165.485
1100	20.786	141.852	222.806	159.781
1200	20.786	143.660	223.345	154.028
1300	20.786	145.324	223.864	148.230
1400	20.786	146.864	224.360	142.393
1500	20.786	148.298	224.835	136.522

T(K)	C_P^o(J/K-mol)	S^o(J/K-mol)	ΔH_f^o(kJ/mol)	ΔG_f^o(kJ/mol)

22. Dihydrogen H_2 (g)

T(K)	C_P^o(J/K-mol)	S^o(J/K-mol)	ΔH_f^o(kJ/mol)	ΔG_f^o(kJ/mol)
298.15	28.836	130.680	0.000	0.000
300	28.849	130.858	0.000	0.000
400	29.181	139.217	0.000	0.000
500	29.260	145.738	0.000	0.000
600	29.327	151.078	0.000	0.000
700	29.440	155.607	0.000	0.000
800	29.623	159.549	0.000	0.000
900	29.880	163.052	0.000	0.000
1000	30.204	166.217	0.000	0.000
1100	30.580	169.113	0.000	0.000
1200	30.991	171.791	0.000	0.000
1300	31.422	174.288	0.000	0.000
1400	31.860	176.633	0.000	0.000
1500	32.296	178.846	0.000	0.000

23. Water H_2O (g)

T(K)	C_P^o(J/K-mol)	S^o(J/K-mol)	ΔH_f^o(kJ/mol)	ΔG_f^o(kJ/mol)
298.15	33.598	188.832	-241.826	-228.582
300	33.606	189.040	-241.844	-228.500
400	34.283	198.791	-242.845	-223.900
500	35.259	206.542	-243.822	-219.050
600	36.371	213.067	-244.751	-214.008
700	37.557	218.762	-245.620	-208.814
800	38.800	223.858	-246.424	-203.501
900	40.084	228.501	-247.158	-198.091
1000	41.385	232.792	-247.820	-192.603
1100	42.675	236.797	-248.410	-187.052
1200	43.932	240.565	-248.933	-181.450
1300	45.138	244.129	-249.392	-175.807
1400	46.281	247.516	-249.792	-170.132
1500	47.356	250.746	-250.139	-164.429

24. Diiodine I_2 (g)

T(K)	C_P^o(J/K-mol)	S^o(J/K-mol)	ΔH_f^o(kJ/mol)	ΔG_f^o(kJ/mol)
298.15	36.887	260.685	62.420	19.324
300	36.897	260.913	62.387	19.056
400	37.256	271.584	44.391	5.447
457.67	37.385	276.610	pressure = 1 bar	
500	37.464	279.921	0.000	0.000
600	37.613	286.765	0.000	0.000
700	37.735	292.573	0.000	0.000
800	37.847	297.619	0.000	0.000
900	37.956	302.083	0.000	0.000
1000	38.070	306.088	0.000	0.000
1100	38.196	309.722	0.000	0.000
1200	38.341	313.052	0.000	0.000
1300	38.514	316.127	0.000	0.000
1400	38.719	318.989	0.000	0.000
1500	38.959	321.668	0.000	0.000

T(K)	C_P^o(J/K-mol)	S^o(J/K-mol)	ΔH_f^o(kJ/mol)	ΔG_f^o(kJ/mol)

25. Hydrogen Iodide HI (g)

T(K)	C_P^o(J/K-mol)	S^o(J/K-mol)	ΔH_f^o(kJ/mol)	ΔG_f^o(kJ/mol)
298.15	29.157	206.589	26.500	1.700
300	29.158	206.769	26.477	1.546
400	29.329	215.176	17.093	-6.289
500	29.738	221.760	-5.481	-9.946
600	30.351	227.233	-5.819	-10.806
700	31.070	231.965	-6.101	-11.614
800	31.807	236.162	-6.323	-12.386
900	32.511	239.950	-6.489	-13.133
1000	33.156	243.409	-6.608	-13.865
1100	33.735	246.597	-6.689	-14.586
1200	34.249	249.555	-6.741	-15.302
1300	34.703	252.314	-6.775	-16.014
1400	35.106	254.901	-6.797	-16.723
1500	35.463	257.336	-6.814	-17.432

26. Dinitrogen N_2 (g)

T(K)	C_P^o(J/K-mol)	S^o(J/K-mol)	ΔH_f^o(kJ/mol)	ΔG_f^o(kJ/mol)
298.15	29.124	191.608	0.000	0.000
300	29.125	191.788	0.000	0.000
400	29.249	200.180	0.000	0.000
500	29.580	206.738	0.000	0.000
600	30.109	212.175	0.000	0.000
700	30.754	216.864	0.000	0.000
800	31.433	221.015	0.000	0.000
900	32.090	224.756	0.000	0.000
1000	32.696	228.169	0.000	0.000
1100	33.241	231.311	0.000	0.000
1200	33.723	234.224	0.000	0.000
1300	34.147	236.941	0.000	0.000
1400	34.517	239.485	0.000	0.000
1500	34.842	241.878	0.000	0.000

27. Nitric Oxide NO (g)

T(K)	C_P^o(J/K-mol)	S^o(J/K-mol)	ΔH_f^o(kJ/mol)	ΔG_f^o(kJ/mol)
298.15	29.862	210.745	91.277	87.590
300	29.858	210.930	91.278	87.567
400	29.954	219.519	91.320	86.323
500	30.493	226.255	91.340	85.071
600	31.243	231.879	91.354	83.816
700	32.031	236.754	91.369	82.558
800	32.770	241.081	91.386	81.298
900	33.425	244.979	91.405	80.036
1000	33.990	248.531	91.426	78.772
1100	34.473	251.794	91.445	77.505
1200	34.883	254.811	91.464	76.237
1300	35.234	257.618	91.481	74.967
1400	35.533	260.240	91.495	73.697
1500	35.792	262.700	91.506	72.425

T(K)	C_P^o(J/K-mol)	S^o(J/K-mol)	ΔH_f^o(kJ/mol)	ΔG_f^o(kJ/mol)

28. Ammonia NH$_3$ (g)

T(K)	C_P^o	S^o	ΔH_f^o	ΔG_f^o
298.15	35.630	192.768	-45.940	-16.407
300	35.678	192.989	-45.981	-16.223
400	38.674	203.647	-48.087	-5.980
500	41.994	212.633	-49.908	4.764
600	45.229	220.578	-51.430	15.846
700	48.269	227.781	-52.682	27.161
800	51.112	234.414	-53.695	38.639
900	53.769	240.589	-54.499	50.231
1000	56.244	246.384	-55.122	61.903
1100	58.535	251.854	-55.589	73.629
1200	60.644	257.039	-55.920	85.392
1300	62.576	261.970	-56.136	97.177
1400	64.339	266.673	-56.251	108.975
1500	65.945	271.168	-56.282	120.779

29. Dioxygen O$_2$ (g)

T(K)	C_P^o	S^o	ΔH_f^o	ΔG_f^o
298.15	29.378	205.148	0.000	0.000
300	29.387	205.330	0.000	0.000
400	30.109	213.873	0.000	0.000
500	31.094	220.695	0.000	0.000
600	32.095	226.454	0.000	0.000
700	32.987	231.470	0.000	0.000
800	33.741	235.925	0.000	0.000
900	34.365	239.937	0.000	0.000
1000	34.881	243.585	0.000	0.000
1100	35.314	246.930	0.000	0.000
1200	35.683	250.019	0.000	0.000
1300	36.006	252.888	0.000	0.000
1400	36.297	255.568	0.000	0.000
1500	36.567	258.081	0.000	0.000

30. Sulfur Dioxide SO$_2$ (g)

T(K)	C_P^o	S^o	ΔH_f^o	ΔG_f^o
298.15	39.842	248.219	-296.810	-300.090
300	39.909	248.466	-296.833	-300.110
400	43.427	260.435	-300.240	-300.935
500	46.490	270.465	-302.735	-300.831
600	48.938	279.167	-304.699	-300.258
700	50.829	286.859	-306.308	-299.386
800	52.282	293.746	-307.691	-298.302
900	53.407	299.971	-362.075	-295.987
1000	54.290	305.646	-362.012	-288.647
1100	54.993	310.855	-361.934	-281.314
1200	55.564	315.665	-361.849	-273.989
1300	56.033	320.131	-361.763	-266.671
1400	56.426	324.299	-361.680	-259.359
1500	56.759	328.203	-361.605	-252.053

Table 5. Enthalpy of Combustion of Selected Organic Compounds

$$-\Delta H^o_{c,298.15}\text{(kJ/mol)}$$

Molecular Formula	Name	Crystal	Liquid	Gas
CH_2O	Formaldehyde			570.7
CH_4	Methane			890.8
$C_2H_2O_4$	Oxalic Acid	251.1		349.1
C_2H_4	Ethylene			1411.2
C_2H_4O	Acetaldehyde		1166.9	1192.5
C_2H_5NO	Acetamide	1184.6		1263.3
C_2H_6	Ethane			1560.7
C_3H_4	Allene			1942.7
C_3H_4	Cyclopropene			2029.3
C_3H_8O	1-Propanol		2021.3	2068.8
C_3H_8O	2-Propanol		2005.8	2051.1
C_3H_8O	Ethyl Methyl Ether			2107.5
$C_4H_{10}O$	1-Butanol		2675.9	2728.2
$C_4H_{10}O$	2-Butanol		2660.6	2710.3
$C_4H_{10}O$	2-Methyl-2-Propanol		2644.0	2690.7
$C_4H_{10}O$	2-Methyl-1-Propanol		2668.5	2719.3
$C_4H_{10}O$	Diethyl Ether		2723.9	2751.1
$C_4H_{10}O$	Methyl Propyl Ether		2737.2	2765.0
$C_4H_{10}O$	Isopropyl Methyl Ether		2724.5	2751.2
$C_6H_5NO_2$	Nitrobenzene		3088.1	3143.1
C_6H_6	Benzene		3267.6	3301.2
C_7H_5N	Benzonitrile		3632.3	3684.8
C_7H_6O	Benzaldehyde		3525.1	3575.4
$C_7H_6O_2$	Benzoic Acid	3226.9		3318.0
C_7H_9N	o-Methyl Aniline		4034.5	4097.2
C_7H_9N	m-Methyl Aniline		4032.7	4095.4
C_7H_9N	p-Methyl Aniline	4017.3		4096.1

Table 6. Bond Strengths in Diatomic Molecules

Molecule	D^o_{298}(kJ/mol)
Br-Br	192.807
Br-C	280 ± 21
Br-Cl	217.53 ± 0.29
Br-H	366.35
Br-I	179.1 ± 21
C-C	607 ± 21
C-Cl	397 ± 29
C-F	552
C-H	338.32
C-I	209 ± 21
C-N	754.3 ± 10
C-O	1076.5 ± 0.4
Cl-Cl	242.580 ± 0.004
Cl-F	256.23
Cl-H	431.62
Cl-I	211.3 ± 0.4
F-F	158.78
F-H	570.3
F-Si	552.7 ± 2.1
H-H	435.990
H-I	298.407
H-N	≤339
H-O	427.6
I-I	151.088
N-N	945.33 ± 0.59
N-O	630.57 ± 0.13
O-O	498.36 ± 0.17

Table 7. Enthalpies of Formation of Gaseous Atoms from Elements in Their Standard States

Atom	$\Delta H^o_{f,298}$(kJ/mol)
Br	111.859
C	716.68 ± 0.46
Ca	178.2 ± 1.7
Cl	121.290 ± 0.008
F	79.41
H	217.995 ± 0.004
I	106.767 ± 0.042
K	89.62 ± 0.21
N	472.67 ± 0.42
O	249.170 ± 0.100
P	332.2 ± 4.2
S	276.98 ± 0.25
Sc	377.8 ± 4
Ti	469.9 ± 2.1

Table 8. Enthalpies of Formation of Free Radicals

Radical	$\Delta H^o_{f,298}$ (kJ/mol)
CH	596.35±4
CH_2(triplet)	392.5±2.1
CH_2(singlet)	430.1±4.2
CH_3	146±1.0
CH≡C	566.1±2.9
CH_2=CH	300.0±3.4
C_2H_5	118.5±1.7
CH_2=CHCH$_2$	163.6±6.3
n-C_3H_7	94.7
i-C_3H_7	89±3
t-C_4H_9	48.6±1.7
$(CH_3)_3CCH_2$	36.4±8
C_6H_5	328.9±8.4
$C_6H_5CH_2$	200.0±6.3
CH_2CN	244.8±10.5
CN	435±8
CHO	37.2±5.0
CH_3CO	-10.0±1.2
C_6H_5CO	109.2±8
CF_2	-194.1±9.2
CF_3	-467.4±15.1
CH_2Cl	121.8±4.2
CHFCl	-60.7±10.0
CF_2Cl	-269.0±8.4
$CHCl_2$	98.3±5.0
$CFCl_2$	-89.1±10.0
CCl_2	239
CCl_3	71.1±2.5
C_6F_5	-547.7±8
NH_2	185.4±4.6
NF_2	34±4
CH_3NH	177.4±8
$(CH_3)_3N$	145.2±8
C_6H_5NH	237.2±8
OH	39.3
CH_3O	17.6
C_2H_5O	-17.2
C_6H_5O	47.7
HO_2	14.6
CH_3CO_2	-207.5±4
$C_2H_5CO_2$	-228.5±4
HS	143.0±2.8
CH_3S	124.6±1.8

Table 9. Bond Strengths in Polyatomic Molecules

Bond	D_{298}^{0}(kJ/mol)
H-CH	421.7
H-CH$_2$	464.8
H-CH$_3$	438.5$\pm$1.5
H-C$_2$H$_5$	420.5$\pm$2.0
H-n-C$_3$H$_7$	417.1
H-C$_6$H$_5$	464.0$\pm$8.4
H-CH$_2$CN	389$\pm$10.5
H-C$_6$F$_5$	476.6
H-SiH$_3$	377.8
H-NH$_2$	449.4$\pm$4.6
H-NHCH$_3$	418.4$\pm$10.5
H-N(CH$_3$)$_2$	382.8$\pm$8
H-NHC$_6$H$_5$	368.2$\pm$8
H-N(CH$_3$)C$_6$H$_5$	366.1$\pm$8
H-OH	498$\pm$4
H-OCH$_3$	436.8$\pm$4
H-O$_2$H	369.0$\pm$4.2
H-ONO	327.6$\pm$2.1
H-ONO$_2$	423.4$\pm$2.1
H-SH	381.6$\pm$2.9
HC$\equiv$CH	965$\pm$8
H$_2$C=CH$_2$	733$\pm$8
CH$_3$-CH$_3$	376.0$\pm$2.1
CH$_3$-t-C$_4$H$_9$	425.9$\pm$8
CH$_3$-CN	509.6$\pm$8
CN-CN	536$\pm$4
CF$_2$=CF$_2$	319.2$\pm$13
CH$_3$-CF$_3$	423.4$\pm$4.6
CF$_3$-CF$_3$	413.0$\pm$10.5
O=CO	532.2$\pm$0.4
F-CH$_3$	472
Cl-CF$_3$	360.2$\pm$3.3
Cl-CF$_2$Cl	346.0$\pm$13.4
Cl-CFCl$_2$	305$\pm$8
Cl-CH$_2$Cl	350.2$\pm$0.8
Cl-CHCl$_2$	338.5$\pm$4.2
Cl-CCl$_3$	305.9$\pm$7.5
Cl-C$_2$F$_5$	346.0$\pm$7.1
I-CF$_3$	223.8$\pm$2.9
I-CH$_2$CF$_3$	235.6$\pm$4
I-C$_2$F$_5$	218.8$\pm$2.9
I-C$_6$H$_5$	273.6$\pm$8
O$_2$N-NO$_2$	56.9
O-SO	552$\pm$

Table 10. Average Single Bond Energies (kJ/mol)

Bond	ε	Bond	ε
H–H	435	N-Cl	154
H-C	414	N-O	
H-N	351	O-O	138
H-O	464	O-F	188
H-F	564	O-Si	443
H-P	317	O-Cl	209
H--S	338	F-F	154
H-Cl	430	F-Si	569
H-Se	276	Si-Si	209
H-Br	368	Si-S	251
H-Te	238	Si-Cl	377
H-I	297	Si-Br	305
C-C	334	Si-I	222
C-N	267	P-P	213
C-O	338	P-Cl	326
C-F	426	P-Br	268
C-Si	313	P-I	205
C-S	255	S-S	205
C-Cl	330	S-Cl	255
C-Br	276	S-Br	213
C-I	217	Cl-Cl	243
N-N	133	Br-Br	192
N-F	234	I-I	151

Table 11. Heat Capacity

$$C_p = a + bT + cT^2, \text{ in J/K-mol}$$

Substance	State	a	$b \times 10^3$	$c \times 10^7$	Range
CO	g	26.861	6.966	-8.20	273 - 1500 K
N_2	g	27.296	5.23	5.188	273 - 1500 K
H_2	g	29.065	-0.8363	20.116	273 - 1500 K
NH_3	g	29.748	25.104	-1.673	273 - 1500 K
O_2	g	25.723	12.978	-38.618	273 - 1500 K
CO_2	g	25.999	43.496	-148.322	273 - 1500 K
H_2O	g	30.359	9.614	11.840	273 - 1500 K
C_2H_2	g	50.751	16.066	-10.292	273 - 1500 K
Br_2	g	35.240	4.0747	-14.874	273 - 1500 K
Cl_2	g	31.695	10.143	-40.375	273 - 1500 K

Table 12. Bond Lengths of Diatomic Molecules

Molecule	Bond Distances (angstroms)
H_2	0.741
F_2	1.417
Cl_2	1.988
Br_2	2.290
I_2	2.667
HI	1.604
O_2	1.207
N_2	1.094
HBr	1.414
HCl	1.275
HF	0.917
HO	0.971

Table 13. Fundamental Vibrational Frequencies

Molecule	Formula	Frequencies (cm $^{-1}$)
Hydrogen	$^1H_2$	4395.2
	$^1H^2H$	3817.1
	$^2H_2$	3118.4
Iodine	I_2	214.2
Hydrogen Iodide	HI	2309.5
Bromine	$^{79}Br^{81}Br$	323.2
Chlorine	$^{35}Cl_2$	564.9
Fluorine	F_2	892.
Hydrogen Chloride	HCl	2989.7
Hydrogen Bromide	HBr	2849.7
Oxygen	O_2	1580.4
Nitrogen	N_2	2359.6
Hydroxyl Radical	OH	3735.2

INDEX